Experimental Approaches to Conservation Biology

Experimental Approaches to Conservation Biology

Edited by

MALCOLM S. GORDON AND SORAYA M. BARTOL

University of California Press

BERKELEY   LOS ANGELES   LONDON

University of California Press
Berkeley and Los Angeles, California

University of California Press, Ltd.
London, England

Library of Congress Cataloging-in-Publication Data

Experimental approaches to conservation biology / edited by Malcolm S.
Gordon and Soraya M. Bartol.
    p.   cm.
  Includes bibliographical references and index.
  ISBN 0–520–24024–3 (cloth : alk. paper).
   1. Conservation biology.   I. Gordon, Malcolm S., 1933–   II. Bartol,
Soraya M., 1970–

QH75.E96   2004
639.9—dc22   2003018997

Manufactured in the United States of America

13   12   11   10   09   08   07   06   05   04
10 9 8 7 6 5 4 3 2 1

The paper used in this publication is both acid-free and totally chlorine-
free (TCF). It meets the minimum requirements of ANSI/NISO z39.48–
1992 (R 1997) (Permanence of Paper).

Contents

Acknowledgments

The coeditors wish to thank the following people for reviewing various chapters of this book: Karen A. Bjorndal, Cynthia S. Brown, Rochelle Buffenstein, Gordon M. Burghardt, Douglas P. Chivers, Pierre Deviche, J. Whitfield Gibbons, Richard Goeden, Karen Goodrowe, John Hanks, William K. Hayes, Graeme Hays, William Holt, Stephen T. Jackson, Lee B. Kats, Robin Liley, Molly Lutcavage, John Maron, James Murray, Norman Owen-Smith, Petr Pyšek, George Roderick, Barney A. Schlinger, Ulie S. Seal, Lesca Strickland, Una Swain, Roy Van Driesche, and David E. Wildt.

In addition to the authors represented in this volume, many other people made important and useful contributions to the development of the scientific program for the conference at which these papers were presented and to the logistical and administrative activities that were involved. The coeditors were cochairs of the local organizing committee and wish to thank the members of the scientific program group: Gary Burness, Donald G. Buth, Pamela Mueller, Kenneth A. Nagy, Philip W. Rundel, Anthony Steyermark, and Hartmut S. Walter. We also wish to thank the members of the administrative staff at the UCLA Institute of the Environment (IoE): Dorothy Fletcher, Marci Green, Evelyn Leon, Mikki Parham, and Joanne Sanger.

Dean V. Lauritzen performed an array of important tasks in the preparation of the manuscript. Publication of the book would not have been possible without the active participation and support of the staff at the University of California Press. Two editors played major roles: Doris Kretschmer and Blake Edgar.

The conference was sponsored by the International Commission on Comparative Physiology of the International Union of Physiological Sciences (M. S. Gordon, chair) and by the UCLA IoE (R. P. Turco, director). Funding was provided by the IoE, the U.S. National Science Foundation (grant number IBN-0120208, to M. S. Gordon), the UCLA Lida Scott Brown Ornithology Trust, and the offices of the dean of life sciences (F. Eiserling) and of the vice-chancellor for research (R. Peccei) at UCLA.

Contributors

E-mail addresses (current as of September 2003) are given for corresponding authors.

ADAMS, DOMINIC C.
Conservation Endocrinology Research Group
Institute of Veterinary, Animal, and Biomedical Sciences
Massey University
Palmerston North, New Zealand

ALBERTS, ALLISON C.
aalberts@sandiegozoo.org
Center for Reproduction of Endangered Species
Zoological Society of San Diego
P.O. Box 120551
San Diego, California, 92112

ANDREWS, RUSSEL D.
russ_andrews@alaskasealife.org
Institute of Marine Science
University of Alaska
Fairbanks, Alaska, 99775, and
Alaska SeaLife Center
Seward, Alaska, 99664

AVERY, HAROLD W.
School of Environmental Science, Engineering, and Policy
Drexel University
Philadelphia, Pa. 19104

BAINBRIDGE, SUSAN J.
Jepson Herbarium
University of California
Berkeley, Calif. 94720-2465

BARTOL, SORAYA M.
sbartol@whoi.edu
Woods Hole Oceanographic Institution
9 Maury Lane, MS #44
Woods Hole, Mass. 02543-1119

BELDEN, LISA K.
Department of Zoology
3029 Cordley Hall
Oregon State University
Corvallis, Oregon, 97331-2914

BENNETT, ELLEN J.
Conservation Endocrinology Research Group
Institute of Veterinary, Animal, and Biomedical Sciences
Massey University
Palmerston North, New Zealand

BLAUSTEIN, ANDREW R.
blaustea@bcc.orst.edu
Department of Zoology
3029 Cordley Hall
Oregon State University
Corvallis, Oregon, 97331-2914

BRAACK, LEO
braack@mweb.co.za
P.O. Box 2550
Brooklyn Square 0075
South Africa

BURNESS, GARY
garyburnessg@trentu.ca
Department of Biology
Trent University
Petersburough, ON K9J7B8
Canada

CANDY, E. JANE
Conservation Endocrinology Research Group
Institute of Veterinary, Animal, and Biomedical Sciences
Massey University
Palmerston North, New Zealand

COCKREM, JOHN F.
j.f.cockrem@massey.ac.nz
Conservation Endocrinology Research Group
Institute of Veterinary, Animal, and Biomedical Sciences
Massey University
Palmerston North, New Zealand

CORBIN, JEFFREY D.
corbin@socrates.berkeley.edu
Department of Integrative Biology
University of California
Berkeley, Calif. 94720-3140

D'ANTONIO, CARLA M.
Department of Integrative Biology
University of California
Berkeley, Calif. 94720-3140

GIBSON, LESLEY A.
Institute of Wildlife Research
School of Biological Sciences A08
University of Sydney
Sydney, NSW 2006
Australia

GORDON, MALCOLM S.
msgordon@ucla.edu
Department of Organismic Biology, Ecology and Evolution and
Institute of the Environment
University of California
Box 951606
Los Angeles, Calif., 90095-1606

HADFIELD, MICHAEL G.
hadfield@hawaii.edu
Department of Zoology and Kewalo Marine Laboratory
Pacific Biomedical Research Center
University of Hawaii at Manoa
41 Ahui St.
Honolulu, Hawaii 96813

HATCH, AUDREY C.
Department of Zoology
University of Washington
Seattle, Washington 98195

HAWKE, EMMA J.
Conservation Endocrinology Research Group
Institute of Veterinary, Animal, and Biomedical Sciences
Massey University
Palmerston North, New Zealand

HENARE, SHARON J.
Conservation Endocrinology Research Group
Institute of Veterinary, Animal, and Biomedical Sciences
Massey University
Palmerston North, New Zealand

HODDLE, MARK S.
mark.hoddle@ucr.edu
Department of Entomology
University of California
Riverside, Calif. 92521

HOLLAND, BRENDEN S.
Department of Zoology and Kewalo Marine Laboratory
Pacific Biomedical Research Center
University of Hawaii at Manoa
41 Ahui St.
Honolulu, Hawaii 96816

HUME, IAN D.
ianhume@bio.usyd.edu.au
Institute of Wildlife Research
School of Biological Sciences A08
University of Sydney
Sydney, NSW 2006
Australia

JONES, DAVID R.
jones@zoology.ubc.ca
Department of Zoology and
Peter Wall Institute for Advanced Studies
University of British Columbia
6270 University Blvd.
Vancouver, B.C. V6T 1Z4
Canada

KIESECKER, JOSEPH M.
Department of Biology
Pennsylvania State University
208 Mueller Lab
University Park, Pa. 16802

LAPIDGE, STEVEN J.
Institute of Wildlife Research
School of Biological Sciences A08
University of Sydney
Sydney, NSW 2006
Australia

MUELLER, PAMELA J.
caperingk9@msn.com
Department of Organismic Biology, Ecology, and Evolution
University of California
Box 951606
Los Angeles, Calif. 90095-1606

OLIVAL, KEVIN J.
Department of Zoology and Kewalo Marine Laboratory
Pacific Biomedical Research Center
University of Hawaii at Manoa
41 Ahui St.
Honolulu, Hawaii 96816

PHILLIPS, JOHN A.
Center for Reproduction of Endangered Species
Zoological Society of San Diego
P.O. Box 120551
San Diego, Calif., 92112

POTTER, MURRAY A.
Conservation Endocrinology Research Group
Institute of Veterinary, Animal and Biomedical Sciences
Massey University
Palmerston North, New Zealand

REJMÁNEK, MARCEL
Section of Ecology and Evolution
University of California
Davis, Calif. 95616

RICHARDSON, DAVID M.
rich@botzoo.uct.ac.za
Institute for Plant Conservation
Botany Department
University of Cape Town
Rondebosch 7701
South Africa

RODGER, JOHN C.
John.Rodger@newcastle.edu.au
Cooperative Research Centre for the Conservation and Management
    of Marsupials
School of Environmental and Life Sciences
University of Newcastle, NSW
Australia 2308

ROUGET, MATHIEU
Institute for Plant Conservation
Botany Department
University of Cape Town
Rondebosch 7701
South Africa

SOUTHWOOD, AMANDA L.
Department of Zoology
University of British Columbia
6270 University Boulevard
Vancouver, B.C. V6T 1Z4
Canada

SPOTILA, JAMES R.
spotilajr@drexel.edu
School of Environmental Science, Engineering, and Policy
Drexel University
Philadelphia, Pa. 19104

STEYERMARK, ANTHONY C.
acsteyermark@stthomas.edu
Department of Biology
University of St. Thomas
St. Paul, Minn. 55105

WILDT, DAVID E.
dewildt@shentel.net
Conservation and Research Center
Smithsonian's National Zoological Park
1500 Remount Road
Front Royal, Virginia 22630

SECTION 1

Introduction

1 Experimental Biology in Conservation Science

Malcolm S. Gordon and Soraya M. Bartol

We are living in the relatively early stages of a worldwide biological extinction event. There is abundant evidence that the rate of extinctions of living species larger than protists is near the highest level since the asteroid collision that occurred 65 million years ago (Wilson 1988, 2002; Reaka-Kudla et al. 1997). An unprecedented feature of these extinctions is that the vast majority are anthropogenic—that is, the direct or indirect result of human activities. Massive modifications in habitats plus global climate changes are combining with comparably massive introductions of exotic, often invasive, species (some introductions are intentional, but most are inadvertent or accidental [Pimentel 2002]) to destroy naturally evolved native floras and faunas on all continents and most major islands. The consequences of these changes can be dramatic, ranging in scale from local disruptions of ecosystem structures and functions up to the destruction of human societies (Diamond 1997; Weiss and Bradley 2001). These destructive changes will continue unless large-scale efforts are made to at least slow if not halt the process (Reid 2000; Western 2000; Little et al. 2001; Schmitz and Simberloff 2001).

GOALS OF THIS BOOK

In the scientific biological literature, many of the most visible of the efforts directed toward slowing this destruction of major parts of the planet's biological infrastructure fall into the disciplinary areas of population and community ecology, population genetics, systematic biology, and evolutionary biology. Organism-centered experimental biology, however, is also making a wide range of important contributions. Hypothesis-driven and mechanism-oriented studies provide perspectives that are often different from those

3

deriving from more observational, descriptive, and historical investigations. All types of studies are essential.

Many useful approaches to the protection of biodiversity are based upon experimental studies. Results from the research fields of molecular genetics, molecular biology, biochemistry, endocrinology, physiology, immunology, nutrition, ecology, and behavior can all be applied to both endangered and exotic species. Examples of such applications may be found in captive breeding programs for many endangered species, management of agricultural pests using biological control agents, and repression of diseases and parasites using molecular biological methods.

Other experimental efforts attempt to vertically integrate data from the molecular level all the way up to the ecosystem level. These efforts can have direct consequences both for policy development and for management decisions relating to forestry, fisheries and wildlife management, and protection and preservation of natural reserves, refuges, parks, and so on.

To deal effectively with this array of scientific and technical advances, policymakers and regulators should be prepared to implement guiding actions based on experimental results. Gaps between researchers and policymakers must be narrowed so that the laws and policies enacted in behalf of biodiversity are rooted in the most-current knowledge developed using the most-advanced methods.

This book has several goals. The first is to present a selection of current case studies relevant to the protection of world biodiversity that illustrate important issues in both basic and applied experimental biology, emphasizing phenomena and processes at the organismic level. Subject matter coverage is, by choice, not encyclopedic. The accessibility of the content to interested students, researchers, and nonscientists is an important consideration. The organisms considered in these studies occur in many parts of the world. They live in a wide range of habitats, both aquatic (marine and freshwater) and terrestrial, and they represent a diversity of taxa: plant and animal, invertebrate and vertebrate. Many of these studies can serve as models for additional work on related matters involving other organisms in other places. The second goal is to raise the visibility of organism-centered experimentation as a valid and useful approach to understanding and, sometimes, resolving major problems in conservation biology. The third goal is to narrow the gaps separating research scientists from managers and policymakers involved in protecting world biodiversity. Describing the lessons learned from recent efforts at integrating scientific knowledge about endangered and exotic species with economic, social, cultural, political, and legal considerations should help in this effort.

ORGANIZATION OF THIS BOOK

The book is composed of a selection of papers that were originally prepared for a three-day international conference having the same title as the book. That conference was scheduled to take place at the University of California, Los Angeles, from September 11 to September 14, 2001. As a result of the tragic events that occurred in New York City and Washington, D.C., on September 11, the conference was truncated and extensively modified. Seventeen papers were presented on September 13 and 14. Nine of those papers are included here. They are supplemented by five papers that were scheduled for delivery but were not presented, since the authors were unable to travel to Los Angeles.

The book is organized into four sections. Sections 2, 3, and 4 begin with short overviews that highlight and synthesize general issues raised by the papers in the section. Section 2 focuses on the conservation of endangered species. Section 3 considers the control of introduced, and often invasive, exotic species. Section 4 illustrates multiple aspects of both:

(1) Applications of experimental results involving endangered and exotic species to attempted resolutions for selected real-world economic, social, cultural, political, legal, and policy controversies; and

(2) The use, in the context of the protection of biodiversity, of management and policy initiatives that are in themselves experimental.

The chapter sequences in each section are based upon the relative phylogenetic positions of the main subject organisms in the chapters.

As stated earlier, the intent of this book is to provide clear, high-quality examples of both the effectiveness and the utility of organism-centered experimental approaches to many important and active aspects of conservation biology. The emphasis is on autecological, rather than synecological, investigations. Despite their importance and relevance, subject areas such as habitat fragmentation and habitat-conservation planning are deemphasized to maintain the focus on organism-centered experimentation.

We define experimental studies broadly. Well-controlled field and laboratory experiments predominate, but "natural" experiments of several kinds are also discussed. The papers vary substantially in form, style, length, and theoretical approaches. The authors were encouraged to exercise their individuality.

WHAT CONSTITUTES AN EXPERIMENT?

Life is an ongoing series of uncontrolled experiments. From this perspective, everything that happens is, in some way, an experimental result. In our view, this definition of an experiment is so broad that it is almost meaning-less—and it contributes nothing to our understanding of how best we can make informed decisions and choices relating to significant issues. An ordinary dictionary definition is better. An *experiment* is defined as "a trial made to confirm or disprove something doubtful; an operation undertaken to discover some unknown principle or effect, or to test some suggested truth, or to demonstrate some known truth" (*Webster's New Collegiate Dictionary*, 1949).

Most of the information summarized in this book derives from studies fitting this definition. Parts of the book, however, deal with yet another category of experiments, termed *natural experiments*. These are also complex, multivariate, and uncontrolled, and they take an alternative approach. Natural experiments take advantage of real-world complexities and variability in ways that permit us to use carefully selected and replicated field observations to test hypotheses and develop pseudocontrols. Such experiments are probably more susceptible to being affected adversely by hidden variables and missing treatments than are many laboratory and microcosm experiments, but they provide ecology and other field-based sciences with tools and information not otherwise obtainable. Cook and Campbell (1979) give a useful discussion of natural experimental approaches.

SOME OBSERVATIONS

The studies described in the following chapters constitute a limited and not necessarily representative sampling from a research enterprise that is international, vibrant, and growing. There is great need for additional work along these lines and for additional people who are willing to commit their creativity and energy to similar efforts. Continuing large-scale biodiversity loss seems inevitable, given long-term trends in human population numbers, patterns of resource use, rates of food production, and so on. The loss of large numbers of species will inevitably erode the functional abilities of the worldwide ecosphere to sustain life as we know it. Unforeseen non-linearities in the affected systems may well produce catastrophic negative changes. The recent (August–September 2002) United Nations World Summit on Sustainable Development was a hopeful sign that national governments and humanity as a whole are beginning to face these and many related problems. We hope that this book will contribute to the effort to avoid devastating losses of biodiversity.

ELECTRONIC RESOURCES

To enhance the value of this book for both reference and teaching purposes, we provide a list of selected Web sites relating to the themes of the three sections. This list is intended to indicate resources available electronically, but it is not exhaustive (Table 1.1).

REFERENCES

Cook, T. D., and Campbell, D. T. 1979. *Quasi-experimentation: Design and Analysis Issues for Field Settings.* Boston: Houghton-Mifflin Co.

Diamond, J. 1997. *Guns, Germs, and Steel: The Fates of Human Societies.* New York: W. W. Norton & Co.

Little, M. D., Badgley, C., Beall, C. M., Balick, M., Munstermann, L. E., Weiss, K. M., Bert, T. M., and Chernoff, B. 2001. A framework for a program in the human dimensions of biodiversity. *Biology International,* no. 42:3–15.

Pimentel, D., ed. 2002. *Biological Invasions: Economic and Environmental Costs of Alien Plant, Animal, and Microbe Species.* Boca Raton, Fla.: CRC Press.

Reaka-Kudla, M. L., Wilson, D. E., and Wilson, E. O., eds. 1997. *Biodiversity II: Understanding and Protecting Our Biological Resources.* Washington, D.C.: National Academy Press.

Reid, W. V. 2000. Ecosystem data to guide hard choices. *Issues in Science and Technology* 16:37–44.

Schmitz, D. C., and Simberloff, D. 2001. Needed: A national center for biological invasions. *Issues in Science and Technology* 17:57–62.

Weiss, H., and Bradley, R. S. 2001. What drives societal collapse? *Science* 291:609–10.

Western, D. 2000. Conservation in a human-dominated world. *Issues in Science and Technology* 16:53–60.

Wilson, E. O. 2002. *The Future of Life.* New York: Knopf.

————, ed. 1988. *Biodiversity.* Washington, D.C.: National Academy Press.

TABLE 1.1 *Selected Web sites relevant to the conservation of biodiversity*

Government Agencies

COSEWIC (Committee on the Status of Endangered Wildlife in Canada)	www.cosewic.gc.ca/index.htm
National Invasive Species Council, a gateway to U.S. federal invasive species programs	www.invasivespecies.gov/
U.S. Geological Survey, National Biological Information Infrastructure	www.nbii.gov/about/
U.S. Geological Survey, Status and Trends of the Nation's Biological Resources	http://biology.usgs.gov/s+t/SNT/index.htm
U.S. Global Change Research Program	www.usgcrp.gov/usgcrp/

Nongovernmental Organizations

Center for Applied Biodiversity Science	www.biodiversityhotspots.org/xp/Hotspots
Center for Conservation Biology, Stanford University	www.stanford.edu/group/CCB
Conservation International, Washington, D.C.	www.conservation.org
CREO Extinctions Database, American Museum of Natural History, New York	http://creo.amnh.org/pdi.html# access
DIVERSITAS (a program sponsored by the International Council for Science)	www.diversitas-international.org/
Global Biodiversity Forum, Switzerland	www.gbf.ch/
International Commission for the Scientific Exploration of the Mediterranean Sea (CIESM) Atlas of Exotic Species in the Mediterranean Sea	www.ciesm.org/atlas
International Union for the Conservation of Nature	www.iucn.org
International Union of Biological Sciences	www.iubs.org
National Council for Science and the Environment, Washington, D.C.	www.ncseonline.org
The Nature Conservancy	http://nature.org
Society for Conservation Biology	www.conbio.net/
World Biodiversity Database, Expert Center for Taxonomic Identification	www.eti.uva.nl/Database/WBD.html
World Conservation Union, 2003 IUCN Red List of Threatened Species	www.redlist.org
World Wildlife Fund	www.worldwildlife.org/

Conservation of Endangered Species

2 Overview

Gary Burness

> To admit that species generally become rare before they become
> extinct—to feel no surprise at the rarity of a species, and yet to
> marvel greatly when it ceases to exist, is much the same as to admit
> that sickness in the individual is the forerunner of death—to feel
> no surprise at sickness, but when the sick man dies, to wonder and
> to suspect that he died by some unknown deed of violence.
>
> CHARLES DARWIN, 1859

Extinction is a natural process. In fact, the number of species currently alive is estimated to be only 2–4% of the total number that have existed over geologic time (May et al. 1995). Although extinction is inevitable for all species, the rate at which species worldwide are currently disappearing is of concern to conservation biologists. We urgently need to understand the factors contributing to species declines, and to identify methods that can be used to mitigate those factors (Mace et al. 2001). The success of this endeavor will, however, depend on a detailed understanding of each species' biology.

In the seven chapters in this section, the authors present case studies of high-profile endangered species. The authors identify factors that are threatening each species' long-term survival, outline approaches currently used as part of management and recovery programs, and identify approaches that may prove successful in the future. The research agendas of the various chapters differ, and although some chapters are very much applied in their focus, others take a more basic-research approach. The common thread that runs through the chapters is the emphasis on experimental rather than purely observational or "non-invasive" studies.

As part of a concerted strategy to protect biodiversity, conservationists often attempt to breed endangered species in captivity. The ultimate goal of captive breeding is usually the reintroduction of species into the field, either to augment existing populations or to start new ones. For example, in Chapter 3, Michael Hadfield, Brenden Holland, and Kevin Olival describe their long-term research on endemic Hawaiian tree snails, a group of invertebrates that have suffered massive extinction in the past 50–70 years. The authors have established captive-breeding programs, and to date, experimental reintroductions of a single species of snail suggest that when predators are excluded, reintroductions will likely be successful. In addition to describing their ecological work, the authors outline recent molecular studies that will be used to guide the genetic management of field populations. Hadfield, Holland, and Olival suggest that captive propagation, predator exclusion, genetic management, and initiation of new field populations provide hope for the few remaining *Achatinella* snail species.

As the general public knows well, populations of many amphibian species are in decline as part of a global loss of biodiversity. In Chapter 4, Andrew Blaustein and colleagues provide an extensive review of experimental studies that have investigated the factors believed to contribute to global amphibian declines. The authors conclude that no single factor appears to be responsible for amphibian declines in all regions; the declines are more likely due to multiple stressors, acting either independently or through complex interactions with each other. Multiple stressors may be more difficult for the general public to understand than is a single widespread stressor. Understanding how these multiple stressors interact will be an important challenge for conservation biologists.

Most chapters in this section discuss terrestrial systems. This reflects the fact that marine conservation biology is a relatively young science, at least in part because of the size and inaccessibility of the oceans. In Chapter 5, David Jones, Amanda Southwood, and Russel Andrews outline their approaches to gathering basic information on the ecology and physiology of endangered leatherback turtles, a species that spends 99% of its life at sea. Studying leatherbacks and other free-ranging marine species poses special problems and has required the development of novel remote-sensing technologies (see also Chapter 9). Fortunately, female leatherbacks must come to land to lay their eggs. This requirement has allowed Jones and his colleagues to attach, to individual females, miniature computers capable of gathering physiological and behavioral information following the individual's return to the sea. Using these data, the authors have been able to estimate the turtles' at-sea metabolic rates, which are crucial to calculating an individual's food requirements. Such information, which is available only through the use of such novel technologies, will prove valuable to management agencies attempting to halt the population decline of leatherbacks and other marine turtle species.

As Hadfield and his colleagues emphasized in their work on tree snails, a major difficulty in working with an endangered species is that performing experimental studies that may harm individuals is both illegal and unethical. This fact is implicit in many other chapters in this section. In Chapter 6, Allison Alberts and John Phillips discuss studies in which they use the Cuban iguana as a model to develop management strategies for related but more critically endangered species. The authors discuss two conservation techniques they have employed: temporarily altering iguana social structure by removing high-ranking males, and headstarting, a technique whereby individuals are raised in captivity until they reach a larger body size that makes them less vulnerable to predation. Altering social structure increases mating opportunities for less-dominant males and will likely prove useful in species with dominance polygyny or in genetically compromised populations. Meanwhile, headstarting may prove particularly effective in populations, or taxa, in which low juvenile recruitment is the major threat to population recovery. Neither technique had long-term negative effects on the behavior or physiology of individual animals; consequently, when combined with other measures such as predator control, both techniques will likely prove valuable in efforts to conserve critically endangered iguana species.

Because reproduction is key to a species' persistence, methods that stimulate breeding or improve the fertility of endangered species will prove particularly valuable. In Chapter 7, John Cockrem and his colleagues suggest that in addition to field programs that control or remove predators, endocrine studies can also significantly contribute to conservation. The authors describe studies of two high-profile endemic New Zealand birds: the northern brown kiwi and the critically endangered kakapo. By analyzing kiwi corticosterone levels, the authors have provided objective data showing that kiwi held in captivity for breeding and public display are not chronically stressed (chronic stress has previously been a concern of many environmentalists). In addition, by measuring fecal steroid levels in free-living kakapo and by developing methods for using exogenous hormones to stimulate breeding, the authors demonstrate that conservation endocrinology can be applied to the management of critically endangered species.

Control or eradication of introduced species is often essential to endangered species recovery programs. In Chapter 8, Ian Hume, Lesley Gibson, and Steven Lapidge argue that one way to control introduced species is by exploiting the evolutionary differences in physiology that exist between native and introduced species. Using isotopic techniques, Hume and his coworkers measured the energy and water requirements of free-living bilbies and yellow-footed rock-wallabies, two of Australia's endangered marsupials. On the basis of the two native species' water requirements, the authors concluded that neither bilbies nor rock-wallabies require access to free water.

Because the introduced, non-arid-adapted species (e.g., feral goats, rabbits, and cats) that contribute to marsupial population declines do require free water, the authors propose that one way to reduce the negative impact of the introduced species on bilbies and rock-wallabies is simply to reduce the number of artificial water supplies. This simple solution will no doubt create a conflict with ranchers who also rely on artificial water supplies; however, closing artesian bores may be possible in nonranching areas, such as national parks. Such a policy may allow threatened desert-adapted native species to begin recovery in areas inhospitable to introduced species.

Uncovering why a species' numbers are declining often requires multifactorial experimental studies (see Chapter 4). In Chapter 9, Russel Andrews discusses the contribution of such studies to our understanding of declining Steller sea lion populations in Alaska. Andrews, who is part of a large consortium of researchers investigating the declines, outlines his comparative work on two distinct stocks of sea lions, one that has decreased in size by 75% over the past 30 years and a second whose size has remained relatively stable. Andrews deployed remote-sensing devices on breeding female sea lions, which allowed him to monitor the physiology and foraging ecology of individuals while they were at sea. Information recovered from the data loggers, coupled with data from fisheries surveys conducted around sea lion rookeries, suggested that females from declining populations were nutritionally stressed. Whether nutritional stress is the sole cause for declining populations will be established only through further experimental work. As Andrews notes, population declines are more likely a consequence of many interacting factors. Identifying the factors contributing to Steller sea lion declines, and understanding how these factors interact with each other, will best be accomplished though the kinds of multidisciplinary research efforts that Andrews is conducting.

Considering the seven chapters in this section as a whole clearly indicates that most species declines are the result not of a single large perturbation but of the cumulative effect of numerous, often unrelated stressors—"death by a thousand wounds" (Soulé and Orians 2001). Understanding how multiple stressors interact at varied spatial and temporal scales will be crucial in halting species declines. Studying the impacts of many unrelated stressors, however, requires researchers to cross traditional disciplinary boundaries and integrate knowledge across various levels of analysis. Many of the authors do precisely this, and as many of them emphasize in their chapters, successful recovery programs for endangered species will depend on a thorough understanding of each species' ecology, life history, physiology, and genetics. As the authors demonstrate, such understanding is often best gained by augmenting the frequently used and more-traditional non-invasive approach of conservation biology with an experimentally driven research agenda. In this way, experimental biology will continue to make important theoretical and practical contributions to conservation efforts.

REFERENCES

Darwin, C. 1859. *On the Origin of Species by Means of Natural Selection.* London: John Murray.

Mace, G. M., Baillie, J.E.M., Beissinger, S. R., and Redford, K. H. 2001. Assessment and management of species at risk. In *Conservation Biology: Research Priorities for the Next Decade,* ed. M. E. Soulé and G. H. Orians, 11–29. Washington, D.C.: Island Press.

May, R. M., Lawton, J. H, and Stork, N. E. 1995. Assessing extinction rates. In *Extinction Rates,* ed. J. H. Lawton and R. M. May, 1–24. New York: Oxford University Press.

Soulé, M. E., and G. H. Orians. 2001. Conservation biology research: Its challenges and contexts. In *Conservation Biology: Research Priorities for the Next Decade,* ed. M. E. Soulé and G. H. Orians, 271–85. Washington, D.C.: Island Press.

3 Contributions of *Ex Situ* Propagation and Molecular Genetics to Conservation of Hawaiian Tree Snails

Michael G. Hadfield, Brenden S. Holland, and Kevin J. Olival

SUMMARY

The Oʻahu tree snail is the poster child for the massive declines that have occurred among nearly all of Hawaiʻi's approximately 800 species of non-marine snails. More than 75% of Oʻahu tree snails of the genus *Achatinella* have become extinct in the last 50 to 70 years, owing to habitat degradation or destruction, shell collecting, and introduced predators. In our efforts to conserve the remaining species and populations, we have undertaken intense field-demographic studies to better understand the immediate causes of decline; laboratory propagation studies to conserve varieties and species that are rapidly vanishing from native habitats; and molecular-genetic investigations to clarify conservation priorities in terms of within-species variation. Field-demographic studies revealed that the achatinellines have slow growth to late maturity (e.g., 5 years) and low fecundity (approximately five offspring per year) and thus depend on longevity to maintain population numbers. To conserve *Achatinella* species in the laboratory, methods were developed to maintain them in conditions approximating the temperature, humidity, rainfall, and day length in their native habitats. Additionally, an epiphytic mold upon which the snails feed was propagated on agar and provided to the captive populations. The developed protocols succeeded for most, but not all, species. Results from an intensive genetic study of one species revealed sufficiently high within-species variation in some DNA sequences (e.g., cytochrome *c* oxidase I) for use in devising field and laboratory strategies to maximally conserve intraspecific biodiversity.

INTRODUCTION

A major challenge in conducting research on federally listed endangered species is that one may not legally or ethically conduct experiments that might kill any specimens. The restriction in the U.S. Endangered Species Act against "take" specifically forbids such experiments. Working with endangered invertebrates adds other problems, including small body sizes and a dearth of information on pathogens, diseases, and methods for captive husbandry. For most endangered invertebrates, the risk of extinction is exacerbated by the limited information on their life histories, distributions, and abundances. Thus, the conservation biologist faces formidable obstacles in attempting to apply the usual methods and approaches of experimental biology to gain knowledge that is often essential.

We describe here both the approaches we have taken for nearly 30 years to understand the causes of serious population declines and species extinctions among Hawai'i's once-abundant endemic tree snails and the steps we have taken to conserve the remaining species. Beginning in the early 1970s, we carried out the first field-demographic study on any of the nearly 800 species of nonmarine gastropods in Hawai'i (Hadfield and Mountain 1980). This mark–recapture investigation of *Achatinella mustelina*—which, like all members of the genus, is endemic to the island of O'ahu—revealed life-history characteristics that were decidedly non-adaptive in the face of the numerous late-twentieth-century challenges to native Hawaiian environments. Pristine native forests are almost gone on O'ahu; nearly every elevation is characterized by some degree of plant invasions, ranging from nearly 100% of plant species at lower elevations to perhaps 5–10% on the least-disturbed summit ridges (Smith 1985). It is likely that the Hawaiian tree snails faced no predators during the long history of their evolutionary radiation, but they are now eaten wholesale by three rat species and a predatory snail that was introduced from the southeastern United States (Hadfield 1986). Low birth rates, slow growth, and late maturity, which are typical of all achatinelline snails that we have studied, coupled with high natural rates of juvenile mortality in many populations, predispose populations to extinction when predators invade. By killing reproductive animals early in what should be a 10–20-year life span, predators (including human shell-collectors) can quickly eradicate entire populations and species (Hadfield 1986). More than 70% of the *Achatinella* species are already extinct, a figure that is comparable to estimates for the entire Hawaiian nonmarine molluscan fauna (Cowie, Evenhuis, and Christensen 1995).

The data from the first study on *Achatinella mustelina* (Hadfield and

Mountain 1980) contributed toward the effort to place the entire genus on the list of endangered species in 1982. The listing was a victory for conservation, but it also severely limited the kinds of research that could be done to conserve remaining species. Thus our efforts have been constrained to completely nonsacrificial methods in three areas: field-demographic studies to determine life-history characteristics and to serve as controls for efforts to propagate snails in captivity; captive propagation in environmental chambers; and molecular-genetic studies to guide field management practices. We have also monitored climatic conditions at field sites to gather data for setting up environmental chambers in the laboratory.

FIELD STUDIES

We have studied populations of *Achatinella mustelina*, a species restricted to the Wai'anae Range of the island of O'ahu, for 3–18-year periods in three different locations using standard mark–recapture techniques (Hadfield 1986; Hadfield and Mountain 1980; Hadfield, Miller, and Carwile 1993). While small between-site differences were found in traits such as maximum size and fecundity, a general pattern was confirmed in the three studies (Table 3.1). Newly born snails are relatively large (shell length, 4–5 mm), but they grow slowly to a reproductive size of approximately 22 mm over 4–5 years. Once reproductive, the hermaphroditic snails have a birth rate of about four to five offspring per year, or fewer. Survivorship of newly born snails is low, at least at the Pahole site, for which the data are best (Hadfield, Miller, and Carwile 1993), with few of the spring- and summer-born snails surviving the first year. In the absence of predators, populations of *A. mustelina* show a relatively even age-frequency distribution after the first year. Snails that survive for 2–3 years tend to be very long-lived, with life-spans up to 10 years or more. Studies of species in the closely related genus *Partulina* on the island of Moloka'i reveal similar life-history patterns (Hadfield and Miller 1989). Life-span estimates for *Partulina proxima* exceed 15 years (Hadfield and Miller 1989); individuals of *P. redfieldi* marked when already mature were recaptured repeatedly over a 13-year period and provided a solid measurement of life-spans of a minimum of 17 years (assuming a minimum age at maturity of 4 years; the snails could have been much older at first marking) (Hadfield, unpublished data).

The extant achatinelline snails (i.e., species in the four genera of the subfamily Achatinellinae, family Achatinellidae) once lived from coastal lowlands to elevations of about 2000 m on the islands of O'ahu, Moloka'i, Lana'i,

TABLE 3.1 *Life-history characteristics of* Achatinella mustelina *estimated from field data*

Birth size (shell length)	4.5 mm
Mean growth rate	4–5 mm/yr
Mean maximum size	21–23 mm
Survivorship in 1st year	21–35%
Age at 1st reproduction	3–5 yr
Fecundity	4–5 offspring/yr
Life-span	Less than 10 yr

SOURCE: Hadfield, Miller, and Carwile 1993.

Maui, and Hawai'i. No living populations of any achatinelline snail were ever found on Kaho'olawe, Kauai, Ni'ihau, or the Northwestern Hawaiian Islands (Pilsbry and Cooke 1912–1914). Lowland populations were extinct by the time Europeans first arrived in the islands and began giving scientific names to Hawaiian snails, although it is difficult to know at what elevations the lowest populations still existed. Information gleaned from field notes made during the second half of the nineteenth century and up through the 1930s indicate that dense populations of the *Achatinella* species still existed relatively near the growing city of Honolulu at elevations less than 150–180 m. That situation was drastically changed by the time we began our research in 1973. The lowest populations found then were at about 450 m on the leeward slopes of the Ko'olau Range, and, to the best of our knowledge, they have vanished in the last 10–15 years. All populations of the surviving seven to eight *Achatinella* species (the indecision here is a taxonomic one) are found at elevations above 550 m in both mountain ranges of O'ahu, extending up to the peak of Konahuanui in the Ko'olau Range, an elevation of about 960 m. No *Achatinella* species were ever recorded at the summit of Ka'ala, at 1220 m, the highest point in the Wai'anae Range. This narrow elevational range indicates a restricted habitat and climatic range for the remaining O'ahu tree snails. Some achatinelline species we have studied on Hawai'i Island *(Partulina physa)*, Moloka'i *(P. proxima* and *P. redfieldi)*, and Lana'i *(P. variabilis* and *P. semicarinata)* live at elevations up to 1500 m.

Using temperature- and humidity-recording apparatus (hygrothermometers and, recently, Hobo electronic temperature-and-humidity recorders), we have accumulated meteorological data at study sites on three islands. At an elevation of about 800 m in West Maui, the average annual temperature

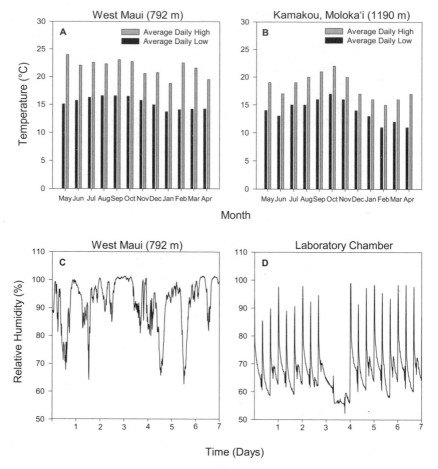

Figure 3.1. Average monthly temperatures for *(A)* the west Maui field site (792 m elevation, 1999–2000) and *(B)* the Kamakou, Moloka'i, field site (1190 m elevation, 1991–1992); relative humidity data for *(C)* the west Maui field site (data collected for 7 days in 1999) and *(D)* a laboratory environmental chamber. Note that relative humidity values in the field fluctuate daily; however, the range shown (60–100%) is representative of values observed throughout the year. Relative humidity data in the laboratory lack the variation seen in the field. Regular spikes correspond to the automated "rainfall" set for three times a day. Once a week there is a simulated dry day when no water is sprayed in the terraria (Day 3). Although the laboratory chamber lacks much of the variation seen in the field, the *range* of humidity is closely approximated.

range was from about 16 °C (nighttime) to about 20 °C (daytime; Figure 3.1A). At an elevation of about 1,200 m on Moloka'i, temperatures averaged somewhat lower but were also moderate (Figure 3.1B). Relative humidity, on the average, varies from about 70% to saturation (Figure 3.1C). On very warm and windy days, relative humidity may drop as low as 50%, but rarely if ever goes lower at these elevations. Thus, the environments of extant Hawaiian tree snails are relatively constant with respect to temperature and humidity. This habitat also supports active growth of the molds that the snails graze from the surfaces of leaves. These environmental data were used as guides in establishing conditions in our laboratory environmental chambers.

In situ conservation efforts have been augmented in recent years for two populations of *Achatinella mustelina*. The Natural Area Reserve System (Hawaii Department of Forestry and Wildlife) constructed predator-repelling fences around the two populations. These fences are 1.2 m high, and their lower edges are buried about 15 cm into the ground. Each fence bears two barriers against incursions by the predatory snail *Euglandina rosea*: a two-wire electric fence, with the wires mounted about 8 mm apart, and a shallow 10-cm-wide trough filled with coarse salt (NaCl). At one of the exclosures, boxes filled with rat bait (diphacinone) are placed both inside and outside. Our research team continues to monitor a population within a 5 m by 5 m quadrat, which we have been studying since 1983 (see Hadfield, Miller, and Carwile 1993) and which is now located within an exclosure of about 400 m². After about 24 months, it appears that such fences can protect field populations, although they require frequent maintenance to fill the salt trough, refill the rat-bait boxes, remove fallen limbs that can bridge the fence, and cut off limbs from nearby trees that are growing across the barrier.

EX SITU CONSERVATION (CAPTIVE PROPAGATION)

The goal of *ex situ* conservation efforts is to remove individuals from seriously endangered populations and bring them into a laboratory setting for propagation. It is anticipated that laboratory populations will increase in size and provide individuals for eventual return to natural habitats. To carry out *ex situ* propagation of endangered species, it is typically necessary to create environments that closely simulate the habitats from which the species were removed. The first part of such an effort must be recreating, in an artificial setting, the physical environment of the species, including day length, substratum, and climate. The second challenge is providing the typical food of the species of concern. We were thus challenged to set up a moist habitat wherein the average temperature varied daily between about 16 °C and 20 °C, where day length averaged 12 hours, and where the snails could feed on their natural diet of molds growing on the surfaces of the leaves of a small number of tree species native to Hawai'i.

In three commercial environmental chambers with controls for both temperature and day length (Precision Scientific Instruments, Revco Scientific, and American Scientific Products), we created a laboratory environment that matched the average for an elevation of approximately 600–650 m in the mountains of Hawai'i. Inside the environmental chambers (capacities 0.5, 0.61, and 1.35 m^3), the snails are kept in terraria of several sizes. A fine spray of water, simulating rainfall, is directed into the terraria for about 2 minutes every 8 hours. The tops and bottoms of the terraria are replaced with nylon window screen so that they do not hold water and so that air flow through them is good. The environmental chambers are equipped with strong fans that maintain relatively even temperatures throughout; this strong air circulation also keeps the terraria from remaining very wet between "rainfalls." The light cycle is set to a constant program of 12 hours of light and 12 hours of dark. Within the terraria, the snails live on small leafy branches from native tree species that are among their common host trees in the field *(Metrosideros polymorpha* and *Freycinetia arborea)*. Branches are collected from the field and replaced in the terraria every 2 weeks, at which time the snails are all censused and the terraria are cleaned. Both plant species remain relatively fresh throughout this period, owing to the frequent watering and relatively low temperatures. In addition to the fungal food occurring naturally on the leaves provided, we propagate a black mold (*Cladosporium* sp.), originally isolated from leaves of a native tree, on potato-dextrose agar and add it to the terraria at each change of leaves. The snails quickly devour the agar-grown mold after it is put into the terraria.

The temperature regime we have established in our environmental chambers compares favorably with that in the field (Figure 3.1A, B). The main difference is, of course, in variability. The daytime highs and nighttime lows in the environmental chambers are close to the *average* temperatures in the field habitats (i.e., 16 °C and 20 °C), although the laboratory chambers do not include seasonal shifts in the means (which are small at Hawaiian latitudes) or the extremes.

The same is true for relative humidity (Figure 3.1C, D). Fans in the chambers, provided to maintain uniform temperatures, produce declines in relative humidity between misting, which is set by timers to occur every 8 hours, 6 days per week. Although the relative-humidity profiles provided by recorders in the chambers indicate low relative humidity most of the time (Figure 3.1D), in fact, humidities remain higher inside the terraria, which are filled with leafy branches.

Kobayashi and Hadfield (1996) compared snail growth when the diet of native molds growing on leaves was and was not augmented with cultured *Cladosporium* sp. in two laboratory populations of Hawaiian tree snails, *Achatinella mustelina* and *Partulina redfieldi*. They also compared growth rates in both laboratory populations with individuals in field populations selected to be close in size to laboratory snails at the beginning of the study. Although laboratory snails of both species grew faster with the mold-augmented diet, the laboratory snails of only one of the species, *P. redfieldi,* grew faster than their counterparts in the field. Thus it may be that laboratory-maintained individuals of *A. mustelina* are food limited in growth rate and in age at maturation, although we have no firm data tying reproductive maturity to size. In fact, the range of sizes of mature snails (those with lipped shells) in the field suggests that maturation is not strictly a size function (Hadfield, Miller, and Carwile et al. 1993). Given the wide variance in growth rate among individuals and within individuals over time observed in field populations of all achatinelline snails studied (Hadfield and Mountain 1980; Hadfield and Miller 1989; Hadfield, Miller, and Carwile 1993), the discrepancy between laboratory and field growth rates seen in *A. mustelina* in Kobayashi and Hadfield's study (1996) may have been simply a sampling effect.

The captive-rearing facility has been used to maintain seven federally listed *Achatinella* species, two candidate species of *Partulina*, five *Partulina* species labeled species of concern, and an additional species of concern, *Newcombia cumingi* (U.S. Fish and Wildlife Service designations), all members of the endemic Hawaiian subfamily Achatinellinae. The success of these efforts has varied greatly: some populations have grown fairly rapidly from the time they were collected (e.g., *Achatinella fuscobasis,* Figure 3.2A), and one was almost nonreproductive over the first 6 years that it was in the laboratory (*A. apexfulva,* Figure 3.2B). Of twenty-two laboratory populations established between 1991 and 2000, none has become extinct, although three have been reduced to only two to four individuals and are unlikely to survive. We have additional, growing populations of two of these species in our facility. By contrast, some populations of some achatinelline species have burgeoned, to the point where our facility is nearing maximum capacity. On October 26, 2001, there were 868 achatinelline snails in fourteen species in our laboratory.

The laboratory population of *A. apexfulva,* a species thought to be

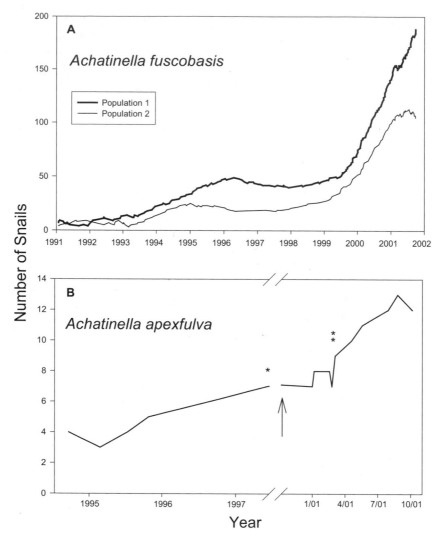

Figure 3.2. *(A)* Two laboratory populations of *Achatinella fuscobasis* showing successful population growth over an 11-year period. *(B)* Laboratory population of *Achatinella apexfulva* showing minimal population growth for the first 5 years (despite the presence of reproductively mature individuals). The axis is interrupted from May 1997 to October 2000, during which time no births occurred. The arrow indicates the date (October 2000) when the temperature was increased by 4 °C day and night. Three months later, the birth rate increased significantly. The asterisks indicate dates when three new adults were added to the population from the field.

extinct in the wild, was a major concern up through 2000. Although it included several mature animals and survived well in the environmental chamber, the population increased only from four to seven snails between 1994 and early 2001. A very low birth rate was offset by mortality through most of this period. In October 2000, the lab population of *A. apexfulva* was moved to a separate environmental chamber, and the temperature cycle was gradually increased to a 20 °C nighttime temperature and a 24 °C daytime temperature. By January 2001, the snails had begun to reproduce more frequently, so that the population in October 2001 included twelve snails (see Figure 3.2B). The birth rate changed significantly after the habitat was warmed; four snails were born to this population between 1994 and the end of 2000, and six were born during 2001. Low snail numbers precluded division of the laboratory population into controls maintained on the 16/20 °C cycle and an experimental group in a warmer regime. This problem will obviously occur in many attempts to apply the approaches of experimental biology to the conservation of severely endangered animals.

The success of a captive-propagation experiment is judged by comparison of the dynamics of laboratory populations, which are the experiment, with those of field populations, which represent a control. When a concerted effort is made to duplicate field-climate conditions in the lab, the main experimental variable becomes the absence of the living, biologically complex forest. In our case, we bring some of this complexity to the laboratory in the form of fresh leafy branches for the snails to live upon in their terraria. Another major variable in the laboratory terraria is the absence of predators, which are active at nearly all field sites for remaining Hawaiian tree snails, at least periodically. To examine our success in duplicating field conditions, we need to compare the experimental regime with the field control in terms of environmental parameters and then examine the population dynamics of the snails as a measure of our success.

In the field, achatinelline tree snails are known to remain in a single tree for long periods of time, even for life. They will rarely, if ever, crawl down from one tree to ascend another, and migration probably occurs most often when high winds knock snails from their home trees and they wander across the ground until they find another tree (Pilsbry and Cooke 1912–1914). Thus the tree species that any snail lives upon will be constant throughout a typical life-span. Most of the snails in our lab were taken from a single tree species in the field, *Metrosideros polymorpha*, and this species has provided the branches and leaves for our lab populations since we began this effort in the late 1980s. One snail species, *A. fuscobasis,* was taken from a mixed vegetation of *M. polymorpha* and *Freycinetia arborea,* so leaves from the latter vinelike pandanus are also routinely added to the terraria. *F. arborea* was once a common host plant for many *Achatinella* species (e.g., Pilsbry and Cooke 1912–1914). Altogether, the major difference between the laboratory and field habitats is the constancy in the former (although perhaps one should also consider the continuous low hum and slight vibration from the motors that cool and warm the chambers and the fans that circulate air in them). How do snail demographies compare?

Figure 3.3A graphically illustrates the growth trend of a one-tree population of *P. redfieldi* at about 1200 m elevation on the island of Moloka'i. This population was experiencing no predation during the illustrated 4-year interval of the mark–recapture study, and it grew from around 40 snails to about 120 (population sizes estimated by the Manly–Parr multiple-recapture method; Begon 1979). In the same graph is plotted a line illustrating the growth of a laboratory population of *P. redfieldi*, begun with individuals taken from trees close to those harboring the field population illustrated. The laboratory population underwent virtually no growth for approximately 2.5 years, principally owing to the absence of adults. Fortunately, mortality was very low. By the beginning of the third year in captivity, ten individuals had matured and the birth rate began to climb. Once the population began to grow, it increased at a rate similar to that measured for the same species in the field, which demonstrates that the lab population is currently demographically equivalent to predator-free field populations and is thus "successful."

The *Achatinella* species that has performed best in our laboratory facility, *A. fuscobasis*, may be extinct in the field (Figure 3.2A). The two laboratory populations of *A. fuscobasis* were started from 11 snails brought from the Ko'olau Mountains on O'ahu in 1991. These snails, ten mature individuals and one subadult, were separated into two terraria, seven in one (labeled Population 1), and the remaining four in the other (Population 2). For the first 2 years, despite births and initial growth, the populations declined, to four snails in Population 1 and three in Population 2. However, by 1994, both populations were growing, although not rapidly. A stable or slightly declining period of about 3 years for both populations ended in 1999, when the snails began to increase in numbers dramatically. Since early 1999, the total number of *A. fuscobasis* has more than doubled, from 127 to 293. Populations of three additional endangered species, *Achatinella lila, A. livida,* and *A. decipiens,* and the candidate species *Partulina variabilis* from Lana'i are also showing significant increases in size often after several years of little or no growth (Figure 3.4). An "adaptation period" may be a real characteristic of captive populations of achatinelline tree snails.

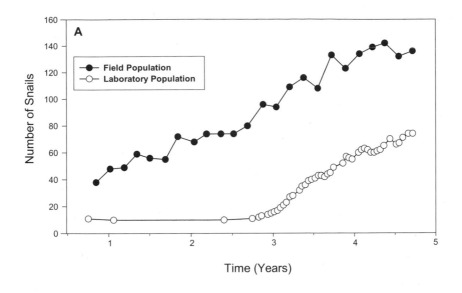

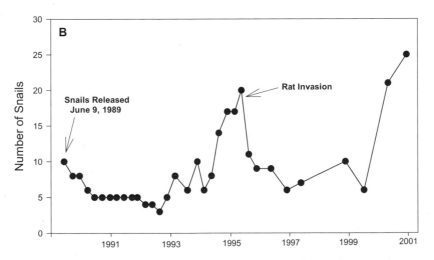

Figure 3.3. *(A)* Growth of field (Manly–Parr estimate based on multiple mark–recapture data; Begon 1979) and laboratory populations of *Partulina redfieldi* from Kamakou, Molokaʻi. Note that after an initial 2-year period during which some of the field-collected snails became reproductively mature, the growth rate of the laboratory population was as fast as or faster than the field population. *(B)* Population dynamics of a single-tree population of *P. redfieldi* derived from ten small, laboratory-born snails in 1989. The parents of these snails had been collected from nearby trees and taken to the laboratory in 1986 and 1987.

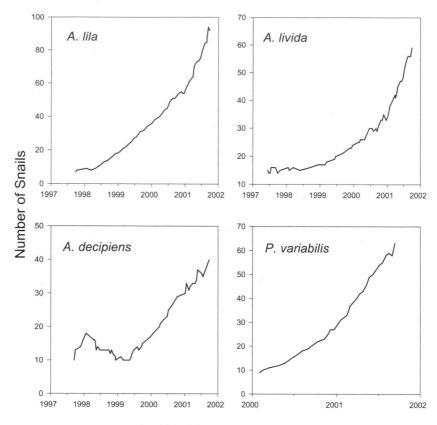

Figure 3.4. Positive growth of four laboratory populations (three *Achatinella* spp. from Oʻahu and one *Partulina* sp. from Lanaʻi). The captive population of *P. variabilis* was begun only in February 2000, whereas the other populations were started in mid-1997.

REINTRODUCTION

The ultimate experiment in captive rearing is the attempt to reintroduce laboratory-born snails to the field. Because the goal of captive propagation is not to build a large snail zoo but rather to provide animals to augment field populations or to start new ones where snails had once existed, this experiment is critical. To determine if such an effort could succeed, we attempted to establish a field population of *Partulina redfieldi* in the Kamakou meadow on Molokaʻi. During 1987 and 1988, adult *P. redfieldi* were brought from the field site—but not the specific study trees—to the laboratory for the purpose of collecting their progeny. As soon as these snails had produced one or two young, the adults were returned to the field. The progeny were held in the laboratory until ten were available for reintroduction. These snails ranged in age from 6 to 18 months and in size from 5.93 mm to 14.80 mm. On June 9, 1989, they were taken to the Moloka'i meadow and placed in a small ohi'a tree that had been searched and found to harbor no snails at that time.

The introduced population of *P. redfieldi* has been followed from its initiation in 1989 to the present (the most recent census was in December 2000); its progress is shown graphically in Figure 3.3B. During the first 4 years, seven of the snails disappeared. Because all were color marked, it would have been possible to recognize these snails as dead ground shells, but none were located in extensive searches beneath the release tree on each visit. In 1993, the three remaining captive-bred snails became reproductively mature, and the population began to grow. In late 1995, 20 snails, all progeny of the three survivors, were present. At that time, rats invaded the Kamakou meadow and decimated the tree snails in all the study trees. The rat-chewed shells of these victims were mostly recovered from the ground beneath the tree. Subsequently, an intensive rat-baiting program maintained by staff with the Nature Conservancy succeeded in bringing the predators under control. Since 1999, the population of *P. redfieldi* in the study tree has grown again and is now larger than at any previous time. Additionally, all age classes are present, as one would expect of a demographically healthy population. While only a single effort for a single species, this experiment provides evidence that it is possible to introduce captive-bred achatinelline snails into the wild to establish new populations. We intend to do this only with genetic stock originally taken from near the site of reintroduction.

MOLECULAR-GENETIC STUDIES FOR FIELD MANAGEMENT

Data obtained using neutral molecular markers can be vital to conservation biology because they provide information regarding the distribution of genetic variation among populations or taxa (Harvey and Steers 1999). In situations where a choice among populations for conservation efforts is necessary, information regarding the partitioning of genetic variation may be a critical criterion in making that choice (Baverstock and Moritz 1996). Therefore, data concerning the phylogeographic structure of fragmented Hawaiian tree snail populations are needed for the formulation of strategies for conservation and management of the remaining species (Holland and Hadfield 2002).

In a conservation-genetics study of the O'ahu tree snail *Achatinella mustelina*, population structure was inferred for seventy specimens from eighteen populations using mitochondrial DNA (mtDNA) sequences (Hol-

TABLE 3.2 *Phylogeographic analysis of* Achatinella mustelina

Pairwise genetic distance summary, also expressed as uncorrected pairwise sequence divergence (%). Values were determined using a Kimura 2-parameter substitution model and the program MEGA 2.1 (Kumar et al. 2001). ESUs represent natural clusters of closely related populations.

Hierarchical Level of Pairwise Comparison	Range	Mean
Among all individuals (70 individuals)	0.0–0.053 (0.0%–5.3%)	0.029 (2.9%)
Within populations (18 populations)	0.0–0.0194 (0.0%–1.94%)	0.0048 (0.48%)
Among populations (18 populations)	0.00–0.044 (0.0%–4.4%)	0.029 (2.9%)
Within ESUs (6 ESUs)	0.001–0.014 (0.1%–1.4%)	0.006 (0.6%)
Among ESUs (6 ESUs)	0.017–0.044 (1.7%–4.4%)	0.034 (3.4%)

SOURCE: Holland and Hadfield 2002.

land and Hadfield 2002). Field sampling of tissues from the tree snails was conducted using a nonlethal technique, as described by Thacker and Hadfield (2000). Tissue samples were placed in 80% ethanol in the field and transported to the laboratory for nucleic acid extraction. Polymerase chain reaction (PCR) amplification of target gene fragments was performed using universal primers for cytochrome *c* oxidase subunit I (COI) (Folmer et al. 1994) and verified by agarose gel electrophoresis. Six-hundred and seventy-five base pair COI fragments were purified with QIAquick-spin columns (QIAGEN Inc., Valencia, Calif.) according to the manufacturer's protocol. PCR-amplified gene fragments were cycle sequenced using PCR primers. Molecular evolutionary genetic analysis (MEGA, version 2.1; Kumar et al. 2001) was used with a Kimura two-parameter model to generate genetic distance matrices and to evaluate intra- and interpopulation level relationships (Table 3.2).

The concept of evolutionarily significant units (ESUs) provides a phylogenetic framework for prioritization of conservation decisions (Avise 2000). A wide range of operational criteria for defining intraspecific ESUs exists in the relevant scientific literature. In this study, we have adhered to the explicit recommendation of Moritz (1994) that ESUs be defined by groups of populations that exhibit reciprocal monophyly based on mtDNA data. The six clusters of populations depicted in Figure 3.5 were found to be reciprocally monophyletic in COI gene trees. Also in accordance with Moritz

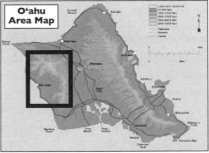

Figure 3.5. Map of the Waiʻanae Range on the island of
Oʻahu (inset), with the positions of eighteen populations
of *Achatinella mustelina* that were sampled. *(A–F)* Six
genetically defined evolutionarily significant units (ESUs)
corresponding to six distinct mitochondrial haplotypes.
Maximum within-ESU sequence divergence was 1% in
all instances.

(1994), we distinguish between ESUs and management units (MUs) in that ESUs are concerned with historical population structure and long-term conservation needs. In that MUs are used to address short-term management issues, such groupings are not being employed in our efforts. Each of the six ESUs depicted is restricted to a distinct mtDNA haplotype. The tree snail populations were found to be highly structured geographically, which indicates long-term separation among these units. The inferred mtDNA phylogeny was supported by genetic distance values calculated using the Kimura two-parameter substitution model (Table 3.2). The among-ESU mean genetic distance was nearly six times the within-ESU mean interpopulation genetic distance. The six designated ESUs are being used to formulate long-term management strategies and prioritize populations for focused conservation efforts, including *in situ* protection and more extensive *ex situ* conservation to preserve intraspecific biodiversity.

For example, as conservation and recovery strategies for *A. mustelina* are developed, one of the long-term goals will include construction of additional predator-exclusion structures. Such structures surround natural stands of native trees that harbor tree snail populations. Decisions regarding placement of new predator exclosures will be based in part on preserving maximum genetic diversity and on maintaining independent lineages of *A. mustelina*, as defined by data on ESUs. Genetically based ESUs may also serve to guide captive propagation, future geographic translocations, and eventual release of captive-bred tree snails.

DISCUSSION AND CONCLUSIONS

Tree snails of the family Partulidae—native to islands spanning almost the entire tropical Pacific Ocean, from the Mariana Islands to the Marquesas— have also been the subject of intense *ex situ* conservation efforts at a number of locations in the United States and the United Kingdom (Mace, Pearce-Kelly, and Clarke 1998). The laboratory rearing effort was begun to conduct genetic studies of *Partula* and *Samoana* species from French Polynesia but became the paramount conservation effort when the predatory snail *E. rosea* was introduced to Tahiti, Moorea, and other islands in the region, again in a terribly misguided attempt to use biological control to eradicate the giant African snail, *Achatina fulica*, a garden pest (Clarke, Murray, and Johnson 1984; Murray et al. 1988). At this time, captive populations of seventeen partulid species represent the sole survivors, as they are extinct in the wild (P. Pearce-Kelly, Zoological Society of London, personal communication, October 2001).

Although most of the forty-one species of *Achatinella,* all endemic to the island of Oʻahu, have already passed into extinction, there is hope for the remaining seven to eight species through captive propagation, genetic management of field populations including translocation of snails between populations to increase intrapopulational heterogeneity, rigid protection of field populations by construction of predator-excluding fences, and initiation of new field populations in suitable areas now empty of native snails and their predators. Only through the combination of field-demographic studies coupled with meteorological monitoring of field sites, laboratory propagation guided by field habitat data and comparisons of demographies of field and lab populations, and the application of modern molecular genetics with non-lethal tissue sampling have we been able to devise such strategies.

ACKNOWLEDGMENTS

Funds to support these researches have come from the U.S. Fish and Wildlife Service for the captive-rearing efforts and field studies; the Hawaii Department of Forestry and Wildlife for supplies and constructing field exclosures for *in situ* conservation; and the U.S. Army Garrison, Hawaiʻi, for molecular-genetic studies of *A. mustelina.* Genetic sampling of eighteen field populations of *A. mustelina* would have been impossible without the assistance of V. Costello, K. Kawelo, and J. Rohrer of the U.S. Army's Department of Public Works, Hawaii, and T. Takahama of the Hawaii Department of Forestry and Wildlife's Natural Area Reserve System. Far too many colleagues, graduate students, and friends have assisted in fieldwork through the last 30 years for us to name them all, but we are grateful for their help. J. Murray and G. Roderick provided useful suggestions for improving the manuscript, for which we are most grateful

REFERENCES

Avise, J. C. 2000. *Phylogeography: The History and Formation of Species.* Cambridge: Harvard University Press.

Baverstock, P. R., and Moritz, C. 1996. Project design. In *Molecular Systematics,* ed. D. M. Hillis, C. Moritz, and B. K. Mable, 17–27. Sunderland, Mass.: Sinauer Associates.

Begon, M. 1979. *Investigating Animal Abundance: Capture-Recapture for Biologists.* Baltimore: University Park Press.

Clarke, B., Murray, J., and Johnson, J. S. 1984. The extinction of endemic species by a program of biological control. *Pacific Sci.* 38:97–104.

Cowie, R. H., Evenhuis, N. L., and Christensen, C. C. 1995. *Catalog of the*

Native Land and Freshwater Molluscs of the Hawaiian Islands. Leiden: Backhuys.

Folmer, O., Black, M., Hoeh, W., Lutz, R., and Vrijenhoek, R. 1994. DNA primers for amplification of mitochondrial cytochrome *c* oxidase subunit I from diverse metazoan invertebrates. *Mol. Mar. Biol. Biotechnol.* 3:294–99.

Hadfield, M. G. 1986. Extinction in Hawaiian achatinelline snails. *Malacologia* 27:67–81.

Hadfield, M. G., and Miller, S. E. 1989. Demographic studies on Hawaii's endangered tree snails: *Partulina proxima. Pacific Sci.* 43:1–16.

Hadfield, M. G., Miller, S. E., and Carwile, A. H. 1993. Decimation of endemic Hawai'ian tree snails by alien predators. *Am. Zool.* 33:610–22.

Hadfield, M. G., and Mountain, B. S. 1980. A field study of a vanishing species, *Achatinella mustelina* (Gastropoda, Pulmonata), in the Waianae Mountains of Oahu. *Pacific Sci.* 34:345–58.

Harvey, P. H., and Steers, H. 1999. One use of phylogenies for conservation biologists: Inferring population history from gene sequences. In *Genetics and the Extinction of Species,* ed. L. V. Landwebber and A. P. Dobson, 101–20. Princeton, N.J.: Princeton University Press.

Holland, B. S., and Hadfield, M. G. 2002. Islands within an island: Phylogeography and conservation genetics of the endangered Hawaiian tree snail *Achatinella mustelina. Mol. Ecol.* 11:365–76.

Kobayashi, S. R., and Hadfield, M. G. 1996. An experimental study of growth and reproduction in the Hawaiian tree snails *Achatinella mustelina* and *Partulina redfieldi* (Achatinellinae). *Pacific Sci.* 50:339–54.

Kumar, S., Tamura, K., Jakobsen, I. B., and Nei, M. 2001. MEGA2: Molecular Evolutionary Genetics Analysis software. Tempe: Arizona State University.

Mace, G. M., Pearce-Kelly, P., and Clarke, D. 1998. An integrated conservation programme for the tree snails *(Partulidae)* of Polynesia: A review of captive and wild elements. *J. Conchol.* (special publication) 2:89–96.

Moritz, C. 1994. Defining "evolutionarily significant units" for conservation. *TREE* 9:373–75.

Murray, J., Murray, E., Johnson, M. S., and Clarke, B. 1988. The extinction of *Partula* on Moorea. *Pacific Sci.* 42:150–53.

Pilsbry, H. A., and Cooke, C. M., Jr. 1912–1914. Achatinellidae. In *Manual of Conchology,* 2d ser., vol. 22. Philadelphia: Academy of Natural Sciences.

Smith, C. W. 1985. Impact of alien plants on Hawai'i's native biota. In *Hawai'i's Terrestrial Ecosystems: Preservation and Management,* ed. C. P. Stone and J. M. Scott, 180–250. Honolulu: Cooperative National Park Resources Studies Unit, University of Hawaii.

Thacker, R. W., and Hadfield, M. G. 2000. Mitochondrial phylogeny of extant Hawaiian tree snails. *Mol. Phylogenet. Evol.* 16:263–70.

4 Multiple Causes for Declining Amphibian Populations

Andrew R. Blaustein, Audrey C. Hatch, Lisa K. Belden, and Joseph M. Kiesecker

SUMMARY

Amphibian population declines are part of an overall decline in biodiversity. Populations of amphibians are declining in all regions where they are found. No single cause for these declines has been documented. Rather, increasing evidence suggests that amphibian population declines are the result of multiple insults interacting with one another. Habitat destruction, global climate changes (including changes in temperature and precipitation), increasing levels of ultraviolet radiation, contaminants, disease, and introduced nonnative species all appear to play a role in amphibian population declines. These factors can act alone or in concert with one another. Moreover, amphibian populations may be declining for different reasons in different regions. The amphibian population decline phenomenon appears to be influenced by a complex array of dynamic interactions. In some cases experimental evidence clearly shows a cause contributing to a particular decline. In other cases, the causes remain unclear. Here, we briefly review studies addressing the causes for amphibian population declines. Whenever possible, we discuss studies that incorporated experimental evidence to address these declines.

INTRODUCTION

Large losses in biodiversity are being documented around the world in almost all classes of plants and animals, mainly due to habitat destruction (Lawton and May 1995). Though the exact number of species being lost is not known, the current rate of extinction is estimated to be greater than any known in the last 100,000 years (Wilson 1992; Eldridge 1998). As part of this overall biodiversity crisis, populations of many amphibian species are in decline, and some species appear to have gone extinct (Blaustein, Wake, and Sousa 1994a; Alford and Richards 1999; Houlahan et al. 2000). In fact, one estimate is that more than 500 populations of frogs, toads, and salamanders on five continents are in decline (Alford and Richards 1999). A number of factors have contributed to these declines, including habitat destruction, pollution, disease, introduced species, and global environmental changes (Alford and Richards 1999; Blaustein and Kiesecker 2002; Blaustein, Belden, Olsen, et al. 2001; Reaser and Blaustein forthcoming). There does not seem to be a single cause for amphibian population declines. Moreover, single agents may act alone or in combination with other factors (Blaustein and Kiesecker 2002).

Amphibians are considered to be bioindicators of environmental change and contamination for a number of reasons (Blaustein 1994). They have permeable, exposed skin (not covered by scales, hair, or feathers) and eggs (not covered by hard or leathery shells) that may readily absorb substances from the environment. The complex life cycles of many species expose them to both aquatic and terrestrial toxicants. Moreover, these attributes and the fact that amphibians are ectotherms may make them especially sensitive to changes in temperature and precipitation and to increases in ultraviolet radiation. Environmental changes affecting amphibians may also affect other organisms.

Habitat destruction is an obvious cause for many amphibian population declines and has been discussed in detail elsewhere (see review in Alford and Richards 1999). Therefore, in this chapter, we summarize the other major causes for amphibian population declines. We focus on studies that used experimental approaches to address the amphibian population decline phenomenon and on those studies in which overt habitat destruction does not seem to play a major role in the decline.

CLIMATE CHANGE

The diversity of locations where amphibian populations have declined has led to investigations concerning global environmental change. Changes in global temperature, precipitation, and levels of ultraviolet radiation may contribute to amphibian population declines. Several investigators have applied a wide array of tools to remote-sensing information databases to examine the interrelationships among climate, ultraviolet radiation, and amphibian population declines. For example, Alexander and Eischeid (2001) examined the relationship between amphibian declines and climate variations in Colorado (U.S.), Puerto Rico, Costa Rica, Panama, and Queensland (Australia) using information gathered from airplanes, land stations, satellites, ships, and weather balloons, plus outputs from a weather forecast model. They concluded that although amphibian declines occurred concurrently with temperature and precipitation anomalies, these anomalies were within the range of normal variability. Moreover, they stated that unusual climate measured as regional estimates of temperature and precipitation is unlikely to be a direct cause for amphibian population declines in the areas they examined.

Stallard (2001) measured time series data sets for Puerto Rico that extended into the 1980s. These data included forest cover; annual mean, minimum, and maximum daily temperatures; annual rainfall; rain and stream chemistry; and atmospheric dust transport. He also used satellite imagery and air chemistry samples from an aircraft flight across the Caribbean. None of the data sets showed changes extreme enough to directly cause amphibian population declines. Stallard called for more experimental research to further examine the amphibian population decline problem.

Several investigators have assessed the breeding biology of amphibians to determine whether they have been affected by climate change. Climate change may have obvious effects on breeding biology, including time of breeding, length of breeding, and hatching success of eggs. Any changes in these parameters may eventually affect an amphibian population. For example, Beebee (1995), plotting the start of breeding activities for six amphibian species in southern England over 16 years, provided evidence that amphibians in temperate countries might be responding to climate change by breeding earlier. Similarly, Gibbs and Breisch (2001) showed that daily temperatures near Ithaca, New York, increased over the twentieth century and that several species of anurans have shifted their breeding patterns accordingly. Gibbs and Breisch studied differences in calling dates during two time periods, 1900–1912 and 1990–1999. They found that during the later period, four species of anurans called 10–13 days earlier than they had during the earlier period, two were unchanged, and none called later.

Not all studies found the earlier-breeding patterns described above. For example, although the main arrival to breeding sites of common toads *(Bufo bufo)* in England was highly correlated with the mean daily temperatures over the 40 days immediately preceding the main arrival, Reading (1998) found no significant trend toward earlier breeding in recent years compared with previous years. At a site in Oregon, western toads *(Bufo boreas)* showed no significant trend toward breeding increasingly early, despite increasing temperatures (Blaustein, Belden, Olson, et al. 2001). At four other sites, neither western toads nor Cascades frogs *(Rana cascadae)* showed statistically significant positive trends toward earlier breeding, even though at three of these four sites breeding time was associated with warmer temperatures. The spring peeper *(Pseudacris crucifer)* in Michigan did not show a statistically significant trend toward breeding earlier but did show a significant positive relationship between breeding time and temperature. Fowler's toads *(Bufo fowleri)* in eastern Canada did not show a trend toward breeding earlier, nor was there a positive relationship between breeding time and temperature.

The general pattern suggested from breeding studies is that some temperate-zone amphibian populations show a trend toward breeding earlier, whereas others do not. The reasons for this variation are not known.

Pounds, Fogden, and Campbell (1999) illustrated the complex interrelationships among global environmental changes and amphibian population declines. They found that changes in water availability associated with changes in large-scale climate processes, such as the El Niño/Southern Oscillation (ENSO) may have significantly affected amphibian, reptile, and bird populations in a Costa Rican cloud forest. They showed that extended dry periods associated with global warming are correlated with amphibian and reptile losses and changes in the bird community. In Costa Rica and possibly in other high-altitude tropical sites, global warming appears to have resulted in a decrease in the amount of mist precipitation received in the forest, owing to the increased altitude of the cloud bank.

A recent experimental field study in Oregon (U.S.) illustrates the complex interrelationships among environmental change, UV radiation, pathogen infection, and amphibian egg mortality (Kiesecker, Blaustein, and Belden 2001). Kiesecker, Blaustein, and Belden linked ENSO events with decreased winter precipitation in the Oregon Cascade Range and suggested that less winter snowpack resulted in lower water levels when western toads *(Bufo boreas)* breed in early spring. Toad embryos developing in shallower water are exposed to higher levels of UV-B radiation, which results in increased mortality from a pathogenic oomycete, *Saprolegnia.*

Amphibians may be particularly sensitive to climatic change because they are ectotherms. Changes in ambient temperature may influence amphibian behaviors, including those related to reproduction. Thus, changes in global ambient temperature could disrupt the timing of breeding, periods of hibernation, and the ability to find food (Donnelly and Crump 1998; Blaustein, Belden, Olson, et al. 2001). Moreover, climate change may affect the types of predators and competitors amphibians come in contact with.

ULTRAVIOLET RADIATION

Just as climate change has varied effects on numerous organisms, so might increasing UV-B (280–315 nm) radiation. At the terrestrial surface, UV-B radiation is extremely important biologically. Critical biomolecules absorb light of higher wavelength (UV-A; 315–400 nm) less efficiently, and stratospheric ozone absorbs most light of lower wavelength (UV-C; 200–280 nm) (Cockell and Blaustein 2001). UV-B radiation can cause mutations and cell death. At the individual level, UV-B radiation can slow growth rates, cause immune dysfunction, and induce sublethal damage (Tevini 1993).

Over evolutionary time, UV radiation has been a ubiquitous stressor on living organisms and has probably exerted selection pressure resulting in the evolution of defenses against its effects (Cockell 2001). Natural events such as comet and asteroid impacts, volcanic activity, and cosmic events such as supernova explosions and solar flares can cause large-scale ozone depletion with accompanying increases in UV radiation (Cockell 2001). However, these natural events are transient and may have significant effects only for a few years.

These events are clearly different from human-induced production of chlorofluorocarbons (CFCs) and other chemicals that continuously deplete stratospheric ozone, inducing long-term increases in UV-B radiation at the surface. Decreases in stratospheric ozone, climate warming, and lake acidification (which leads to decreases in dissolved organic carbon concentrations, e.g., Schindler et al. 1996) all result in increasing levels of UV radiation. In fact, information from several sources indicates that levels of UV-B radiation have recently risen significantly (especially since 1979) in both tropical and temperate regions (Kerr and McElroy 1993; Herman et al. 1996; Middleton et al. 2001).

Spectral measurements made at Toronto, Canada, indicate a 35% increase in UV-B radiation per year in winter and a 7% increase in summer from 1989 to 1993 (Kerr and McElroy 1993). These increases were caused by a downward trend in ozone that was measured at Toronto during the same time. Moreover, increases in UV-B in late spring may have a disproportionately larger effect on some species if they occur at critical phases of the species' development (Kerr and McElroy 1993). This suggestion is consonant with the results of several studies showing the adverse effects of UV-B radiation in spring-conducted field experiments on developing amphibian embryos (discussed below).

Middleton et al. (2001) assessed trends in solar UV-B radiation over twenty sites in Central and South America derived from the Total Ozone Mapping Spectrometer satellite data. They showed that annually averaged UV-B doses, as well as the maximum values, have been increasing in both regions since 1979. The UV index was consistently higher for Central America, where many amphibian declines have been reported (Lips 1998; Pounds, Fogden, and Campbell 1999).

UV-B radiation adversely affects a wide variety of plants, animals, and microorganisms (e.g., Tevini 1993; Cockell and Blaustein 2001). For example, recent studies have demonstrated that UV radiation affects photosynthesis in plankton, contributes to coral bleaching, and is lethal to a number of species of invertebrates, fishes, and amphibians (van der Leun, Tang, and Tevini 1998; Cockell and Blaustein 2001). Moreover, UV radiation may interact with other stressors to affect a wide array of organisms and even entire ecological communities (Tevini 1993; van der Leun, Tang, and Tevini 1998; Cockell and Blaustein 2001).

Living organisms have had less than 100 years to cope with a significant human-induced rise in UV radiation. Moreover, the combined effects of UV radiation with pollutants and other agents that have been on Earth for a relatively short time may be especially damaging to organisms that have not had time to develop mechanisms to survive their effects.

UV-B Radiation: Experimental Studies

Investigators at numerous sites worldwide have shown that ambient UV-B radiation decreases the hatching success of some amphibian species, in the field, at natural oviposition sites (reviewed in Blaustein et al. 1998; Blaustein, Belden, Hatch, et al. 2001). These investigators typically placed fertilized eggs in enclosures with filters that either removed UV-B radiation or allowed UV-B to penetrate (control filters) (Blaustein et al. 1998). In some studies, enclosures with no filters were used as an additional control. Researchers compared the hatching success of eggs under each regime.

These studies have demonstrated that the embryos of some species are more susceptible to UV-B radiation than others (Blaustein et al. 1998; Table 4.1 in this chapter). For example, in the Pacific Northwest (U.S.), the hatching success of Cascades frogs *(R. cascadae)*, western toads *(B. boreas)*, and long-toed *(Ambystoma macrodactylum)* and northwestern *(A. gracile)* salamanders was lower when the eggs were exposed to ambient UV-B radiation than when eggs were shielded from UV-B (Blaustein et al. 1998). However, the hatching success of spotted *(R. pretiosa* and *R. luteiventris)*, red-legged *(R. aurora)*, and Pacific tree *(Hyla regilla)* frogs was not significantly different in the UV-shielded and UV-exposed treatments (Blaustein et al. 1998). In California, the hatching success of Pacific tree frogs was not affected by ambient levels of UV-B radiation, but hatching success was lower in California tree frogs *(Hyla cadaverina)* and California newts *(Taricha torosa)* exposed to UV-B (Anzalone, Kats, and Gordon 1998). The hatching success of common toads *(B. bufo)* in Spain was lower in UV-B-exposed eggs than in those shielded from UV-B, whereas UV-B had no effect on the hatching success of the natterjack toad *(B. calamita)* (Lizana and Pedraza 1998). In Finland, the hatching success of moor frogs *(Rana arvalis)* increased when embryos were shielded from UV-B, but shielding the embryos of common toads *(B. bufo)* and common frogs *(Rana temporaria)* from UV-B had no effect (Häkkinen, Pasanen, and Kukkonen 2001).

As the studies described above illustrate, species cope differently with UV-B radiation (Table 4.1). For some species, in field experiments, hatching success is lower in eggs exposed to UV-B radiation compared with shielded controls. In other species, hatching success is not affected by UV-B exposure. These studies are not contradictory, as some investigators have erroneously reported (Corn 1998; Carey 2000). Rather, these studies illustrate clear interspecific differences in tolerance to UV-B radiation at early life stages. In fact, within the same study, conducted at the same time and same site, embryos of some species have been shown to be sensitive to UV-B, whereas the embryos of other species were resistant (e.g., Blaustein, Hoffman, et al. 1994; Anzalone, Kats, and Gordon 1998; Lizana and Pedraza 1998).

UV-B Radiation Causes Sublethal Damage

Even though hatching rates of some species appear unaffected by ambient UV radiation in field experiments, an increasing number of studies illustrate a variety of sublethal effects due to UV exposure (Blaustein et al. 2003). For example, when exposed to UV-B radiation, amphibians may change their behavior (Nagl and Hofer 1997; Blaustein et al. 2000; Kats et al. 2000); and growth and development may be slowed (e.g., Belden, Wildy, and Blaustein 2000; Pahkala, Laurila, and Merilä 2000, 2001; Smith, Waters, and Rettig 2000). UV can induce developmental and physiological malformations (e.g., Worrest and Kimeldorf 1976; Hays et al. 1996; Blaustein, Kiesecker, Chivers, et al. 1997; Fite et al. 1998). Sublethal effects may become evident even in species whose embryos appeared to be resistant in field experiments.

Several experimental studies illustrate that early exposure to UV-B radiation causes delayed effects in later stages. For example, UV-B radiation did not influence hatching success of plains leopard frogs *(Rana blairi)*, but growth and development are slower in tadpoles when they were exposed to the highest levels of UV-B radiation as embryos (Smith, Waters, and Rettig

TABLE 4.1 *Experimental studies in the field investigating the effect of UV-B radiation on hatching success of amphibians*

Species	UV Effect on Hatching Success	Location	Population Status	Investigators
Rana cascadae Cascades frog	Decreases	Oregon, U.S.	Declining	Blaustein, Hoffman, Hokit, et al. 1994
Rana arvalis moor frog	Decreases	Finland	Declining	Häkkinen, Pasanen, and Kukkonen 2001
Rana temporaria common frog	None	Finland	Declining	Häkkinen, Pasanen, and Kukkonen 2001
Rana aurora red-legged frog	None	Oregon, U.S.	Declining	Blaustein et al. 1996
Rana aurora	None	Western Canada	Declining	Ovaska, Davis, and Flamarique 1997
Rana pretiosa Western spotted frog	None	Washington, U.S.	Declining	Blaustein et al. 1999
Rana luteiventris Columbia spotted frog	None	Washington, U.S.	Declining	Blaustein et al. 1999
Rana sylvatica wood frog	None	North Carolina, U.S.	Unknown	Starnes et al. 2000
Bufo boreas western toad	Decreases	Oregon, U.S.	Declining	Blaustein, Hoffman, Hokit, et al. 1994
Bufo boreas	None	Colorado, U.S.	Declining	Corn 1998
Bufo bufo common toad	Decreases	Spain	Declining	Lizana and Pedraza, 1998
Bufo bufo[1]	None	Finland	Declining	Häkkinen et al. 2001
Bufo calamita natterjack toad	None	Spain	Declining in some areas	Lizana and Pedraza 1998
Hyla regilla Pacific tree frog	None	Oregon, U.S.; California, U.S.; Western Canada	Persistent	Blaustein et al. 1994b; Anzalone. et al. 1998; Ovaska et al. 1997

TABLE 4.1 *(continued)*

Species	UV Effect on Hatching Success	Location	Population Status	Investigators
Hyla cadaverina California tree frog	Decreases	California, U.S.	Unknown	Anzalone et al. 1998
Hyla chrysoscelis gray tree frog	None	North Carolina, U.S.	Unknown	Starnes et al. 2000
Pseudacris triseriata chorus frog	None	North Carolina, U.S.	Unknown	Starnes et al. 2000
Crineria signifera common froglet	Decreases	Australia	Persistent	Broomhall et. al. 2000
Litoria verreauxii alpine tree frog	Decreases	Australia	Declining	Broomhall et. al. 2000
Litoria aurea bell frog	Uncertain[2]	Australia	Declining	van de Mortel and Buttemer 1996
Litoria dentata Keferstein's tree frog	None	Australia	Persistent	van de Mortel and Buttemer 1996
Litoria peronii Peron's tree frog	None	Australia	Persistent	van de Mortel and Buttemer, 1996
Crinia signifera eastern froglet	Decreases	Australia	Persistent	Broomhall et al. 2000
Ambystoma macrodactylum long-toed salamander	Decreases	Oregon, U.S.	Unknown	Blaustein et al. 1997a
Ambystoma maculatum spotted salamander	None	North Carolina, U.S.	Unknown	Starnes et al. 2000
Ambystoma gracile northwestern salamander	Decreases	Oregon, U.S.	Unknown	Blaustein et al. 1995
Taricha torosa California newt	Decreases	California, U.S.	Unknown	Anzalone et al. 1998

[1] Significant mortality of larvae exposed to UV-B compared with controls shielded from UV-B
[2] Possible effects (see van de Mortel and Buttemer 1996)

2000). Exposure of *Rana temporaria* embryos to UV-B radiation showed no effects on survival rates, frequency of developmental anomalies, or hatching size. However, larvae exposed to UV-B radiation as embryos displayed an increased frequency of developmental anomalies, metamorphosed later, and were smaller than larvae shielded from UV-B as embryos (Pahkala, Laurila, and Merilä 2001). Similarly, UV-B radiation had no effects on hatching success in red-legged *(R. aurora)* frogs (Blaustein et al. 1996). However, larvae exposed to UV-B radiation as embryos were smaller and less developed than those shielded from UV-B radiation (Belden and Blaustein 2002a).

Delayed growth and development after exposure to UV-B radiation may significantly affect populations of amphibians that live in ephemeral habitats. For example, if growth and development are slowed significantly and amphibians cannot metamorphose and move to land before a pond dries or freezes, significant mortality may occur (Blaustein, Wildy, et al. 2001).

Some species of amphibians may be more resistant to UV-B radiation than others because they evolved mechanisms to counteract its harmful effects. Thus, behavioral, physiological, anatomical, and molecular adaptations can minimize damage caused by UV-B radiation (Blaustein and Kiesecker 1997; Blaustein, Belden, Hatch, et al. 2001). For example, embryos of some amphibian species may be more resistant to UV-B because they can repair UV-induced DNA damage more efficiently than others (Blaustein, Hoffman, et al. 1994; Blaustein, Belden, Hatch, et al. 2001; van de Mortel et al. 1998). One important repair process, enzymatic photoreactivation, uses the enzyme CPD-photolyase to remove the most frequent UV-induced lesion in DNA, cyclobutane pyrimidine dimers (CPDs) (Blaustein, Belden, Hatch, et al. 2001). CPD-photolyase appears to be the first level of defense against CPDs for many organisms exposed to sunlight (Blaustein, Belden, Hatch, et al. 2001). Because photoreactivation is probably the most important repair mechanism in amphibians, those species with the highest photolyase activities are likely to be the most resistant to UV damage. In fact, the amount of CPD-photolyase in eggs is positively correlated with survival of embryos in field experiments (Blaustein, Belden, Hatch, et al. 2001). For example, eggs of the most-resistant species in field experiments (e.g., *H. regilla, Litoria peronii, L. dentata, R. aurora, R. pretiosa,* and *R. luteiventris*) have higher CPD-photolyase activity than eggs of more susceptible species (e.g., *H. cadaverina, L. aurea, R. cascadae, B. boreas, A. macrodactylum,* and *A. gracile*).

Even if eggs are laid in the open at high altitudes (where UV levels may be high) and have long developmental periods during which they are subjected to prolonged UV-B exposure, they may not be adversely affected by

UV-B radiation if they can repair DNA damage efficiently. However, species with low photolyase levels may be especially sensitive to UV-B radiation even if they live at very low altitudes and receive relatively low doses of UV-B radiation (Blaustein et al. 1998). Furthermore, within a species, individuals from one population may differ from members of another population in their sensitivity to UV-B radiation (Belden et al. 2000; Belden and Blaustein 2002a,b). This variation may be due to differences in their ability to repair DNA damage; differences in their physiology, behavior, or both; or differences in various ecological parameters.

Although DNA repair is one parsimonious explanation for the resistance of some species to UV-B radiation and the susceptibility of others, it is becoming increasingly clear that behavioral and ecological attributes may limit UV-B damage to amphibians (Blaustein and Belden 2003). For example, some species wrap their eggs in leaves, which protect them from UV-B radiation (Marco et al. 2001). Many salamander species, some frogs, and some toads lay their eggs under leaf litter, in crevices, or under logs where they are not exposed to high levels of sunlight. Many amphibians lay their eggs in muddy water, where light penetration may not be great enough to damage eggs. Thus, in some clear high-mountain lakes of Oregon, where many UV studies have taken place, the potential for UV penetration is greater than in some relatively turbid sea-level ponds, where many species in the eastern United States lay their eggs (Stebbins 1954).

The effects of UV-B radiation on amphibians are a significant cause for concern. UV-B radiation may act alone or in concert with contaminants, pathogens, or changes in climate to adversely affect amphibians. As described above, the hatching success of many species is lower when their eggs are exposed to UV-B radiation. Sublethal effects may develop after exposure to UV-B radiation. Sublethal damage may eventually lead to death if it significantly alters the physiology, anatomy, and behavior of an individual. For example, sublethal damage may affect an individual's ability to find food, avoid predators, or seek shelter (e.g., Fite et al. 1998; Kats et al. 2000). Continued mortality of embryos or significant sublethal damage in many individuals may eventually contribute to a population decline.

CONTAMINANTS

A wide array of contaminants and low pH can affect amphibians. Contaminants include pesticides, herbicides, fungicides, fertilizers, and numerous pollutants (Blaustein, Kiesecker, Hoffman, et al. 1997; Sparling et al. 2000; Boone and Bridges 2003). Toxic substances acting alone can severely affect amphibians in a variety of ways. They can kill amphibians, affect their behavior, reduce their growth rates, act as endocrine disruptors, or induce immunosuppression (Alford and Richards 1999).

Contaminants transported atmospherically have the potential to harm amphibians. In remote, relatively undisturbed environments, atmospheric transport may be the major route of exposure. Although levels of pollutants from atmospheric deposition are typically low, these levels may still significantly impair developing amphibians under complex ecological conditions (e.g., Relyea and Mills 2001). Pounds and Crump (1994) suggested that atmospheric scavenging of contaminants by clouds might concentrate contaminants and release them in remote areas such as Monteverde, Costa Rica, where numerous amphibian populations have declined (Pounds, Fogden, and Campbell 1999). This concentrating effect may be particularly important under unusually hot, dry conditions (Pounds and Crump 1994; Pounds, Fogden, and Campbell 1999). Atmospheric deposition of organophosphate pesticides from California's highly agricultural Central Valley may have contributed to frog declines (Aston and Seiber 1997; Sparling et al. 2001). Pesticides may adhere to foliage, where they threaten amphibian species (Aston and Seiber 1997). Activity of the enzyme cholinesterase in Pacific tree frogs *(H. regilla)* was impaired in areas where ranid frogs were in decline in California. Sparling et al. (2001) suggested that this impairment might be linked to the presence of organophosphate pesticides. Davidson, Shaffer, and Jennings (2001) concluded that patterns of decline in red-legged frogs in California were most likely caused by pesticides carried downwind from the Central Valley.

Contaminants can also act locally. For example, recent experiments have shown that nitrate and nitrite fertilizers affect survival, behavior, and development in a number of amphibians in North America (e.g., Hecnar 1995; Marco and Blaustein 1999; Marco, Quilchano, and Blaustein 1999; Hatch et al. 2001; Hatch and Blaustein 2003). Fertilizers can affect both terrestrial- and aquatic-stage amphibians.

Contaminants can interact with other agents synergistically. For example, a number of investigators have examined the combined effects of UV radiation and contaminants on amphibians (e.g., Kagan, Kagan, and Buhse 1984; Ankley et al. 1998; Hatch and Burton 1998; Walker, Taylor, and Oris 1998; Zaga et al. 1998). Synergism may be especially important when developing amphibians have reduced ability to respond to one stressor in the presence of another. For example, increasing levels of UV-B radiation may reduce the ability of developing amphibians to cope with other environmental insults such as acidification. Because acidification alone can adversely affect amphibians (Pierce 1985; Harte and Hoffman 1989; Dunson, Wyman, and Corbett 1992; Kiesecker 1996), there may be synergistic effects between low pH and UV-B radiation in regions where acid pollution is a concern. Thus, in one study when UV intensity was experimentally increased to levels expected at high elevations and combined with low pH, amphibian survival was decreased (Long, Saylor, and Soule 1995). No significant effects were attributed to either factor alone.

Synergism between UV radiation and pollutants may also occur when the former enhances the toxicity of the latter. Chemical contaminants that absorb strongly in a portion of the UV spectrum are most likely to be phototoxic. Because UV may penetrate freshwater to significant depths, phototoxicity is especially relevant.

There are different general mechanisms by which UV radiation can enhance the toxicity of contaminants (Blaustein, Belden, Hatch, et al. 2001). For example, UV can directly alter the chemical structure of a contaminant to a more toxic form (e.g., Zaga et al. 1998). Alternatively, toxicity can result when chemicals accumulate in biological tissue and undergo transformation when an organism is exposed to UV radiation (e.g., Bowling et al. 1983).

Polycyclic aromatic hydrocarbons (PAHs) are multiple-ringed hydrocarbons that may contaminate small bodies of water (e.g., ponds, streams) by way of road runoff, direct industrial discharge, or atmospheric deposition. PAHs absorb UV-A (315–400 nm) and are acutely toxic because they cause singlet oxygen to form within the cell. In the presence of sunlight, some PAHs (such as anthracene, benzo[a]pyrene, and fluoranthene) can be highly toxic to aquatic animals (Bowling et al. 1983), including amphibians (Hatch and Burton 1998; Monson et al. 1999; Walker, Taylor, and Oris 1998).

In laboratory experiments, UV-A enhanced the toxicity of anthracene, benzo[a]pyrene, and 1,12-benz[a]anthraquinone to larval newts *(Pleurodeles waltl)* exposed for 6 days (Fernandez and l'Haridon 1992). Similarly, exposure to UV-A after newts were exposed to benzo[a]pyrene resulted in greater toxicity than did exposure to benzo[a]pyrene alone (Fernandez and l'Haridon 1994). The toxic mechanism involves a reaction that occurs within biological tissue, rather than an external alteration of PAH chemical structure by UV-A.

Hatch and Burton (1998) experimentally investigated the effects of fluoranthene with and without UV on amphibians in both laboratory and outdoor exposures. In the laboratory, African clawed frogs *(Xenopus laevis)* exhibited the most deformities (including abdominal edema and gut malformations) due to the combination of UV and fluoranthene, whereas the spotted salamander *(Ambystoma maculatum)* exhibited no deformities.

Outdoor experiments demonstrated that newly hatched larvae were more sensitive than embryos to phototoxic fluoranthene. The mortality risk was correlated with fluoranthene and UV-intensity levels. Full sunlight exposures and increasing fluoranthene levels greatly increased mortality rates for *A. maculatum* and *X. laevis*.

Walker, Taylor, and Oris (1998) found behavioral and histological effects of fluoranthene in the presence of UV on bullfrog *(Rana catesbeiana)* larvae. A 2-day exposure to low levels of fluoranthene in the presence of UV caused structural disorganization at the microscopic level in the skin. After 2 days of high fluoranthene exposure with UV, larvae exhibited hyperactivity.

Zaga et al. (1998) investigated the synergism between the insecticide carbaryl and UV-B on *X. laevis* and grey tree frog *(Hyla versicolor)* embryos and larvae using a solar simulator. Carbaryl strongly absorbs UV-B radiation. Swimming activity of larvae was decreased by UV-B alone and by UV-B in combination with carbaryl. *Xenopus laevis* demonstrated increased swimming with increasing carbaryl concentration. This response was reversed with concurrent exposure to UV-B: the larvae reduced their activity.

Zaga et al. (1998) further investigated the photoactivation of carbaryl and the photosensitization of *X. laevis* by irradiating carbaryl before exposure to amphibians. Mortality increased significantly when the chemical was irradiated, which suggests that the mechanism of toxicity involves photoproducts of carbaryl. In the photosensitization experiment, *X. laevis* were exposed to nonirradiated carbaryl without UV and then transferred to clean water with concurrent UV exposure. Mortality was correlated with increasing levels of carbaryl and UV-B.

Interactions among several agents simultaneously can affect amphibians in a variety of ways that can contribute to population declines. For example, Hatch and Blaustein (2000) showed that survival and activity levels of frog larvae were significantly reduced when the larvae were exposed to low pH, high nitrate levels, and UV-B radiation together.

These studies demonstrate potentially striking interactions between UV radiation and environmental contaminants. Understanding the mechanisms of the UV–chemical interactions is vital to predicting the potential effects of phototoxic chemicals on amphibians in the environment.

PATHOGENS

Although little is known about the effects of pathogens on wild amphibians, a number of studies show that pathogens contribute significantly to declining amphibian populations. A variety of pathogens may affect wild amphibian populations, including viruses, bacteria, trematode parasites, protozoans, oomycetes, and fungi (e.g., Smith et al. 1986; Blaustein, Hokit, et al. 1994; Drury, Gough, and Cunningham 1995; Jancovich et al. 1997; Berger et al. 1998; Daszak et al. 1999; Longcore, Pessier, and Nichols 1999; Pessier et al. 1999; Kiesecker and Skelly 2001; Johnson et al. 2002). These pathogens can be the proximate causes of mortality, or they can cause sublethal damage such as severe developmental and physiological deformities.

In some cases, massive mortality of amphibians has been attributed to pathogens. For example, substantial mortality attributable to the bacterium *Aeromonas hydrophilum* was documented in *Rana sylvatica* tadpoles in Rhode Island (Nyman 1986), in a population of *R. muscosa* in California (Bradford 1991), and perhaps in populations of *B. boreas* in Colorado (Carey 1993). Several reports suggest that disease of unknown origin may be playing a role in the decline of tropical amphibian populations (e.g., Laurance, McDonald, and Speare 1996; Lips 1998; Pounds, Fogden, and Campbell 1999). Two pathogens have received a great deal of recent attention with regard to amphibian population declines; a chytridiomycete fungus and a funguslike oomycete, *Saprolegnia.*

The chytridiomycete *Batrachochytrium dendrobatidis*, which is found in several areas where population declines have occurred, is fatal to frogs under certain experimental conditions (e.g., Berger et al. 1998). Chytrids appear to affect both tadpoles and adults (Berger et al. 1998; Longcore, Pessier, and Nichols 1999). Berger et al. (1998) suggested that *B. dendrobatidis* is fatal only after metamorphosis, when the skin becomes keratinized and infection results in cutaneous chytridiomycosis. However, recent experimental evidence suggests that tadpoles of the western toad *(Bufo boreas)* are highly susceptible to *Batrachochytrium* and that mortality occurs rapidly after exposure (Blaustein, unpublished data). A number of investigators are examining the hypothesis that this fungus is the proximate cause for amphibian population declines in various regions.

Saprolegnia has long been known to affect amphibian eggs, embryos, and larvae and, ultimately, whole populations of amphibians (Bragg 1958, 1962; Blaustein, Hokit, et al. 1994). Recent experimental evidence suggests that *Saprolegnia* may affect amphibians in several ways and that its infection rate and virulence depend upon a number of factors. For example, in field experiments in which the spatial position and time of egg laying were manipulated, the highest mortality from *Saprolegnia* infection was found in eggs laid later and in closer proximity to communal masses (Kiesecker and Blaustein 1997a). Susceptibility to *Saprolegnia* is enhanced when developing eggs are exposed to UV-B radiation (Kiesecker and Blaustein 1995). One of the main vectors of *Saprolegnia* infection appears to be introduced fishes (Blaustein, Hokit, et al. 1994; Kiesecker, Blaustein, and Miller 2001). Laboratory experiments have shown that hatchery-reared rainbow trout *(Oncorhynchus mykiss)* can transmit *S. ferax* to developing amphibians and to soil substrate. Amphibian embryos exposed to either infected fish or infected soil were more likely to develop *Saprolegnia* infections and had higher mortality rates compared to embryos exposed to control conditions. Furthermore, different strains of *Saprolegnia* may have different virulence (Kiesecker, Blaustein, and Miller 2001).

Other studies suggest that introduced species act as vectors for pathogen transmission to amphibians. For example, in the San Rafael Valley in southern Arizona, populations of the Sonora tiger salamander *(Ambystoma tigrinum stebbinsi)* have experienced decimating epizootics (Jancovich et al. 1997). An iridovirus *(Ambystoma tigrinum* virus, ATV) was isolated from diseased tiger salamanders and determined to be the cause of the epizootics. Jancovich et al. (1997) speculated that the origin of ATV at their sites may have been introduced fish, bullfrogs, or possibly introduced salamanders that are used as fish bait.

Declines of amphibian populations in Australia have been attributed to viruses and fungi (Laurance, McDonald, and Speare 1996; Alford and Richards 1999). Laurence, McDonald, and Speare (1996) suggested that a pathogen might have been introduced from aquarium fish.

It is likely, in some cases, that pathogens have been introduced by fishes or other amphibian species or with other introduced vectors. It is important to note that introduced species (e.g., fish) are often found at sites where disease outbreaks and catastrophic population declines of amphibians have been observed (e.g., the mountains of western Oregon [Blaustein, Hokit, et al. 1994]; the rainforest of eastern Australia [Laurance, McDonald, and Speare 1996]; montane streams of Central and South America [LaMarca and Reinthaler 1991]). Clearly more work is needed to assess the role of introduced species as vectors for amphibian disease.

Besides causing mortality in amphibians, *Saprolegnia* and other pathogens may have significant effects in ecological communities. For example, *Saprolegnia* differentially affects larval recruitment in different amphibian species. Thus, Kiesecker and Blaustein (1999) showed that larval recruitment of *R. cascadae* was reduced by almost half in the presence of *Saprolegnia*, whereas larval recruitment in *H. regilla* was not affected. In the absence of *Saprolegnia*, *R. cascadae* had strong negative effects on growth, development, and survival in *H. regilla*. However, in the presence of *Saprolegnia*, the outcome of competition between *R. cascadae* and *H. regilla* was reversed. These results suggest that pathogens can strongly affect species interactions and that they may significantly affect community structure.

INTRODUCED SPECIES

Introduced species may have a number of effects on native amphibians. A few representative studies illustrate how introduced species may affect amphibian populations. Population declines of ranid frogs native to the western United States have been reported by a number of investigators (e.g., Moyle 1973; Fisher and Shaffer 1996). One hypothesis often invoked to explain these declines is the introduction of nonnative bullfrogs *(Rana catesbeiana)*, which may compete with and prey upon other frog species (e.g., Moyle 1973; Kiesecker and Blaustein 1998). Bullfrogs are native to eastern North America, occurring naturally as far west as the Great Plains (Stebbins 1966). However, they have been widely introduced to the western United States (Kiesecker and Blaustein 1998). The effects of bullfrogs on native frog populations are often unclear because at many locations bullfrogs have been introduced in areas where the habitat has been altered or where fishes have also been introduced (Hayes and Jennings 1986; Adams 1999; Kiesecker, Blaustein, and Miller 2001). However, several recent experimental studies have generated information on the mechanisms of interactions between bullfrogs and native species (Kiesecker and Blaustein 1997b, 1998; Kupferberg 1997; Adams 1999; Kiesecker, Blaustein, and Miller 2001).

In the Willamette Valley of Oregon, the impact of bullfrogs on native red-legged frogs *(Rana aurora)* is complicated. Both direct and indirect interactions occur between these species. Moreover, these interactions are compounded by habitat modifications that appear to promote the success of introduced species and intensify interactions between bullfrogs and red-legged frogs (Kiesecker, Blaustein, and Miller 2001). Surveys of forty-two historical red-legged frog breeding sites in the Willamette Valley indicate that red-legged frogs are absent from about 69% of their historical breeding sites (Kiesecker, unpublished data). Breeding populations of bullfrogs are found at about 83% of the sites presently unoccupied by red-legged frogs. Competition between red-legged frogs and bullfrogs or predation of bullfrogs on red-legged frogs may account for their inverse relationship in Oregon.

In the presence of bullfrog larvae and adults, red-legged frogs appear to alter their use of microhabitat, making them more susceptible to fish predation (Kiesecker and Blaustein 1998). Laboratory experiments illustrate that red-legged frog tadpoles from populations that are syntopic with bullfrogs display antipredator behaviors when presented with chemical cues of bullfrogs (Kiesecker and Blaustein 1997b). In contrast, red-legged frog tadpoles from populations that are allotopic to bullfrogs do not. Field and laboratory experiments illustrate that these behavioral differences result in higher rates of predation in tadpoles from allotopic populations (Kiesecker and Blaustein 1997b, 1998). Thus, individuals (e.g., red-legged frogs in Oregon) that are unfamiliar with novel introduced predators (e.g., bullfrogs in Oregon) may not possess adaptations that would prevent a negative encounter. However, even red-legged frogs that are syntopic with bullfrogs accrue costs when displaying antipredator behavior. Thus, in field experiments, *R. aurora* tadpoles that shifted their microhabitat use in response to bullfrog presence grew more slowly and were at increased risk of predation by introduced fish. Moreover, survivorship of red-legged frogs was significantly affected only when they were exposed to the combined effects of bullfrog larvae and adults or bullfrog larvae and smallmouth bass. The interaction between stages (larval and adult) or species (bullfrog and smallmouth bass) produced indirect effects that were greater than when each factor was considered separately.

Habitat alteration, which frequently favors bullfrogs, influences the interaction of red-legged frogs and bullfrogs. For example, in the western United States, large, ephemeral wetlands are often converted to smaller permanent ponds that become good breeding sites for bullfrogs (Richter et al. 1997; Adams 1999; Kiesecker, Blaustein, and Miller 2001). These ponds may contain less shallow water and emergent vegetation, which tends to be clumped along a narrow area at the edge of the pond. Reducing and clumping vegetation seems to increase competition between larval bullfrogs and red-legged frogs (Kiesecker, Blaustein, and Miller 2001) and may intensify predation of adult bullfrogs on larval and juvenile red-legged frogs (Kiesecker and Blaustein 1998).

Amphibian population declines associated with introduced bullfrogs have also occurred in California. For example, yellow-legged frogs *(Rana boylii)* in California were nearly an order of magnitude less abundant in stream reaches where bullfrogs were found (Kupferberg 1997). Furthermore, bullfrogs had negative effects on the growth and survival of *R. boylii*. Kupferberg (1997) found that competitive interactions between these species were mediated by bullfrog-induced changes in algal resources. Bullfrogs also had negative effects on growth of native Pacific tree frogs *(Hyla regilla)* (Kupferberg 1997). Bullfrogs have been introduced into other regions of the world, including parts of Central and South America, Italy, and the Netherlands, where they may have similar affects on other native species (e.g., Albertini and Lanza 1987; Stumpel 1992).

Introduced fish may also have significant direct effects on amphibian populations. Numerous studies (observational, experimental, or both) have shown that fish introductions significantly impact amphibian populations. In the western United States, high-elevation lakes are of particular concern because most were historically fishless, yet they are regularly stocked with nonnative salmonid fishes (rainbow trout, *O. mykiss*, native to western North America; brook trout, *Salvelinus fontinalis*, native to eastern North America; and brown trout, *Salmo trutta*, native to Europe and Asia). In high-elevation lakes and ponds of the Sierra Nevada of California, strong evidence suggests that introduced salmonids have negatively affected populations of the mountain yellow-legged frog *(R. muscosa)*. Population declines and disappearances of *R. muscosa* have been reported from many locations in the Sierra Nevada, including sites within Yosemite, Sequoia, and Kings Canyon National Parks and from wilderness areas outside these parks (Bradford, Tabatabai, and Graber 1993; Knapp, Mathews, and Sarnelle 2001).

In observational studies, Bradford (1989) and Bradford et al. (1998) suggested that declines of *R. muscosa* may have resulted from tadpole predation by introduced salmonids because of the strong overlap between sites that can successfully receive fish and sites that meet the habitat requirements of *R. muscosa*. Even more evidence of the negative impacts of fishes on amphibians has been gathered from recent experimental studies that have incorporated fish-removal regimes with subsequent observations of amphibian populations. Thus, Knapp and Mathews (1998) have removed entire trout populations from select lakes within the Sierra Nevada and are assessing the recolonization of amphibian populations. Initial results suggest that after removal of predatory fish, populations of mountain yellow-legged frogs recover. These results add to the growing body of studies that evaluate the recovery of mountain lake ecosystems following the removal of nonnative fishes (e.g., Parker, Wilhelm, and Schindler 1996; Frank and Dunlap 1999; Drake and Naiman 2000; Knapp, Mathews, and Sarnelle 2001).

Amphibian population declines associated with introduced predatory fish have also been found in other parts of the western United States. For example, Tyler, Liss, Ganio, et al. (1998) found reductions in larval long-toed salamander *(Ambystoma macrodactylum)* densities in lakes with introduced trout in North Cascades National Park, Washington. In a laboratory experiment, Tyler, Liss, Hoffman, et al. (1998) demonstrated that survivorship and size of long-toed and northwestern salamander *(Ambystoma gracile)* larvae were lower in tanks with trout. Furthermore, both species restricted their habitat use when in the presence of trout.

Amphibian population declines in Australia may also occur with the introduction of fishes into historically fishless waters. In New South Wales, Gillespie (2001) suggested that the geographic pattern of spotted tree frog *(Litoria spenceri)* declines may be related to the introduction of brown *(S. trutta)* and rainbow trout *(O. mykiss)*. Results of field experiments support this hypothesis. Native fishes consumed no or few spotted tree frog tadpoles, whereas introduced trout significantly reduced survivorship of tadpoles, even though alternative prey and tadpole refugia were available.

Because of their broad tolerances (Otto 1973) and their effectiveness as mosquito predators, mosquitofish *(Gambusia affinis)* have been introduced throughout warm regions of the world. However, they appear to have detrimental effects on amphibian populations (Gamradt and Kats 1996; Goodsell and Kats 1999; Komak and Crossland 2000). Gamradt and Kats (1996) compared streams in Los Angeles County, which in the 1980s had populations of the California newt *(Taricha torosa)* but not populations of introduced predators. In this system, the impacts of mosquitofish seem to working simultaneously with impacts of the introduced crayfish *(Procambarus clarkii)*. Laboratory and field experiments confirmed that both introduced species were effective predators of larval newts and that crayfish were successful egg predators as well (Gamradt and Kats 1996; Gamradt, Kats, and Anzalone 1997). Moreover, in Australia, where mosquitofish have been introduced to control mosquitoes, they appear to be detrimental to native amphibian populations (Komak and Crossland 2000).

COMPLEX INTERACTIONS

It is evident from many of the studies discussed above that amphibian population declines are the result of extremely complex interactions, which we are just beginning to understand. For example, one potential consequence of global warming is the increased spread of infectious disease (Cunningham et al. 1996; Epstein 1997). Wider-spread disease may occur, for example, when rising temperatures affect the distribution of the vectors of a pathogen, making additional hosts susceptible, or when an environmental agent renders a host's immune system more susceptible. Changes in precipitation, acidification, and contaminants and increased UV-B radiation are some of the stressors that may affect the immune and endocrine systems of amphibians (Figure 4.1). Damage to these systems by multiple stressors could affect amphibians directly or make them more susceptible to pathogens. New

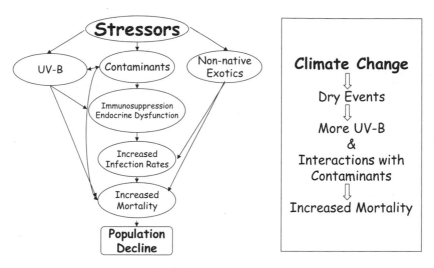

Figure 4.1. Complex interactions involved in amphibian population declines discussed in this paper. This figure describes several scenarios that could contribute to amphibian population declines. Stressors such as UV-B radiation, contaminants, or nonnative exotic species may directly affect amphibians, or they may act in concert with one another. For example, UV-B radiation or contaminants may kill amphibians alone or they may interact with one another synergistically. Moreover, they could adversely affect the immune and endocrine systems, which in turn could make amphibians more susceptible to disease. Continued mortality could eventually lead to a population decline. The right portion of the figure depicts one possible scenario discussed in this paper that relates to the effects of global climate change on amphibian populations. In this example, dry events lead to lower water levels, which lead to greater exposure to UV-B or potentially to a greater concentration of contaminants that may interact with UV-B radiation. Many other scenarios are possible.

pathogens, whose ranges may change as, for example, temperatures rise globally, may be especially significant. Thus, numerous cofactors may be involved in triggering disease outbreaks. As changes in the environment occur, new diseases may emerge or presently benign pathogens may become virulent.

The parallel studies by Pounds, Fogden, and Campbell (1999) and Kiesecker, Blaustein, and Belden (2001) illustrate the complex interrelationships among environmental parameters and amphibian population declines. These studies suggest that global environmental changes, such as changes in temperature and precipitation, increasing levels of UV-B radiation, and pathogens (often coming from introduced species such as fish), may contribute to amphibian population declines at a single location. Similar events may be mirrored at other locations.

We argue that many amphibian population declines are the result not of a single cause but rather of multiple stressors acting in concert with one another. There are no clearly documented cases of how these interactions affect long-term population dynamics per se. In fact, the agents affecting amphibians may have different effects on populations because those agents impact different life stages (Biek et al. 2002; Vonesh and De la Cruz 2002). Which stages are affected depends upon the life-history characteristics of the species. For example, species especially prone to damage at the egg stage may be more affected at the population level if they lay relatively few eggs. What is clear though is that studies of single stressors alone or studies of a single life-history stage may not provide an overall understanding of how amphibians are affected by environmental stressors in complex environments.

Long-term studies, preferably with experimental components, on the effects of multiple stressors are necessary to more fully understand how amphibian populations are affected so that conservation efforts can be implemented to ameliorate the problem. We suggest that multifactorial experimental tests, conducted in the field, are the most rigorous means for determining the causes of amphibian population declines.

ACKNOWLEDGMENTS

Financial support was provided by NSF grant IBN-9977063 and the Katherine Bisbee II Fund of the Oregon Community Foundation. We thank Rachelle Pascua, Barbara Han, Walter Kurtz, Bill Kilgore, Ben Willard, and Lance Johnson for their help.

REFERENCES

Adams, M. J. 1999. Correlated factors in amphibian decline: Exotic species and habitat change in western Washington. *J. Wildl. Manage.* 63:1162–71.

Albertini, G., and Lanza, B. 1987. *Rana catesbeiana* Shaw, in Italy. *Alytes* 6:117–29.

Alexander, M. A., and Eischeid, J. K. 2001. Climate variability in regions of amphibian declines. *Cons. Biol.* 15:930–42.

Alford, R. A., and Richards, S. J. 1999. Global amphibian declines: A problem in applied ecology. *Annu. Rev. Ecol. System.* 30:133–65.

Ankley, G. T., Tietge, J. E., DeFoe, D. L., Jensen, K. M., Holcombe, G. W., Durham, E. J., and Diamond, S. A. 1998. Effects of ultraviolet light and methoprene on survival and development of *Rana pipiens. Environ. Toxicol. Chem.* 17:2530–42.

Anzalone, C. R., Kats, L. B., and Gordon, M. S. 1998. Effects of solar UV-B radiation on embryonic development in *Hyla cadaverina, Hyla regilla,* and *Taricha torosa. Cons. Biol.* 12:646–53.

Aston, L. S., and Seiber, J. N. 1997. Fate of summertime airborne organophosphate pesticide residues in the Sierra Nevada Mountains. *J. Environ. Qual.* 26:1483–92.

Beebee, T.J.C. 1995. Amphibian breeding and climate. *Nature* 374:219–20.

Belden, L. K., and Blaustein, A. R. 2002a. Exposure of red-legged frog embryos to ambient UV-B in the field negatively affects larval growth and development. *Oecologia* 130:551–54.

———. 2002b. Population differences in sensitivity to UV-B radiation in larval long-toed salamanders. *Ecology* 83:1586–90.

Belden, L. K., Wildy, E. L., and Blaustein, A. R. 2000. Growth, survival, and behaviour of larval long-toed salamanders *(Ambystoma macrodactylum)* exposed to ambient levels of UV-B radiation. *J. Zool.* (London) 251:473–79.

Berger, L., Speare, R., Daszak, P., Green, D. E., Cunningham, A. A., Goggin, C. L., Slocombe, R., Ragan, M. A., Hyatt, A. D., McDonald, K. R., Hines, H. B., Lips, K. R., Marantelli, G., and Parkes, H. 1998. Chytridiomycosis causes amphibian mortality associated with population declines in the rain forests of Australia and Central America. *Proc. Nat. Acad. Sci.* (USA) 95:9031–36.

Biek, R., Funk, C. W., Maxell, B. A., and Mills, S. L. 2002. What is missing in amphibian decline research? Insights from ecological sensitivity analysis. *Cons. Biol.* 16:728–34.

Blaustein, A. R. 1994. Chicken Little or Nero's fiddle? A perspective on declining amphibian populations. *Herpetologica* 50:85–97.

Blaustein, A. R., and Belden, L. K. 2003. Amphibian defenses against UV-B radiation. *Evol. Develop.* 5:89–97.

Blaustein, A. R., Belden, L. K., Hatch, A. C., Kats, L. B., Hoffman, P. D., Hays, J. B., Marco, A., Chivers, D. P., and Kiesecker, J. M. 2001. Ultraviolet radiation and amphibians. In *Ecosystems, Evolution, and Ultraviolet Radiation,* ed. C. S. Cockell and A. R. Blaustein, 63–79. New York: Springer.

Blaustein, A. R., Belden, L. K., Olson, D. H., Green, D. L., Root, T. L., and Kiesecker, J. M. 2001. Amphibian breeding and climate change. *Cons. Biol.* 15:1804–9.

Blaustein, A. R., Chivers, D. P., Kats, L. B., and Kiesecker, J. M. 2000. Effects of ultraviolet radiation on locomotion and orientation in roughskin newts *(Taricha granulosa). Ethology* 108:227–34.

Blaustein, A. R., Edmond, B., Kiesecker, J. M., Beatty, J. J., and Hokit, D. G. 1995. Ambient ultraviolet radiation causes mortality in salamander eggs. *Ecol. Appl.* 5:740–43.

Blaustein, A. R., Hatch, A. C., Belden, L. K., Scheessele, E., and Kiesecker, J. M. 2003. Global change: Challenges facing amphibians. In *Amphibian Conser-*

vation, ed. R. D. Semlitsch, 187–98. Washington, D. C.: Smithsonian Institution Press.

Blaustein, A. R., Hays, J. B., Hoffman, P. D., Chivers, D. P., Kiesecker, J. M., Leonard, W. P., Marco, A., Olson, D. H., Reaser, J. K., and Anthony, R. G. 1999. DNA repair and resistance to UV-B radiation in western spotted frogs. *Ecol. Appl.* 9:1100–1105.

Blaustein, A. R., Hoffman, P. D., Hokit, D. G., Kiesecker, J. M., Walls, S. C., and Hays, J. B. 1994. UV repair and resistance to solar UV-B in amphibian eggs: A link to populations? *Proc. Nat. Acad. Sci.* (USA) 91:1791–95.

Blaustein, A. R., Hoffman, P. D., Kiesecker, J. M., and Hays, J. B. 1996. DNA repair and resistance to solar UV-B radiation in eggs of the red-legged frog. *Cons. Biol.* 10:1398–402.

Blaustein, A. R., Hokit, D. G., O'Hara, R. K., and Holt, R. A. 1994. Pathogenic fungus contributes to amphibian losses in the Pacific Northwest. *Biol. Cons.* 67:251–54.

Blaustein, A. R., and Kiesecker, J. M. 1997. The effects of ultraviolet-B radiation on amphibians in natural ecosystems. In *The Effects of Ultraviolet Radiation on Aquatic Ecosystems,* ed. D. P. Häder, 175–88, Austin, Texas: Landes Co.
———. 2002. Complexity in conservation: Lessons from the global decline of amphibian populations. *Ecol. Lett.* 5:597–608.

Blaustein, A. R., Kiesecker, J. M., Chivers, D. P., and Anthony, R. G. 1997. Ambient UV-B radiation causes deformities in amphibian embryos. *Proc. Nat. Acad. Sci.* (USA) 94:13735–37.

Blaustein, A. R., Kiesecker, J. M., Chivers, D. P., Hokit, D. G., Marco, A., Belden, L. K., and Hatch, A. 1998. Effects of ultraviolet radiation on amphibians: Field experiments. *Am. Zool.* 38:799–812.

Blaustein, A. R., Kiesecker, J. M., Hoffman, P. D., and Hays, J. B. 1997. The significance of ultraviolet-B radiation to amphibian population declines. *Rev. Toxicol.* 1:147–65.

Blaustein, A. R., Wake, D. B., and Sousa, W. P. 1994. Amphibian declines: Judging stability, persistence, and susceptibility of populations to local and global extinctions. *Cons. Biol.* 8:60–71.

Blaustein, A. R., Wildy, E. L., Belden, L. K., and Hatch, A. 2001. The influence of abiotic and biotic factors on amphibians in ephemeral ponds with special reference to long-toed salamanders *(Ambystoma macrodactylum). Israel J. Zool.* 47:333–45.

Boone, M. D., and Bridges, C. M. 2003. Effects of pesticides on amphibian populations. In *Amphibian Conservation,* ed. R. D. Semlitsch, 152–67. Washington, D. C.: Smithsonian Institution Press.

Bowling J. W., Leversee, G. J., Landrum, P. F., and Giesy, J. P. 1983. Acute mortality of anthracene-contaminated fish exposed to sunlight. *Aquat. Toxicol.* 3:79–90.

Bradford, D. F. 1989. Allotopic distribution of native frogs and introduced fishes in high Sierra Nevada lakes of California: Implications of the negative effect of fish introduction. *Copeia* 1989:775–78.

————. 1991. Mass mortality and extinction in a high elevation population of *Rana muscosa. J. Herpetol.* 25:174–77.

Bradford, D. F., Cooper, S. D., Jenkens, T. M., Jr., Kratz, K., Sarnelle, O., and Brown, A. D. 1998. Influences of natural acidity and introduced fish on faunal assemblages in California alpine lakes. *Can. J. Fish. Aquat. Sci.* 55:2478–91.

Bradford, D. F., Tabatabai, F., and Graber, D. M. 1993. Isolation of remaining populations of the native frog *Rana muscosa,* by introduced fishes in Sequoia and Kings Canyon National Parks, California. *Cons. Biol.* 7:882–88.

Bragg, A. N. 1958. Parasitism of spadefoot tadpoles by *Saprolegnia. Herpetologica* 14:34.

————. 1962. *Saprolegnia* on tadpoles again in Oklahoma. *Southwest. Nat.* 7:79–80.

Broomhall S. D., Osborne, W., and Cunningham, R. 2000. Comparative effects of ambient ultraviolet-B (UV-B) radiation on two sympatric species of Australian frogs. *Cons. Biol.* 14:420–27.

Carey, C. L. 1993. Hypothesis concerning the causes of the disappearance of boreal toads from the mountains of Colorado. *Cons. Biol.* 7:355–62.

————. 2000. Infectious disease and worldwide declines of amphibian populations, with comments on emerging diseases in coral reef organisms and in humans. *Environ. Health Perspect.* 108 (supp. 1): 143–50.

Cockell, C. S. 2001. A photobiological history of earth. In *Ecosystems, Evolution, and Ultraviolet Radiation,* ed. C. S. Cockell and A. R. Blaustein, 1–35. New York: Springer.

Cockell, C. S., and Blaustein, A. R., eds. 2001. *Ecosystems, Evolution, and Ultraviolet Radiation.* New York: Springer.

Corn, P. S. 1998. Effects of ultraviolet radiation on boreal toads in Colorado. *Ecol. Appl.* 8:18–26.

Cunningham, A. A., Langton, T.E.S., Bennett, P. M., Drury, S.E.N., Gough, R. E., and MacGregor, S. K. 1996. Pathological and microbiological findings from incidents of unusual mortality of the common frog *(Rana temporaria). Philos. Trans. R. Soc. Lond., ser. B* 351:1529–57.

Daszak, P., Berger, L., Cunningham, A. A., Hyatt, A. D., Green, D. E., and Speare, R. 1999. Emerging infectious diseases in amphibian population declines. *Emerg. Infect. Dis.* 5:735–48.

Davidson, C., Shaffer, H. B., and Jennings, M. R. 2001. Declines of the California red-legged frog: Climate, UV-B, habitat, and pesticides hypotheses. *Ecol. Appl.* 11:464–79.

Donnelly, M. A., and Crump, M. L. 1998. Potential effects of climate change on two neotropical amphibian assemblages. *Clim. Change* 39:541–61.

Drake, D. C., and Naiman, R. J. 2000. An evaluation of restoration efforts in fishless lakes stocked with exotic trout. *Cons. Biol.* 14:1807–20.

Drury, S.E.N., Gough, R. E., and Cunningham, A. A. 1995. Isolation of an iridovirus-like agent from common frogs *(Rana temporaria). Vet. Rec.* 137:72–73.

Dunson, W. A., Wyman, R. L., and Corbett, E. S. 1992. A symposium on amphibian declines and habitat acidification. *J. Herpetol.* 26:349–52.

Eldridge, N. 1998. *Life in the Balance: Humanity and the Biodiversity Crisis.* Princeton, N.J.: Princeton University Press.

Epstein, P. R. 1997. Climate, ecology, and human health. *Consequences* 3:3–19.

Fernandez, M., and l'Haridon, J. 1992. Influence of lighting conditions on toxicity and genotoxicity of various PAHs in the newt in vivo. *Mut. Res.* 298:31–41.

———. 1994. Effects of light on the cytotoxicity and genotoxicity of benzo(a)pyrene and an oil refinery effluent in the newt. *Environ. Mol. Mutagen.* 24:124–36.

Fisher, R. N., and Shaffer, H. B. 1996. The decline of amphibians in California's great Central Valley. *Cons. Biol.* 10:1387–97.

Fite, K. V., Blaustein, A. R., Bengston, L., and Hewitt, H. E. 1998. Evidence of retinal light damage in *Rana cascadae:* A declining amphibian species. *Copeia* 1998:906–14.

Frank, W. C., and Dunlap, W. W. 1999. Colonization of high-elevation lakes by long-toed salamanders *(Ambystoma macrodactylum)* after the extinction of local trout populations. *Can. J. Zool.* 77:1759–67.

Gamradt, S. C., and Kats, L. B. 1996. Effect of introduced crayfish and mosquitofish on California newts. *Cons. Biol.* 10:1155–62.

Gamradt, S. C., Kats, L. B., and Anzalone, C. B. 1997. Aggression by non-native crayfish deters breeding in California newts. *Cons. Biol.* 11:793–96.

Gibbs, J. P., and Breisch, A. R. 2001. Climate warming and calling phenology of frogs near Ithaca, New York, 1900–1999. *Cons. Biol.* 15:1175–78.

Gillespie, G. R. 2001. The role of introduced trout in the decline of the spotted tree frog *(Litoria spenceri)* in south-eastern Australia. *Biol. Cons.* 100:187–98.

Goodsell, J. A., and Kats, L. B. 1999. Effect of introduced mosquitofish on Pacific treefrogs and the role of alternative prey. *Cons. Biol.* 13:921–24.

Häkkinen, J., Pasanen, S., and Kukkonen, J.V.K. 2001. The effects of solar UV-B radiation on embryonic mortality and development in three boreal anurans *(Rana temporaria, Rana arvalis,* and *Bufo bufo). Chemosphere* 44:441–46.

Harte J., and Hoffman, E. 1989. Possible effects of acidic deposition on a Rocky Mountain population of the tiger salamander *Ambystoma tigrinum. Cons. Biol.* 3:149–58.

Hatch, A. C., Belden, L. K., Scheessele, E., and Blaustein, A. R. 2001. Juvenile amphibians do not avoid potentially lethal levels of urea on soil substrate. *Environ. Toxicol. Chem.* 20:2328–35.

Hatch, A. C., and Blaustein, A. R. 2000. Combined effects of UV-B, nitrate, and low pH reduce the survival and activity level of larval Cascades frogs *(Rana cascadae). Arch. Environ. Contam. Toxicol.* 39:494–99.

———. 2003. Combined effects of UV-B radiation and nitrate fertilizer on larval amphibians. *Ecol. Appl.* 13:1083–93.

Hatch, A. C., and Burton, G. A., Jr. 1998. Effects of photoinduced toxicity of fluoranthene on amphibian embryos and larvae. *Environ. Toxicol. Chem.* 17:1777–85.

Hayes, M. P., and Jennings, M. R. 1986. Decline of frog species in western North America: Are bullfrogs *(Rana catesbeiana)* responsible? *J. Herpetol.* 20:490–509.

Hays, J. B., Blaustein, A. R., Kiesecker, J. M., Hoffman, P. D., Pandelova, I., Coyle, A., and Richardson, T. 1996. Developmental responses of amphibians to solar and artificial UV-B sources: A comparative study. *Photochem. Photobiol.* 64:449–56.

Hecnar, S. J. 1995. Acute and chronic toxicity of ammonium nitrate fertilizer to amphibians from southern Ontario. *Environ. Toxicol. Chem.* 14:2131–37.

Herman, J. R., Bhartia, P. K., Ziemke, J., Ahmad, Z., and Larko, D. 1996. UV-B increases (1979–1992) from decreases in total ozone. *Geophys. Res. Lett.* 23:2117–20.

Houlahan, J. E., Findlay, C. S., Schmidt, B. R., Myer, A. H., and Kuzmin, S. L. 2000. Quantitative evidence for global amphibian population declines. *Nature* 404:752–55.

Jancovich, J. K., Davidson, E. W., Morado, J. F., Jacobs, B. L., and Collins, J. P. 1997. Isolation of a lethal virus from the endangered tiger salamander *Ambystoma tigrinum stebbinsi. Dis. Aquatic Organ.* 31:161–67.

Johnson, P.T.J., Lunde, K. B., Thurman, E. M., Ritchie, E. G., Wray, S. N., Sutherland, D. R., Kapfer, J. M., Frest, T. J., Bowerman, J., and Blaustein, A. R. 2002. Parasite *(Ribeiroia ondatrae)* infection linked to amphibian malformations in the western United States. *Ecology* 72:151–68.

Kagan J., Kagan, P. A., and Buhse, H. E., Jr. 1984. Light-dependent toxicity of alpha-terthienyl and anthracene toward late embryonic stages of *Rana pipiens. J. Chem. Ecol.* 10:1115–22.

Kats, L. B., Kiesecker, J. M., Chivers, D. P., and Blaustein, A. R. 2000. Effects of UV-B on antipredator behavior in three species of amphibians. *Ethology* 106:921–32.

Kerr, J. B., and McElroy, C. T. 1993. Evidence for large upward trends of ultraviolet-B radiation linked to ozone depletion. *Science* 262:1032–34.

Kiesecker, J. M. 1996. pH induced growth reduction and its effects on predator-prey interactions between *Ambystoma tigrinum* and *Pseudacris triseriata. Ecol. Appl.* 6:1325–31.

Kiesecker, J. M., and Blaustein, A. R. 1995. Synergism between UV-B radiation and a pathogen magnifies amphibian embryo mortality in nature. *Proc. Nat. Acad. Sci.* (USA) 92:11049–52.

———. 1997a. Egg laying behavior influences pathogenic infection of amphibian embryos. *Cons. Biol.* 11:214–20.

———. 1997b. Population differences in responses of red-legged frogs *(Rana aurora)* to introduced bullfrogs *(Rana catesbeiana). Ecology* 78:1752–60.

———. 1998. Effects of introduced bullfrogs and smallmouth bass on microhabitat use, growth, and survival of native red-legged frogs *(Rana aurora). Cons. Biol.* 12:776–87.

———. 1999. Pathogen reverses competition between larval amphibians. *Ecology* 80:2442–48.

Kiesecker, J. M., Blaustein, A. R., and Belden, L. K. 2001. Complex causes of amphibian population declines. *Nature* 410:681–84.

Kiesecker, J. M., Blaustein, A. R., and Miller, C. L. 2001. Potential mechanisms underlying the displacement of native red-legged frog larvae by introduced bullfrog larvae. *Ecology* 82:1964–70.

Kiesecker, J. M., Miller, C. L., and Blaustein, A. R. 2001. The transfer of a pathogen from fish to amphibians. *Cons. Biol.* 15:1064–70.

Kiesecker, J. M., and Skelly, D. K. 2001. Interactions of disease and pond drying on the growth, development, and survival of the gray treefrog *(Hyla versicolor)*. *Ecology* 82:1956–63.

Knapp, R. A., and Mathews, K. R. 1998. Eradication of nonnative fish by gill netting from a small mountain lake in California. *Restor. Ecol.* 6:207–13.

Knapp, R. A., Mathews, K. R., and Sarnelle, O. 2001. Resistance and resilience of alpine lake fauna to fish introductions. *Ecol. Monogr.* 71:401–21.

Komak, S., and Crossland, M. R. 2000. An assessment of the introduced mosquitofish *(Gambusia affinis holbrooki)* as a predator of eggs, hatchlings, and tadpoles of native and nonnative anurans. *Wildl. Res.* 27:185–89.

Kupferberg, S. J. 1997. Bullfrog *(Rana catesbeiana)* invasion of a California river: The role of larval competition. *Ecology* 78:1736–51.

LaMarca, E., and Reinthaler, H. P. 1991. Population changes in *Atelopus* species of the Cordillera de Merida, Venezuela. *Herpetol. Rev.* 22:125–28.

Laurance, W. F., McDonald, K. R., and Speare, R. 1996. Epidemic disease and the catastrophic decline of Australian rain forest frogs. *Cons. Biol.* 10:406–13.

Lawton, J. H., and May, R. M. 1995. *Extinction Rates*. Oxford, England: Oxford University Press.

Lips, K. R. 1998. Decline of a tropical montane amphibian fauna. *Cons. Biol.* 12:106–17.

Lizana, M., and Pedraza, E. M. 1998. Different mortality of toad embryos *(Bufo bufo* and *Bufo calamita)* caused by UV-B radiation in high mountain areas of the Spanish Central System. *Cons. Biol.* 12:703–7.

Long, L. E., Saylor, L. S., and Soule, M. E. 1995. A pH/UV-B synergism in amphibians. *Cons. Biol.* 9:1301–3.

Longcore, J. E., Pessier, A. P., and Nichols, D. K. 1999. *Batrachochytrium dendrobatidis gen. et sp. nov.,* a chytrid pathogenic to amphibians. *Mycologia* 91:219–27.

Marco, A., and Blaustein, A. R. 1999. Morphological effects of nitrite on Cascades frog *(Rana cascadae)* metamorphosis. *Environ. Toxicol. Chem.* 18:946–49.

Marco, A., Lizana, M., Alvarez, A. and Blaustein, A. R. 2001. Egg-wrapping behaviour protects newt embryos from UV radiation. *Anim. Behav.* 61:639–44.

Marco, A., Quilchano, C., and Blaustein, A. R. 1999. Sensitivity to nitrate and nitrite in pond-breeding amphibians from the Pacific Northwest. *Environ. Toxicol. Chem.* 18:2836–39.

Middleton, E. M., Herman, J. R., Celarier, E. A., Wilkinson, J. W., Carey, C., and

Rusin, R. J. 2001. Evaluating ultraviolet radiation exposure with satellite data at sites of amphibian declines in Central and South America. *Cons. Biol.* 15:914–29.

Monson, P. D., Call, D. J., Cox, D. A., Liber, K., and Ankley, G. T. 1999. Photoinduced toxicity of fluoranthene to northern leopard frogs *(Rana pipiens)*. *Environ. Toxicol. Chem.* 18:308–12.

Moyle, P. B. 1973. Effects of introduced bullfrogs, *Rana catesbeiana*, on the native frogs of the San Joaquin Valley, California. *Copeia* 1973:18–22.

Nagl, A. M., and Hofer, R. 1997. Effects of ultraviolet radiation on early larval stages of the alpine newt, *Triturus alpestris*, under natural and laboratory conditions. *Oecologia* 110:514–19.

Nyman, S. 1986. Mass mortality in larval *Rana sylvatica* attributable to the bacterium *Aeromonas hydrophilum*. *J. Herpetol.* 20:196–201.

Otto, R. G. 1973. Temperature tolerance of the mosquitofish. *J. Fish Biol.* 5:575–85.

Ovaska, K., Davis, T. M., and Flamarique, I. M. 1997. Hatching success and larval survival of the frogs *Hyla regilla* and *Rana aurora* under ambient and artificially enhanced solar ultraviolet radiation. *Can. J. Zool.* 75:1081–88.

Pahkala, M., Laurila, A., and Merilä, J. 2000. Ambient ultraviolet-B radiation reduces hatchling size in the common frog *Rana temporaria*. *Ecography* 23:531–38.

———. 2001. Carry-over effects of ultraviolet-B radiation on larval fitness in *Rana temporaria*. *Proc. Roy. Soc. Lond., ser. B* 268:1699–706.

Parker, B. R., Wilhelm, F. M., and Schindler, D. W. 1996. Recovery of the *Hesperodiaptomus arcticus* populations from diapausing eggs following elimination by stocked salmonids. *Can. J. Zool.* 74:1292–97.

Pessier, A. P., Nichols, D. K., Longcore, J. E., and Fuller, M. S. 1999. Cutaneous chytridiomycosis in poison dart frogs *(Dendrobates* spp.) and White's tree frogs *(Litoria caerulea)*. *J. Vet. Diag. Invest.* 11:194–99.

Pierce, B. J. 1985. Acid tolerance in amphibians. *BioScience* 35:239–43.

Pounds, J. A., and Crump, M. L. 1994. Amphibian declines and climate disturbance: The case of the golden toad and the harlequin frog. *Cons. Biol.* 8:72–85.

Pounds, J. A., Fogden, M.P.L., and Campbell, J. H. 1999. Biological responses to climate change on a tropical mountain. *Nature* 398:611–15.

Reading, C. J. 1998. The effect of winter temperatures on the timing of breeding activity in the common toad *Bufo bufo*. *Oecologia* 117:469–75.

Reaser, J. K., and Blaustein, A. R. In press. Repercussions of global change in aquatic systems. In *Status and Conservation of North American Amphibians*, ed. M. Lanoo. Berkeley: University of California Press.

Relyea, R. A., and Mills, N. 2001. Predator-induced stress makes the pesticide carbaryl more deadly to gray treefrog tadpoles *(Hyla versicolor)*. *Proc. Nat. Acad. Sci.* (USA) 98:2491–96.

Richter, B. D., Braun, D. P., Mendelson, M. A., and Master, L. L. 1997. Threats to imperiled freshwater fauna. *Cons. Biol.* 11:1081–93

Schindler, D. W., Curtis, P. J., Parker, B. R., and Stainton, M. P. 1996. Consequences of climate warming and lake acidification for UV-B penetration in North American boreal lakes. *Nature* 379:705–8.

Smith, A. W., Anderson, M. P., Skilling, D. E., Barlough, J. E., and Ensley, P. K. 1986. First isolation of calicivirus from reptiles and amphibians. *Am. J. Vet. Res.* 8:1718–21.

Smith, G. R., Waters, M. A., and Rettig, J. E. 2000. Consequences of embryonic UV-B exposure for embryos and tadpoles of the plains leopard frog. *Cons. Biol.* 14:1903–07.

Sparling, D. W., Fellers, G. M., and McConnell, L. L. 2001. Pesticides and amphibian population declines in California, USA. *Environ. Toxicol. Chem.* 20:1591–95.

Sparling, D. W., Linder, G., and Bishop, C. A., eds. 2000. *Ecotoxicology of Amphibians and Reptiles.* Pensacola, Fla.: SETAC Press.

Stallard, R. F. 2001. Possible environmental factors underlying amphibian decline in eastern Puerto Rico: Analysis of U.S. government data archives. *Cons. Biol.* 15:943–53.

Starnes, S. M., Kennedy, C. A., and Petranka, J. W. 2000. Sensitivity of southern Appalachian amphibians to ambient solar UV-B radiation. *Cons. Biol.* 14:277–82.

Stebbins, R. C. 1954. *Amphibians and Reptiles of Western North America.* New York: McGraw-Hill.

———. 1966. *A Field Guide to Western Reptiles and Amphibians.* Boston: Houghton Mifflin Company.

Stumpel, A.H.P. 1992. Successful reproduction of introduced bullfrogs, *Rana catesbeiana,* in northwestern Europe: A potential threat to indigenous amphibians. *Biol. Cons.* 60:61–62.

Tevini, M., ed. 1993. *UV-B Radiation and Ozone Depletion: Effects on Humans, Animals, Plants, Microorganisms, and Materials.* Boca Raton, Fla.: Lewis Publishers.

Tyler, T. J., Liss, W. J., Ganio, L. M., Larson, G. L., Hoffman, R. L., Deimling, E., and Lomnicky, G. 1998. Interaction between introduced trout and larval salamanders *Ambystoma macrodactylum* in high elevation lakes. *Cons. Biol.* 12:94–105.

Tyler, T. J., Liss, W. J., Hoffman, R. L., and Ganio, L. M. 1998. Experimental analysis of trout effects on survival, growth, and habitat use of two species of ambystomatid salamanders. *J. Herpetol.* 32:345–49.

van de Mortel, T. F., and Buttemer, W. A. (1996). Are *Litoria aurea* eggs more sensitive to ultraviolet-B radiation than eggs of sympatric *L. peronii* or *L. dentata? Aust. Zool.* 30:150–57.

van de Mortel, T., Buttemer, W., Hoffman, P., Hays, J., and Blaustein, A. 1998. A comparison of photolyase activity in three Australian tree frogs. *Oecologia* 115:366–69.

van der Leun, J. C., Tang, X., and Tevini, M., eds. 1998. *Environmental Effects of Ozone Depletion 1998 Assessment.* Lausanne, Switzerland: Elsevier.

Vonesh, J. R., and De la Cruz, O. 2002. Complex life cycles and density dependence: Assessing the contribution of egg mortality to amphibian declines. *Oecologia* 133:325–33.

Walker S. E., Taylor, D. H., and Oris, J. T. 1998. Behavioral and histopathological effects of fluoranthene on bullfrog larvae *(Rana catesbeiana)*. *Environ. Toxicol. Chem.* 17:734–39.

Wilson, E. O. 1992. *The Diversity of Life.* Cambridge: Harvard University Press.

Worrest, R. D., and Kimeldorf, D. J. 1976. Distortions in amphibian development induced by ultraviolet-B enhancement (290–310 nm) of a simulated solar spectrum. *Photochem. Photobiol.* 24:377–82.

Zaga, A., Little, E. E., Rabeni, C. F., and Ellersieck, M. R. 1998. Photoenhanced toxicity of a carbamate insecticide to early life stage anuran amphibians. *Environ. Toxicol. Chem.* 17:2543–53.

5 Energetics of Leatherback Sea Turtles

A Step toward Conservation

David R. Jones, Amanda L. Southwood,
and Russel D. Andrews

INTRODUCTION

The leatherback sea turtle *(Dermochelys coriacea)* is the largest and certainly the most widespread pelagic reptile, ranging from New Zealand in the south to beyond the Arctic Circle in the north. Although we now have a rough picture of the movements and behavior of adult females, their early lives, from hatching to first nesting, remain "lost years" (Carr 1982). We know little about these large, charismatic, and gentle animals, and unfortunately we may never learn more now that leatherbacks are on the verge of extinction.

Long-time residents of Playa Grande, Costa Rica, one of the largest nesting beaches in the world, recall a time in the recent past when the beach seemed to come alive with hundreds of leatherbacks nesting in a single night. Now on a dusk-to-dawn patrol of the beach at the height of breeding season, one is likely to see only 10 turtles. A recent study by Spotila et al. (2000) reports that the entire Pacific population is now nearing extinction. Twenty years ago there may have been more than 70,000 leatherback turtles in the Pacific, but today there may be as few as 3000. The dramatic decline of the Pacific population may be due to adult mortality from fisheries, with a major role being played by swordfish boats plying the South American coast (Eckert and Sarti 1997). Although 27,000 leatherbacks may remain in the Atlantic population, this population is also severely threatened by fishery interactions as well as by beach development and consequent disruption of nesting behavior or even total destruction of nesting habitats. As a consequence, the 2000 IUCN Red List of Threatened Species has classified leatherbacks as "critically endangered" because they have undergone an estimated population reduction of at least 80% over the last 30 years, and they appear to be facing an extremely high risk of extinction in the immediate future.

Management agencies throughout the world have recently accelerated their attempts to halt the decline and aid the recovery of this species, but they lack a great deal of necessary information, including such basic life-history parameters as age at first reproduction, longevity, daily nutritional needs, and foraging strategies. There is no doubt that a reduction in poaching of adults and eggs has been an important conservation achievement, but nevertheless only one in a hundred eggs will develop into a mature adult. The establishment of hatcheries at the major nesting beaches has increased hatching success, but unless the threats to adults and subadults are addressed, these efforts will be wasted.

Monitoring of both diving and migratory behavior (Morreale et al. 1996; Eckert and Sarti 1997) has highlighted the threat from current fishery practices. Although fishery-related mortality is clearly a major factor in the decline of the Pacific leatherback population, other factors, such as food limitation, may also be involved. Leatherback turtles forage primarily, perhaps exclusively, on jellyfish, salps, and other gelatinous prey. It is clear that both fishing practices and natural climate shifts can dramatically affect the abundance of gelatinous zooplankton (*cf.* Parsons 1992; Brodeur et al. 1999), but determining whether changes in jellyfish abundance might affect leatherbacks requires an estimate of the total biomass necessary to sustain the current population. Such an estimate is also critically important because fishing practices and climate fluctuations are altering food webs (Pauly et al. 1998). The complexities of food webs may lead to an ironic outcome in that fishing-related decreases in leatherback populations may increase the abundance of medusae that forage on ichthyoplankton, including the larvae of fish species that will then not recruit into the adult population targeted by the fishery (Eckert 2000).

Much more information about the physiology and metabolism of leatherbacks at sea will be required before restoration and management of the species can be placed on anything more than a quasi-scientific basis. The approach has to be holistic, spanning from the individual level to the population level. For instance, food requirements at the individual level determine distribution and abundance at the population level. Similarly, individual growth rates are important factors in setting the age structure and other demographics of the population. Individual reproductive output determines the potential recruitment into the population. The total amount of energy required by the population ultimately defines population structure and sustainability and is reflected at the individual level by the field metabolic rate

(FMR), the energetics of daily life integrated over time. Unfortunately, at-sea metabolic rates have never been measured in the leatherback sea turtle, nor have they been determined for any other sea turtle species.

MEASURING FIELD METABOLIC RATE

Leatherbacks spend 99% of their lives at sea, and up to 85% of at-sea time may be spent underwater (Keinath and Musick 1993). Given such elusive behavior, how can we probe their physiology? To study the foraging behavior, energetics, physiology, and movements of animals such as the leatherback turtle that may swim far out of the sight of human observers, sophisticated remote-monitoring devices must be employed. Happily, we can take advantage of the leatherback's homing behavior to deploy and retrieve computers that record and store heart rate, swim velocity, dive time, dive depth, and body temperature (Andrews 1998a, b). Leatherback turtles return to their natal beach every 2–3 years to lay eggs (Steyermark et al. 1996). In a single nesting season a female leatherback may lay ten clutches of eggs with 8–14 days spent at sea between individual nestings. Turtles come ashore at night and remain on the beach for up to an hour to complete the nesting process. Hence, the turtles set the protocol, and researchers have to work rapidly in darkness to deploy instruments during the short quiescent period that the turtle enters during egg deposition. Behavioral and physiological variables are measured while the turtle is freely swimming at sea during the internesting interval. Radio telemetry may be used to relocate turtles on their return to the beach to lay more eggs, so that instruments may be recovered.

One of the most common ways of determining FMR in a free-ranging animal is the doubly-labeled-water method (DLW; Nagy 1989). The DLW method requires intramuscular, intravenous, or intraperitoneal injection of water that has been enriched with stable isotopes of hydrogen (deuterium) and oxygen (oxygen-18). An initial blood sample is taken, before injection of DLW, to establish the background levels of these isotopes in the animal's body. After DLW injection, a series of blood samples is taken over several hours to establish the time course of isotope equilibration. A final sample is used to calculate the concentration of both stable isotopes after the labeled water has reached equilibrium with the turtle's body water pool. The turtle is then released, and another blood sample is taken when she returns for her next nesting event. Labeled water is lost through water efflux processes such as urination, defecation, and evaporation, and new unlabeled water is added by ingestion and metabolic water formation. Although both deuterium and oxygen-18 are lost in the form of water, oxygen-18 is also lost in the form of carbon dioxide, so the difference between the washout slopes of deuterium and oxygen-18 gives a measure of carbon dioxide production, which is proportional to metabolic rate.

The amount of DLW that needs to be injected is crucial and depends on water turnover, metabolic rate, and sampling interval. Leatherbacks forage exclusively on gelatinous organisms (cnidarians and tunicates) that have an extremely low energy density and may be up to 95% water (Davenport 1998). It has been estimated (Lutcavage and Lutz 1986) and observed (Duron 1978) that leatherbacks may consume up to their body mass in jellyfish each day, which is a lot of water to excrete. On the other hand, metabolic rates in leatherbacks may be higher than those normally associated with reptiles, owing to the leatherbacks' high body temperatures. Paladino, O'Connor, and Spotila (1990) suggested that leatherbacks were "gigantotherms" able to maintain elevated body temperatures even in cold waters, owing to their large size, good insulation, and circulatory regulation. Pacific leatherbacks maintain body temperatures several degrees higher than water temperatures (16–27.5 °C) for the majority of the internesting interval in waters off Costa Rica (Southwood, Andrews, Paladino, et al. 1999). Even in waters of the frigid North Atlantic (8 °C), a leatherback was reported to have a body temperature of 25 °C (Frair, Ackman, and Mrosovsky 1972). Perhaps metabolic rates in leatherbacks may be high enough to offset a large water exchange, but it would be reassuring to find support for this assumption before embarking on logistically complex, as well as expensive, field studies.

AN ESTIMATION OF FIELD METABOLIC RATES BASED ON AEROBIC DIVE LIMITS

Is it possible to estimate at-sea FMRs for leatherbacks that might at least give us confidence that pursuing field studies will be worthwhile? Initially, it is best to be parsimonious, and the simplest approach would be to use metabolic rates recorded on the beach to estimate the potential limits for dive times, which can then be compared to actual recordings of dive times at sea.

In studies of leatherback females on a nesting beach, Lutcavage, Bushnell, and Jones (1992) measured both metabolism and heart rate for unrestrained and restrained turtles. Mean oxygen consumption for unrestrained turtles laying eggs was 0.25 mL min^{-1} kg^{-1} with heart rates in the range of 8–29 beats min^{-1}. Mean oxygen consumption for turtles restrained in a cargo net was 1.09 mL min^{-1} kg^{-1}, and heart rates were 43–48 beats min^{-1}. Paladino, O'Connor, and Spotila (1990) reported a mean oxygen consumption of

3.7 mL min^{-1} kg^{-1} for turtles covering nest cavities or walking on the beach, but heart rates were not measured in this study.

These studies point to a range of oxygen consumptions (relative metabolic scope) of fourteen, with the higher exercise values approaching resting metabolic rates of similarly sized mammals. Perhaps we can determine the range of feasible metabolic rates at sea by calculating aerobic dive limits. The current view with respect to voluntary submergence is that diving is aerobic and that the animal surfaces when the onboard oxygen supplies are depleted (Kooyman et al. 1983; Butler and Jones 1997). Hence divers do not routinely rely on anaerobic metabolism to supplement oxygen-based metabolism; that is, oxygen stores, and their utilization, govern aerobic dive times. Any factors modulating metabolism will affect underwater endurance, so that the dive time is inversely proportional to the rate of aerobic metabolism.

Oxygen is stored in the lungs, blood, and tissues, and oxygen levels have been measured by Lutcavage, Bushnell, and Jones (1990). Using the range of values for oxygen consumption recorded from turtles on the beach, Lutcavage, Bushnell, and Jones (1992) reported that the calculated aerobic dive limit (cADL) for an average-size leatherback turtle (340 kg) would be from 5 to 70 min if about 85% of total O_2 stores were available for use. Obviously, this approach tells us nothing more than we already knew: the extremely wide range for aerobic dive limits just mirrors the similarly wide range of oxygen consumptions measured on the beach.

Dives are punctuated by surface intervals, which Fedak, Pullen, and Kanwisher (1988) have characterized as unproductive but necessary, suggesting that surface time should be reduced to the bare minimum in order to improve foraging success. Foraging efficiency can be improved in two ways. One is to enhance ventilation and gas exchange to reduce the interval necessary for repaying oxygen deficits incurred in the dive. The other is to increase dive time by ensuring that aerobic sources (myoglobin-bound oxygen) and anaerobic sources (phosphocreatine) of energy production are fully in place before the dive. Hence, the circulatory work normally required for exercise is completed before the dive, which reduces energy costs during the dive (Fedak, Pullen, and Kanwisher 1988). This energy cost reduction does not necessarily hold, however, for recovery from dives that have been extended into the anaerobic metabolic domain, beyond the aerobic dive limit; in this case, the extended dive time is proportional to the concentration of lactate, the end product of anaerobic metabolism. Recovery from extended dives therefore requires not only preparation for the dive ahead but also "payback time" for costs incurred in the preceding dive, and this

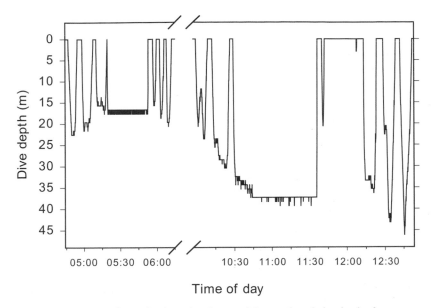

Figure 5.1. Excerpts from the dive-depth record from a female leatherback sea turtle tagged at Playa Grande, Costa Rica. Note the 33-min dive on the left-hand side of the break in the time axis, and the 67-min dive on the right-hand side of the break.

requirement will prolong the surface interval. In fact, we have observed extreme disruptions of the postdive recovery pattern after a 67-min dive (Figure 5.1). After this long dive, the animal submerged after a normal surface interval but immediately resurfaced and then remained at the surface for more than 30 min (Southwood, Andrews, Lutcavage, et al. 1999). We have also noticed more-subtle changes in dive pattern after a 33-min dive (Figure 5.1). After this dive, the animal made a series of dives that were much shorter than usual, which suggests that the 33-min dive had stretched into the anaerobic domain. Unfortunately, long dives are rare events, and close inspection of this turtle's overall dive pattern showed considerable irregularities in that bouts of longer and shorter dives alternated. Consequently, it seems most reasonable to assume that the cADL is higher than 30 min and that metabolic rates during dives are low.

In diving animals, maintenance and regulation of aerobic metabolism are even more crucial than for terrestrial animals, to avoid significant anaerobic episodes and to ensure that when an animal is submerged, sufficient oxygen remains in the oxygen stores to maintain the physiological integrity of tissues (i.e., heart and brain) that cannot survive without it. Hence, cardiac output and tissue perfusion are important regulators of aerobic metabolism, and metabolic rate and perfusion rate indicate oxygen flux under both reduced and elevated metabolic conditions. In diving, circulation is the proximate regulator of metabolism and modulates dive time; heart rate is the single most recorded and useful monitor of metabolic performance. In effect, heart rate is proportional to aerobic metabolism, which is inversely proportional to dive time. This relationship implies that heart rate is variable, decreasing with increased dive time, so tissue perfusion will also be variable in intensity. It seems plausible that in highly active species, the locomotor muscles would be perfused in preference to visceral vascular beds, but in more lethargic animals the situation is less clear.

Examination of heart rate data from turtles on the beach (for which we also have metabolic rates) and turtles at sea should allow us to further refine our estimate of at-sea metabolic rate. Diving heart rates for leatherback turtles ($17-20$ beats min^{-1}) are more similar to heart rates during egg deposition ($8-29$ beats min^{-1}) than to heart rates of turtles restrained in nets on the beach ($43-48$ beats min^{-1}). Assuming there is some positive relation between activity and heart rate in leatherbacks, the use of metabolic rates recorded from nesting turtles on the beach, as opposed to restrained or active turtles, may give a more accurate estimate of the cADL of leatherback turtles. If so, then leatherback turtles have a cADL at the high end of the range of estimates given previously and, therefore, a low FMR. CADL calculated from egg-laying metabolic rates, however, does not correlate with dive times recorded for turtles diving in near-shore waters, where most dives are less than 10 min long.

AN ESTIMATION OF FIELD METABOLIC RATES BASED ON THE SIMILARITY PRINCIPLE

Estimating metabolic rates of leatherbacks at sea requires another approach. Is it possible to make this estimate based on similarities in the behavior and physiology of two disparate species if metabolic rate is known for one of them? This method is an application of the *similarity principle*. Both Claude Bernard and Charles Darwin were strong proponents of the similarity principle, arguing that morphology, physiology, and behavior appear to be similar in different animals because of common evolutionary origins, similar physiological and metabolic control systems, or both. The general idea is that morphological adaptations, physiology, and behavior are inextricably linked, though not necessarily in direct sequence. Therefore, we can approach the question of leatherback FMRs indirectly by comparing the performance of two very different animals that share enough characteristics to make comparison appropriate and meaningful.

Dermochelys is an unusual turtle, sharing characteristics with both the large marine mammals and other sea turtles. For instance, in terms of oxygen transport and storage, the hematocrit, hemoglobin, and myoglobin levels approach mammalian values; whereas in blood oxygen affinity, Hill coefficient, and Bohr effect, the leatherback resembles other sea turtles (Lutcavage, Bushnell, and Jones 1990). However, size sets the leatherback apart from other sea turtles and is probably the most important criterion to be matched in any comparison because large size confers numerous advantages on animals whose diving lifestyle poses major problems for homeostasis. Oxygen supplies are limited during submergence, yet because cellular metabolic rates decrease with increased size, large animals are preadapted for hypoxia tolerance. Furthermore, many vital functions also reflect this cellular preadaptation. The volumes of the respiratory media, air and blood, scale in direct proportion to body size in the face of falling cellular demand for oxygen, so heart and respiratory rates are low.

Fortunately, we have more than just generalities to consider when attempting to justify our selections for comparison. In view of similarities in dive behavior, either the Weddell or elephant seal appears to be the most likely candidate to compare with leatherbacks. Because we have directly determined FMRs for elephant seals but not for Weddell seals, the elephant seal is the animal of choice. We do, however, realize that such a cross-class comparison requires a leap of faith.

Morphologically, both elephant seals and leatherbacks are extremely well adapted, in their own respective ways, for life at sea. But it is in their diving behavior that similarities are most striking. Diving is continuous, interrupted by only short periods at the surface (Eckert et al. 1989; Le Boeuf et al. 1989). Their dives can be exceptionally deep, compared with those of many other reptilian or mammalian divers, reaching depths in excess of 1000 m. At these depths, the lungs, which are small in both species, will be collapsed, and the animals have to rely on blood- and tissue-oxygen stores (Lutcavage, Bushnell, and Jones 1990). In our own studies with leatherbacks, we have monitored dive behavior only on the continental shelf rather than in the open ocean. For leatherbacks, mean dive duration in the first 5 days of the internesting interval is 9.5±1.5 min (Figure 5.2A; Southwood, Andrews, Lutcavage, et al. 1999), which is similar to that of the elephant seal diving on the continental shelf (9.8±2.5 min) (Figure 5.2B; Andrews et al. 1997). When the animals enter deeper water, dive times increase by about 30% in leatherbacks (Eckert et al. 1986) and nearly double in elephant seals.

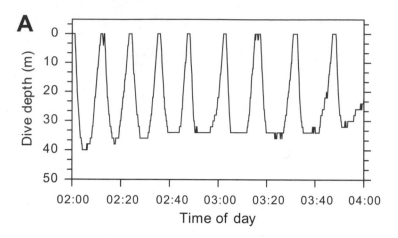

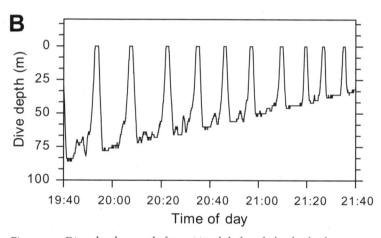

Figure 5.2. Dive-depth records from *(A)* adult female leatherback sea turtles and *(B)* juvenile northern elephant seals diving on the continental shelf.

Even when diving out in the open ocean, elephant seals have exceptionally short surface intervals, seldom exceeding 3 min. In contrast, leatherback surface intervals are much longer, being three to four times those of the elephant seal, even after dives of similar duration. After one exceptionally long dive (67 min) by a leatherback, the subsequent diving pattern was disrupted for more than 30 min (Figure 5.1), whereas elephant seals seem able to take dives of this length with consummate ease, showing no effects on the subsequent dive pattern (Hindell et al. 1992). The upshot of prolonged surface intervals in the leatherback is that they spend only about two-thirds of their time underwater, whereas elephant seals are submerged for 90% of their time at sea. While at sea, both animals swim at similar speeds (around 1 m sec^{-1}) and show clear indications of feeding or drinking (Andrews 1999; Southwood, Andrews, Lutcavage et al. 1999b). In terms of diving behavior, there are enough similarities to indicate that an approach based on the similarity principle might be scientifically viable.

Furthermore, both leatherbacks and elephant seals show reductions in heart rate when diving. The decline in heart rate is much faster and greater in elephant seals: heart rates fall from 107±4.4 beats min^{-1} on the surface to 41.2±4.4 beats min^{-1} within 1 min when the animals are diving on the shelf (Figure 5.3A; Andrews et al. 1997). In leatherbacks, the decline is slower and smaller, with heart rate falling from 25 beats min^{-1} to 17 beats min^{-1} (Figure 5.3B; Southwood, Andrews, Lutcavage, et al. 1999). In both species, heart rate begins to return to surface levels before the animal actually surfaces and breathes. In proportional terms, the reduction is twice as large in the seal as in the turtle. When dives are prolonged more than usual, the inverse relationship between heart rate and dive time is much stronger in the seal than in the turtle. In extremely long dives, or under unusual circumstances, however, both species exhibit heart rates below 7 beats min^{-1} (Figure 5.3C, D).

We don't know how the heart rate–metabolism relationship in reptiles compares to that in mammals. The very high postdive heart rates in elephant seals may aid recovery, accounting for surface intervals much shorter than those observed for leatherbacks after dives of similar durations. The changes in diving heart rates are smaller in the leatherback. For both animals, lower diving heart rates suggest lower diving metabolic rates. However, given the smaller heart size and therefore cardiac stroke volume in the reptile (Schmidt-Nielsen 1990), we do not know the relative magnitudes of metabolic rate changes. In extended dives, when heart rates are below 7 beats min^{-1}, aerobic metabolism is undoubtedly low in both species. Low heart rates are an expression of the classic Irving–Scholander dive response, which is typically evoked when diving animals are prevented from surfacing either by restraint or, when at sea, by some unfortunate event.

Body temperatures have marked effects on tissue metabolism. Body and stomach temperatures of leatherbacks are usually higher than the temperatures recorded from the surrounding water. In Costa Rica, body temperatures are in the 28.7–31.1 °C range. This is not so dissimilar from core body temperatures in diving elephant seals, which hover around 35–36 °C. Interestingly, elephant seals show considerable regional heterogeneity in temperature: subcutaneous and muscle–blubber interface temperatures fall to

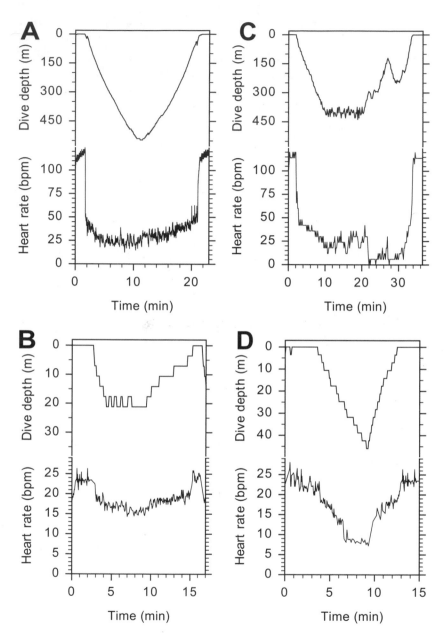

Figure 5.3. Dive-depth and heart rate records from typical dives of *(A)* juvenile northern elephant seals and *(B)* adult female leatherback sea turtles and from unusual dives when heart rate decreased to a very low level: *(C)* elephant seal and *(D)* leatherback.

12 °C and 26 °C, respectively, in 9 °C water. Overall, one might suspect that the grand mean body temperature of the seal is in the same range as that of the leatherback.

Diving and heart-rate profiles allied to similar body temperatures attest to a wide range of behavioral and physiological similarities between leatherbacks and elephant seals. Are these similarities enough to allow extrapolation to metabolic rates? The FMR of juvenile elephant seals diving both on and off the shelf has been determined using doubly-labeled water (Andrews 1999). The at-sea FMR was 2.43 ± 0.42 W kg^{-1}, which is only slightly higher than that of animals on the beach (2.08 ± 0.63 W kg^{-1}). The at-sea metabolic rate is equivalent to an oxygen consumption rate of 7.3 mL min^{-1} kg^{-1}. The total oxygen stores of juvenile elephant seals are in the range of 69 mL kg^{-1} (Thorson and Le Boeuf 1994), so the cADL is given by the quotient of oxygen stores to the oxygen consumption rate, or 9.4 min, which is remarkably close to the average dive time of seals diving on the shelf.

Leatherbacks have less than half the oxygen stores of the elephant seal (27 mL kg^{-1}; Lutcavage, Bushnell, and Jones 1992) and are nearly twice as big. So, adjusting the elephant seal FMR by applying the three-quarter-power law (metabolism = mass$^{0.75}$) reduces the estimated leatherback FMR to around 6.4 mL O$_2$ kg^{-1} min^{-1}. Therefore, these figures give a leatherback cADL of 4.2 min, which is well below that seen when animals are diving on the shelf. This value suggests that the FMR of the leatherback is overestimated by unqualified use of the similarity principle.

Some qualifiers are certainly in order, if only because the animals differ in age, maturity, body mass, and body temperature. The smaller, juvenile elephant seals (100–200 kg) are 1–2 years old, whereas the leatherbacks (250–400 kg) are mature adults. Regional heterothermy in the elephant seal constrains a strict application of a Q_{10} adjustment to our estimate even though stomach (deep body) temperatures in leatherbacks are 4–6 °C below those in elephant seals. In this context, the metabolism of isolated pectoral muscle tissue from adult leatherbacks is reported to be independent of temperature over the range of 5–38 °C ($Q_{10} = 1$) (Pennick et al. 1998). Independence of metabolism from body temperature would be a unique quality of the leatherback, suggesting yet again that attempting to correct metabolism for temperature is inappropriate. On the other hand, metabolic rates in leatherbacks could be much lower than their body temperatures would suggest.

FMRs in the leatherback clearly cannot be in the same range as those of the elephant seal. In fact, the FMR of leatherbacks can be only about half that of the elephant seal to account for the dive durations observed when diving in inshore waters. It should be emphasized that we are not dealing here with maximum possible aerobic dive limits. We suspect these exceed 30 min for the leatherback and may be effectively infinite in the elephant seal if the only indication of breaching the ADL is a significant prolongation of the postdive surface interval. Aerobic dive limits might best be viewed as the period that leads to minimum metabolic disequilibrium and is followed by short surface-recovery times. Hence, in the rather unique situation of the internesting interval, during which the leatherback is faced with producing a large quantity of eggs, diving metabolic rates might be higher than during postnesting migrations to areas where prey is usually more abundant and concentrated than in the tropics. Even so, FMRs of continuously exercising leatherbacks are low by mammalian standards, just approaching the mammalian resting range, despite gigantothermy (Paladino, O'Connor, and Spotila 1990).

CONCLUSIONS

What, if anything, can we conclude from the above exercise? A within-species estimation of diving metabolism, based on aerobic dive performance, yields FMRs well within the reptilian range. On the other hand, comparing the leatherback to a diving mammal yields an FMR of about half that of the elephant seal and at the lower end of the mammalian range. Overall, this exercise suggests that leatherbacks have remained prisoners of their phylo-genetic position, probably having a resting metabolic rate considerably lower than that of a similarly sized mammal. Low metabolic rates in reptiles vis à vis mammals are due to the fact that reptiles have only half the mito-chondrial density and half the mass of metabolically important organs (Else and Hulbert 1981; Hulbert and Else 1981). Also, cell membranes are much leakier in the mammal, so mammals expend considerably more energy in maintaining intracellular homeostasis (Hulbert and Else 1990).

Obviously, direct measurements of metabolic rate at sea would clear up the confusion about leatherback metabolism. In this respect, one might be well advised to attempt a determination of FMR in leatherbacks in three stages. First, measure water turnover using deuterated or tritiated (H-3) water, which is certainly less expensive than using DLW in an all-or-nothing approach. Second, use the data from the water turnover study to calculate what differences might be expected in declines of water and water plus carbon dioxide slopes, using values for FMR in the range of those calculated above or actually measured on the beach ($0.25-3.7$ mL kg^{-1} min^{-1}). The third step, which should be taken only if the prior stages yield a somewhat convincing answer, is to go for broke and attempt a measurement using

DLW in the field. Even if it fails, given the exorbitant cost of injecting DLW into a 350-kg turtle, one can almost certainly guarantee that "broke" status will have been achieved.

The leatherback displays traits that make it an ideal subject for the comparative physiologist, especially one who specializes in diving animals. The leatherback is the largest living turtle, and, while at sea, it dives continuously, sometimes for long periods and to great depths while maintaining body temperature well above ambient without relying on solar radiation. Hence, it is possible to plan both basic and applied research with leatherbacks. Metabolic rates of leatherback turtles on the beach span a wide range, which leads to some important questions about the leatherbacks' ability to manage their oxygen stores during diving and to maintain their body temperatures above ambient temperature. A low at-sea metabolic rate would facilitate long duration dives, but would not explain the ability to maintain a high body temperature. The question of whether a large reptile can be both bradymetabolic and homeothermic has received considerable attention with the recent discovery of an ornithischian dinosaur that is alleged to possess a four-chambered heart and a single systemic aorta (Fisher et al. 2000). Therefore, basic research will have a significant bearing on numerous questions of interest both to comparative physiologists and to members of the public anxious for information on the biology of dinosaurs. Basic research studies notwithstanding, measuring energy expenditure (FMR) during the nesting season will allow a major contribution to be made to ensuring the leatherbacks' survival. This is applied research at its best, for it also yields fundamental information. Extrapolating FMRs to the population level will allow estimates of rates of food intake, growth, and reproduction that are crucial for understanding the leatherbacks' role in the ecosystem, because the energy demand of the population is a key ecological factor in determining animal abundance, distribution, and density.

ACKNOWLEDGMENTS

We would like to thank the Natural Sciences and Engineering Research Council of Canada for funding and Robin Liley, Molly Lutcavage, and Graeme Hays for providing useful comments on the draft manuscript.

REFERENCES

Andrews, R. D. 1998a. Instrumentation for the remote monitoring of physiological and behavioral variables. *J. Appl. Physiol.* 85:1974–81.

———. 1998b. Remotely releasable instruments for monitoring the foraging behaviour of pinnipeds. *Mar. Ecol. Prog. Ser.* 175:289–94.

———. 1999. The cardiorespiratory, metabolic, and thermoregulatory physiology of juvenile northern elephant seals *(Mirounga angustirostris).* Ph.D. diss., University of British Columbia, Vancouver, Canada.

Andrews, R. D., Jones, D. R., Williams, J. D., Thorson, P. H., Oliver, G. W., Costa, D. P., and Le Boeuf, B. J. 1997. Heart rates of northern elephant seals diving at sea and resting on the beach. *J. Exp. Biol.* 200:2083–95.

Brodeur, R. D., Mill, C. E., Overland, J. E., Walters, G. E., and Schumacher, J. D. 1999. Evidence for a substantial increase in gelatinous zooplankton in the Bering Sea, with possible links to climate change. *Fish. Oceanogr.* 8:296–306.

Butler, P. J., and Jones, D. R. 1997. Physiology of diving of birds and mammals. *Physiol. Rev.* 77:837–99.

Carr, A. 1995. Notes on the behavioural ecology of sea turtles. In *Biology and Conservation of Sea Turtles,* ed. K. A. Bjorndal, 19–26. Rev. ed. Washington: Smithsonian Institution Press.

Davenport, J. 1998. Sustaining endothermy on a diet of cold jelly: Energetics of the leatherback turtle *Dermochelys coriacea. Brit. Herp. Soc. Bull.* 62:4–8.

Duron, M. 1978. Contribution à l'étude de la biologie de *Dermochelys coriacea* (Linné) dans les Pertuis Charentais. Ph.D. diss., Université de Bordeaux.

Eckert, S. A. 2000. Letter to the editor. *National Geographic,* October 2000.

Eckert, S. A., Eckert, K. L., Ponganis, P., and Kooyman, G. L. 1989. Diving and foraging behavior of leatherback sea turtles *(Dermochelys coriacea). Can. J. Zool.* 67:2834–40.

Eckert, S. A., Nellis, D. W., Eckert, K. L., and Kooyman, G. L. 1986. Diving patterns of two leatherback sea turtles *(Dermochelys coriacea)* during internesting intervals at Sandy Point, St. Croix., U.S. Virgin Islands. *Herpetologica* 42:381–88.

Eckert, S. A., and Sarti, L. 1997. Distant fisheries implicated in the loss of the world's largest leatherback nesting population. *Mar. Turtle Newsl.* 78:2–7.

Else, P. L., and Hulbert, A. J. 1981. Comparison of the "mammal machine" and the "reptile machine": Energy production. *Am. J. Physiol.* 240:R3–R9.

Fedak, M. A., Pullen, M. R., and Kanwisher, J. 1988. Circulatory responses of seals to periodic breathing: Heart rate and breathing during exercise and diving in the laboratory and open sea. *Can. J. Zool.* 66:53–60.

Fisher, P. E., Russell, D. A., Stoskopf, M. K., Barrick, R. E., Hammer, M., and Kuzmitz, A. A. 2000. Cardiovascular evidence for an intermediate or higher metabolic rate in an ornithischian dinosaur. *Science* 288:503–5.

Frair, W., Ackman, R. G., and Mrosovsky, N. 1972. Body temperature of *Dermochelys coriacea:* Warm turtle from cold water. *Science* 177:791–93.

Hindell, M. A., Slip, D. J., Burton, H. R., and Bryden, M. M. 1992. Physiological implications of continuous, prolonged, and deep dives of the southern elephant seal *(Mirounga leonina). Can. J. Zool.* 70:370–79.

Hulbert, A. J., and Else, P. L. 1981. Comparison of the "mammal machine" and the "reptile machine": Energy use and thyroid activity. *Am. J. Physiol.* 241:R350–R356.

———. 1990. The cellular basis of endothermic metabolism: A role for "leaky" membranes? *NIPS* 5:25–28.

Keinath, J. A., and Musick, J. A. 1993. Movements and diving behavior of a leatherback turtle, *Dermochelys coriacea. Copeia* 1993 (4): 1010–17.

Kooyman, G. L., Castellini, M. A., Davis, R. W., and Maue, R. A. 1983. Aerobic diving limits of immature Weddell seals. *J. Comp. Physiol. B* 151:171–74.

Le Boeuf, B. J., Naito, Y., Huntley, A. C., and Asaga, T. 1989. Prolonged, continuous, deep diving by northern elephant seals. *Can. J. Zool.* 67:2514–19.

Lutcavage, M. E., Bushnell, P. G., and Jones, D. R. 1990. Oxygen transport in leatherback sea turtle *Dermochelys coriacea. Physiol. Zool.* 63:1012–24.

———. 1992. Oxygen stores and aerobic metabolism in the leatherback sea turtle. *Can. J. Zool.* 70:348–51.

Lutcavage, M. E., and Lutz, P. L. 1986. Metabolic rate and food energy requirements of the leatherback sea turtle, *Dermochelys coriacea. Copeia* 1986 (3): 796–98.

Morreale, S. J., Standora, E. A., Spotila, J. R., and Paladino, F. V. 1996. Migration corridor for sea turtles. *Nature* 384:319–20.

Nagy, K. A. 1989. Doubly-labeled water studies of vertebrate physiological ecology. In *Ecological Studies: Stable Isotopes in Ecological Research*, ed. P. W. Rundel, J. R. Ehleringer, and K. A. Nagy, 68:270–87. New York: Springer-Verlag.

Paladino, F. V., O'Connor, P., and Spotila, J. R. 1990. Metabolism of leatherback turtles, gigantothermy, and thermoregulation of dinosaurs. *Nature* 344:858–60.

Parsons, T. R. 1992. The removal of marine predators by fisheries and the impact of trophic structure. *Mar. Poll. Bull.* 25:51–53.

Pauly, D., Christensen, V., Dalsgaard, J., Froese, R., and Torres, F., Jr. 1998. Fishing down marine food webs. *Science* 279:860–63.

Pennick, D. N., Spotila, J. R., O'Connor, M. P., Steyermark, A. C., George, R. H., Salice, C. H., and Paladino, F. V. 1998. Thermal independence of muscle tissue metabolism in the leatherback turtle, *Dermochelys coriacea. Comp. Biochem. Physiol.* 120A: 399–403.

Schmidt-Nielsen, K. 1990. *Animal Physiology: Adaptation and Environment.* Cambridge: Cambridge University Press.

Southwood, A. L., Andrews, R. D., Lutcavage, M. E., Paladino, F. V., West, N. H., George, R. H., and Jones, D. R. 1999. Heart rates and diving behavior of leatherback sea turtles in the eastern Pacific Ocean. *J. Exp. Biol.* 202:1115–25.

Southwood, A. L., Andrews, R. D., Paladino, F. V., and Jones, D. R. 1999. Body temperatures of leatherback sea turtles (*Dermochelys coriacea*) during the internesting interval. *Comp. Biochem. Physiol.* 124A:S21.

Spotila, J. R., Reina, R. D., Steyermark, A. C., Plotkin, P. T., and Paladino, F. V. 2000. Pacific leatherback turtles face extinction. *Nature* 405:529–30.

Steyermark, A. C., Williams, K., Spotila, J. R., Paladino, F. V., Rostal, D. C., Morreale, S. J., Koberg, M. T., and Arauz, R. 1996. Nesting leatherback turtles at Las Baulas National Park, Costa Rica. *Chel. Cons. Biol.* 2:173–83.

Thorson, P. H., and Le Boeuf, B. J. 1994. Developmental aspects of diving in northern elephant seal pups. In *Elephant Seals: Population Ecology, Behavior, and Physiology*, ed. B. J. Le Boeuf and R. M. Laws, 271–89. Berkeley: University of California Press.

6 Experimental Strategies for the Recovery of Depleted Populations of West Indian Rock Iguanas

Allison C. Alberts and John A. Phillips

INTRODUCTION

West Indian rock iguanas (genus *Cyclura*) are the largest extant native vertebrates inhabiting the dry tropical forests of the Greater Antilles and the Bahamas (Schwartz and Henderson 1991). As a group they are threatened with extinction, with six of eight species classified by the IUCN as critically endangered (IUCN 2003). Reasons for their decline include predation on juveniles by introduced species—particularly mongooses, dogs, and feral cats—and free-ranging hoofstock, which not only severely degrade native vegetation but also disturb nesting burrows (Iverson 1978; Henderson 1992; Mitchell 1999; Tolson 2000). The Jamaican iguana, believed extinct until its rediscovery in 1991, may number no more than a hundred adults, and several other species and subspecies have declined to fewer than 500 individuals (Hudson 1994; Vogel, Nelson, and Kerr 1996; Alberts 2000) (Table 6.1).

The loss or reduction of rock iguana populations has serious consequences for Caribbean ecosystems. Hartley et al. (2000) dissected seeds from iguana scat collected in the Dominican Republic and compared their germination rates to those of seeds collected from beneath parent plants. They found that seeds that had passed through the digestive tracts of iguanas germinated more rapidly than seeds that had not, which indicates that iguanas may provide significant benefits to native plants, particularly following hurricane disturbance or in xeric habitats with sporadic rainfall. Seedlings produced from seeds left in Cuban iguana scat have been shown to grow approximately twice as fast as seedlings originating from seeds dissected from iguana scat (Alberts et al., unpublished data). In addition, it has been hypothesized that repetitive cropping of vegetation by iguanas encourages additional shoot and foliage development (Auffenberg 1982; Knapp and Hudson 2004) and that their movement patterns enhance dispersal of seeds into new microhabitats (Iverson 1985). It is possible that grazing by iguanas

TABLE 6.1 *Conservation status of West Indian rock iguana species and subspecies*

Taxon	Range Countries	Estimated Population	IUCN Threat Classification
Turks and Caicos iguana *Cyclura carinata carinata*	Turks and Caicos Islands	30,000	Critically endangered
Bartch's iguana *Cyclura carinata bartschi*	Bahamas	200–300	Critically endangered
Jamaican iguana *Cyclura collei*	Jamaica	100–200	Critically endangered
Rhinoceros iguana *Cyclura cornuta cornuta*	Dominican Republic; Haiti	10,000–17,000	Vulnerable
Mona Island iguana *Cyclura cornuta stejnegeri*	Puerto Rico (Mona Island)	1500–2000	Endangered
Andros Island iguana *Cyclura cychlura cychlura*	Bahamas	2500–5000	Vulnerable
Exuma Island iguana *Cyclura cychlura figginsi*	Bahamas	1000–1200	Endangered
Allen's Cay iguana *Cyclura cychlura inornata*	Bahamas	400–500	Endangered
Cuban iguana *Cyclura nubila nubila*	Cuba	40,000–60,000	Vulnerable
Lesser Caymans iguana *Cyclura nubila caymanensis*	Cayman Islands	1000	Critically endangered
Grand Cayman iguana *Cyclura nubila lewisi*	Cayman Islands	100–175	Critically endangered
Anegada Island iguana *Cyclura pinguis*	British Virgin Islands (Anegada Island)	200	Critically endangered
Ricord's iguana *Cyclura ricordii*	Dominican Republic	2000–4000	Critically endangered
San Salvador iguana *Cyclura rileyi rileyi*	Bahamas	500–1000	Critically endangered
White Cay iguana *Cyclura rileyi cristata*	Bahamas	150–200	Critically endangered
Acklin's iguana *Cyclura rileyi nuchalis*	Bahamas	15,000	Endangered

could increase species diversity of local plant communities, as has been shown for other native herbivores (Collins et al. 1998).

For several reasons, the Cuban iguana *(Cyclura nubila)* provides an ideal model for developing experimental conservation strategies that can be applied to the more highly endangered West Indian iguanas. Healthy populations of Cuban iguanas still exist in their native range (Perera 1985, 2000), data on their ecology and behavior in the wild are available (Berovides 1980; Christian 1986; Cubillas and Berovides 1991; Rodriguez-Schettino 1991), and their captive husbandry is relatively well understood (Duval and Christie 1990; Rehák 1994; Lemm and Alberts 1997). Although interspecific differences exist, West Indian iguanas as a group face common threats and are for the most part very similar in terms of their life-history strategies and resource requirements (Schwartz and Henderson 1991; Alberts 2000).

EXPERIMENTAL ALTERATION OF LOCAL SOCIAL STRUCTURE

In 1993, we initiated a field study of Cuban iguanas inhabiting the U.S. Naval Base at Guantánamo Bay. To gather baseline data on the local social system, we spent a year documenting hormone–behavior interactions in a group of iguanas inhabiting a section of rocky coastline on the windward side of the base (Alberts et al. 2002). Using hand nets, we captured all resident adults once per month throughout the 1993 breeding season. At the time of capture, we made a series of morphological measurements and collected a blood sample from the caudal vein for hormonal analysis (Gorzula, Arocha-Piñango, and Salazar 1976). Animals were permanently marked with a unique crest scale clip and were temporarily marked with a nontoxic painted number on the flank for visual identification from a distance.

Behavioral observations revealed that approximately 80% of adult males engaged in aggressive interactions with other males. Male–male interactions were usually in the form of a chase, but occasionally prolonged stereotypical facial pushing and biting matches ensued. If a male forced his opponent to retreat, he was considered to have won the encounter. Males winning more than 50% of encounters were classified as high-ranking, and those winning less than 50% of encounters were classified as low-ranking. The remaining 20% of males never participated in agonistic interactions and were classified as nonranking. We found that high-ranking males exhibited higher testosterone levels and significantly larger body length, mass, head width, and femoral pore diameter than low-ranking males.

High-ranking males defended small home ranges that each overlapped the ranges of one to four females. Nonranking males occupied peripheral home ranges with very limited access to females, tending to avoid movement to escape aggression. Low-ranking males did not defend territories and instead moved extensively throughout the study area while suffering constant chases by high-ranking males. These three male social classes appear to be functionally equivalent to the territorial male, peripheral/marginal/satellite male, and pseudofemale/sneaker male categories previously documented for green and marine iguanas (Dugan 1982; Rodda 1992; Wikelski, Carbone, and Trillmich 1996).

An analysis of mean distances between pairs of individual Cuban iguanas indicated that each of the resident females on the site was closer to a high-ranking male on average than to a low- or nonranking male. Headbob displays, chases, and mouth gaping, which were generally performed in the context of territorial defense, were exhibited by high-ranking males significantly more often than by low-ranking males. Additionally, there was a trend for courtship to be performed more often by high-ranking males than by other males. Although it is impossible to be certain in the absence of genetic studies, our results suggested that high-ranking males, through their more robust body morphology and behavioral dominance, had better access to potential mates than low- and nonranking males. Only about 30% of all males in our study were classified as high-ranking. If these males contribute disproportionately to the gene pool, then the observed variability in male social behavior may have significant implications for genetic structuring of local populations (Alberts et al. 2002).

During the subsequent breeding season, we conducted an experiment to determine whether temporary alteration of local social structure could increase the probability that sexually mature but genetically underrepresented male iguanas would have enhanced opportunities to mate (Alberts et al. 2002). This type of manipulation represents a unique approach to lizard conservation and, when appropriately applied, has the potential to serve as an important management tool for endangered populations. For the duration of the 1994 breeding season, the five highest-ranked males were temporarily removed from the study site. During the time these males were absent and for 5 weeks after their return, we observed behavior, analyzed hormones, and monitored home-range use as before.

Removal of the high-ranking males immediately and dramatically changed male social structure. Within a few days, the five largest previously low-ranking males began to win more than 50% of their encounters and could be classified as high-ranking. All the previously nonranking males began to move throughout the study site and engage in agonistic interactions with other males, and could be classified as low-ranking. The newly dominant males showed increased rates of headbob display and chases associated with territorial defense, as well as testosterone levels typical of high-ranking males during the breeding season. Active courtship of females was seen in both the newly dominant males as well as in the low-ranking males, which suggests that both classes of males attempted to attract mates in the absence of the previously dominant individuals. Home ranges of the five males that were removed from the study site were oriented such that the ranges each overlapped the home ranges of several females. Once the previously dominant males were removed from the site, the five males that achieved high-ranking status in their absence each defended territories that were strikingly spatially similar to those vacated by the removed individuals.

Resident females did not significantly alter their behavior, hormones, or home-range size in response to removal of the dominant males. As in the previous year, analysis of distances between pairs of individuals showed that females on the site were generally closer to one of the newly high-ranking males than to the low-ranking males. The only observable change in female behavior was a slight increase in aggressive interactions between females, possibly reflecting increased competition for access to limited high-quality mates.

At the close of the breeding season, four of the five previously dominant males were returned to the study site. Unfortunately, one individual escaped from the holding enclosure and could not be retrieved. The other four males each regained their previous territories within 2 days, although the aggressive interactions required for these males to reestablish themselves were among the longest and most intense observed during the entire study. Behavioral observations and home-range mapping for 5 weeks following the return of the dominant males indicated no long-term disruption of behavior or social relationships. On the basis of our results, we conclude that temporary alteration of local social structure may represent a potential management tool for small or otherwise genetically compromised populations by enhancing the chances that a greater percentage of males will have opportunity to mate (Alberts et al. 2002).

EXPERIMENTAL INCUBATION PROTOCOLS

In addition to our research on adult iguanas, we also studied how environmental factors affect egg incubation and the subsequent growth and viability of hatchlings (Alberts et al. 1997). Previous studies on reptile eggs had shown that the physical environment inside the nest affects the survival, metabolism, and growth of the developing embryos (Packard and Packard 1988; Packard and Phillips 1994). Reptile eggs incubated in relatively cool, moist conditions usually consume more of their yolk and hatch at a larger body size than eggs exposed to relatively warm, dry conditions (Packard et al. 1983, 1987). Because environmentally induced variation in size and nutrient reserves of hatchlings may affect their ability to successfully acquire food and escape predators, the physiological responses of reptile embryos to their physical environment could play a key role in their survival (Burger 1989, 1990; Van Damme et al. 1992; Miller 1993; Phillips and Packard 1994).

In April and May 1993, we captured gravid female Cuban iguanas at several different locations on the base and transferred them to a large outdoor enclosure containing twenty artificial nest sites. After oviposition (clutch size two to thirteen eggs), females were released at their site of capture. Eggs were assigned to one of three incubators maintained at 28, 29.5, or 31 °C. Within each incubator, water was combined with vermiculite to create water potentials of −150 kPa (wet), −550 kPa (moist), or −1100 kPa (dry) in three separate containers. These treatments were within the range of environmental conditions encountered in natural nests of Cuban iguanas (Christian and Lawrence 1991; Christian, Lawrence, and Snell 1991). Because eggs from individual females were uniformly distributed among treatments in a three-by-three design, it is unlikely that random clutch effects were confounded with fixed effects arising from the experimental conditions.

Eggs hatched after 89 to 136 days of incubation, with those incubated at higher temperatures hatching sooner. Early-hatching individuals tended to have small amounts of external yolk that were absorbed within a few days, whereas those hatching later emerged with fully internalized yolk sacs. There was a significant negative relationship between female body length and fertility, with larger females producing a higher percentage of infertile eggs. In addition, the initially viable eggs of larger females tended not to survive the incubation period as well as did those of smaller females. That larger, presumably older, female Cuban iguanas showed higher levels of infertility and egg mortality suggests that reproductive senescence may occur in this long-lived species.

The hatchling sex ratio was equal. Whereas water potential had little influence on size of animals at hatching or on their subsequent growth, incubation temperature significantly affected these variables. Although slightly smaller at hatching, iguanas from eggs incubated at higher temperatures had faster growth rates and were larger in body length, mass, and head dimensions than those incubated at lower temperatures beginning at 3 months of age. These higher growth rates persisted through the first year, which resulted in significantly larger body sizes for hatchlings incubated at higher temperatures. In contrast to results for alligators (Lang 1987), our results provided no evidence that the initially faster growth rates of hatchlings incubated at higher temperatures were the result of differences in thermal selectivity (Alberts et al. 1997).

For reptile conservation programs involving artificial egg incubation, it has been proposed that incubation temperatures be maintained in the lower part of the acceptable range to produce hatchlings of larger size and presumably higher quality (Packard and Phillips 1994). However, results of our studies caution that initially larger size is not necessarily indicative of larger size throughout the neonatal period. As has been suggested for turtles (Brooks et al. 1991; McKnight and Gutzke 1993) and alligators (Joanen, McNease, and Ferguson 1987), long-term monitoring of hatchling growth may be essential to determining the ultimate influence of incubation conditions on viability of lizard hatchlings.

EXPERIMENTAL HEADSTARTING STUDIES

Larger reptiles tend to have higher juvenile survival rates than smaller ones because the former are better at avoiding predation and competing for food (Vleck 1988; Ferguson, Brown, and DeMarco 1982; Haskell et al. 1996). This observation provides the basis for headstarting, a conservation strategy in which animals are raised in captivity until they reach a larger, presumably less vulnerable, body size prior to release into the wild. While headstarting has the potential to significantly increase juvenile survival, there are significant concerns associated with it. Headstarted animals may lose their fear of humans and other potential predators, experience problems adapting to natural food sources after months to years on an artificial captive diet, and potentially expose the wild population to novel diseases or parasites acquired in captivity (Campbell 1980; Kleiman 1989; Beck, Cooper, and Griffith 1993; Chiszar, Smith, and Radcliffe 1993; Griffith et al. 1993).

Although headstarting has been employed extensively as a conservation strategy for sea turtles, Frazer (1992) has pointed out that it does not address the fundamental cause of population decline, which appears to be excessive mortality of adult females. He raises the further concern that removal of hatchling turtles for headstarting prevents them from serving important ecological functions in the marine environment. In contrast, most rock iguana populations are depressed because of heavy predation on juveniles by introduced species rather than increased adult mortality or a lack of suitable habitat (Gicca 1980; Iverson 1978; Woodley 1980; Hayes et al. 1995; Gerber and Iverson 2000). Raptors, cuckoos, boas, and racers are the most significant native predators of rock iguanas (Wiewandt 1977; Iverson 1979). These species are all generalist predators, and juvenile iguanas form only one component of their varied diets (Tolson and Henderson 1993; Henderson and Sajdak 1996; Raffaele et al. 1998). As a result, removal of a proportion of juvenile iguanas for headstarting is unlikely to have a significant impact on their population dynamics. Although headstarting may not be appropriate or successful for all reptilian species (Congdon, Dunham, and Van Loben Sels 1993; Heppell, Crowder, and Crouse 1996), for rock iguanas it has the potential to directly address the problem of reduced juvenile recruitment in wild populations (Hudson 2000; Knapp and Hudson 2004) (Table 6.2). Over the last 15 years, headstarting programs for Galápagos land iguanas have proven highly successful in augmenting wild populations (Cayot et al. 1994).

From 1994 to 1997, we conducted an experimental study to investigate the behavioral and physiological effects of an 8- to 20-month headstarting period on juvenile Cuban iguanas (Alberts et al. 2004). At 1 month of age, 45 hatchlings (n = 30 from 1993; n = 15 from 1994) were transported to our off-exhibit lizard research facility at the San Diego Zoo. Each cohort was maintained in a 48-m^3 enclosure equipped with rock and wood climbing structures under UV-transmitting skylights. Because these animals would eventually be returned to the wild, human contact was minimized. During the headstarting period, we carried out a series of experiments to assess how well the juvenile iguanas might be expected to adapt to life in the wild (Alberts et al. 2004).

We conducted monthly antipredator experiments to assess whether a headstarting period altered the response of juvenile iguanas to potential predators. When the juveniles were between 3 and 12 months old, we measured how close individual juveniles would allow a human to approach before fleeing. Loss of the natural flight response in captivity is a particularly important concern if individual variation in wariness is due not to genetic factors but to recent experiences with predators (Snell et al. 1988; but see Schall and Pianka 1980). While mean flushing distances in captivity were only about 80% of those measured in the wild, flushing distances in the captive group increased throughout the first year, indicating that captivity had not significantly decreased fear response over time (Alberts et al. 2004). In addition, we observed that the juveniles often slept together in groups throughout the study, a potential antipredator strategy employed by young green iguanas (Burghardt and Rand 1985).

TABLE 6.2 *West Indian iguanas for which headstarting programs are predicted to be most effective*

Taxon	Feral Predators	Status of Program	Initial Release	Planned Release Schedule	In-Country Partners
Grand Cayman iguana *Cyclura nubila lewisi*	Dogs, cats	Operational since 1990	1996	5 animals per year	National Trust for the Cayman Islands
Jamaican iguana *Cyclura collei*	Dogs, cats, pigs, mongooses	Operational since 1991	1996	15 animals per year	Hope Zoo, University of the West Indies, Natural Resources Conservation Authority
Mona Island iguana *Cyclura cornuta stejnegeri*	Cats, pigs	Operational since 1999	2002	10 animals per year	Puerto Rico Department of Natural and Environmental Resources
Anegada iguana *Cyclura pinguis*	Dogs, cats	Operational since 1997	2003	15 animals per year	British Virgin Islands National Parks Trust
Ricord's iguana *Cyclura ricordii*	Dogs, cats, mongooses	In planning stages	2004	5 animals per year	Parque Zoologico Nacional, Dominican Republic

In a second experiment, we investigated whether headstarted juveniles would be willing to accept natural food sources after several months on an artificial diet (Alberts et al. 2004). At 10 months of age, we offered each of the 1993 hatchlings a simultaneous choice between the freshly collected leaves of a consumed but not highly palatable food plant from their native habitat, the touch-me-not *(Malpighia polytricha)*, and several foods that they had been routinely fed in captivity, including *Hibiscus* leaves, kale leaves, and collard greens. The responses of the juveniles to the various choices showed that although the native plant was not a favored food item, the majority of animals were willing to try it even though it was unfamiliar to them. In addition, there was no difference in the latency to feed on each of the four food types (Alberts et al. 2004). On the basis of these results, we concluded that headstarted juveniles would be unlikely to experience difficulty adapting to natural food sources once released into the wild.

A third concern associated with headstarting is that headstarted animals in captivity could be exposed to novel diseases or pathogens, which could then be transmitted to a naive wild population (Woodford and Kock 1991; Jacobson 1993). To address this issue, we completed an extensive health-screening examination of all juveniles prior to release, including physical examinations, analyses of fecal samples, complete blood counts, and serum chemistry panels (Alberts et al. 1998). We then compared these results to similar data for samples collected from the wild population at Guantánamo Bay. The white blood cell counts and hematocrits of both captive and wild juveniles fell within the expected normal range for healthy reptiles. However, white blood cell counts were somewhat lower in wild individuals, perhaps because they were more heavily parasitized than captive ones. Between 40% and 70% of the wild individuals examined were infected with hemoparasites, as well as an unidentified virus.

Light microscopic evaluation of red blood cells followed by electron microscopy revealed the presence of a single hemoparasite, the piroplasm *Sauroplasma*, in slightly over half of the captive juveniles (Alberts et al. 1998). Given that the juveniles were isolated from other species while in captivity and *Sauroplasma* was also present in 79% of wild juveniles sampled, we assume the captive animals were naturally infected prior to arrival at the zoo. In diverse families of lizards worldwide (Svahn 1976), this organism appears self-limiting and is not associated with any type of pathology or disease. We have also documented *Sauroplasma* in two healthy captive-reared groups of Grand Cayman and Jamaican iguanas, which suggests that it is probably native throughout the Greater Antilles (unpublished data).

Plasma chemistry panels indicated that the calcium, phosphorus, sodium, potassium, and glucose levels of captive and wild iguanas were very similar (Alberts et al. 1998). We did find somewhat higher protein and cholesterol levels in wild juveniles. This may have been a result of higher overall activity levels, as the selected basking temperature of wild iguanas exceeded that of similar-age captives by more than 2 °C. Despite these differences, for both groups all values were within normal ranges and indicative of good nutritional condition. Taken together, the screening results indicated that the headstarted iguanas were healthy and could safely be released.

Prior to release, all juveniles received passive integrated transponder (PIT) tags implanted subcutaneously in the inguinal region for permanent individual identification (Germano and Williams 1993), and a unique combination of colored beads was attached to the nuchal crest for visual identification from a distance (Hayes, Carter, and Mitchell 2000). Although the use of radio transmitters to track movement following release would have been ideal, it was precluded by the presence of radio-controlled ordnance on the base.

In June 1995, we released two groups of juvenile Cuban iguanas—thirty 1993-hatched animals that had been headstarted for 20 months and fifteen 1994-hatched animals that had been headstarted for 8 months—at Guantánamo Bay (Alberts et al. 2004). We chose two release sites, one within a pristine nature reserve and the other in a more disturbed area where contact with people and feral cats would occur. Every 6 months, we returned to Guantánamo to survey the release sites and evaluate the health and status of the released juveniles. Although surveys were conducted over an area ten times the size of the release area, we recaptured only 10% of the released individuals (Alberts et al. 2004). However, given that juvenile rock iguanas show little evidence of philopatry (Wiewandt 1977; Iverson 1979; Christian 1986; Mitchell 1999) and tend to hide in trees and dense brush for the first few years of life (personal observation), it would be premature to conclude that the nonrecaptured animals did not survive. Our recapture rates are similar in magnitude to those reported for other successful reptile reintroduction programs involving crocodilians (King 1990) and turtles (Haskell et al. 1996).

Based on twelve recaptures of headstarted individuals postrelease, we found that growth rates among headstarted and wild iguanas were virtually identical for the first 2.5 years following release (Alberts et al. 2004). Clearly, these headstarted animals did not experience significant difficulty finding, ingesting, and processing natural foods. In addition to measuring growth, we also conducted antipredator experiments on both wild and released juveniles. Although flushing distances had been shorter in captivity, within 2 weeks of release, the behavior of headstarted juveniles matched that of similarly sized wild juveniles in terms of predator avoidance (Alberts et al. 2004). While they had been slightly lower in captivity, basking temperatures selected by headstarted individuals following release were also similar to those of their wild counterparts. Finally, social interactions among the released juveniles appeared to be normal in all respects.

In terms of antipredator, feeding, social, and thermoregulatory behavior, we found that juvenile Cuban iguanas raised in captivity for up to 20 months were not at a disadvantage compared to wild juveniles. Although our results suggest that a headstarting period appears to have no discernable negative effects on the behavior and physiology of juvenile iguanas, we recaptured too few individuals to adequately evaluate the potential positive effects of headstarting on juvenile survival following release. To better address this question, we recommend that future headstarting programs increase the number of release sites and the number of years over which releases occur. The survival and reproduction of released animals should be closely monitored using radiotelemetry, as is currently being attempted for Grand Cayman and Jamaican iguanas (Burton 2000; Wilson et al. 2004). Further studies on the interactions between introduced mammalian predators and juvenile iguanas would be helpful in determining the optimal length for the headstarting period.

CONCLUSIONS

Given their precarious status in the wild, most if not all West Indian rock iguanas are in need of immediate and intensive intervention if they are to survive (Alberts 2000). The existence of a less-threatened species that can serve as a model for the group has allowed us to adopt an experimental approach to designing practical conservation management strategies. The two techniques described here, temporary alteration of local social structure and headstarting, may not be equally appropriate for all species of rock iguanas, and to be fully effective these techniques will need to be combined with other measures, such as predator control, that directly counter the factors responsible for population decline.

In populations for which inbreeding has become a serious threat to genetic integrity, temporary removal of dominant males is likely to be most efficacious for species that show strong dominance polygyny. Because of the possibility that high variance in male reproductive success is naturally maintained through genetic or age-dependent balanced polymorphism, it is important that this strategy be considered only as an emergency interim measure (Alberts et al. 2002). As soon as the effective population size is large enough to insure genetic viability, the social system should be allowed to return to its natural state.

Headstarting appears to be appropriate for those taxa for which reduced juvenile recruitment due to unnatural causes is known to be the overriding threat to wild populations. Currently, headstarting programs are underway for four species of rock iguanas, on Grand Cayman, Jamaica, Anegada Island in the British Virgin Islands, and Mona Island in Puerto Rico (Hudson 2000; Knapp and Hudson 2004). Although these programs range from those in which multiyear survival and reproduction of released animals have been documented (Wilson et al. 2004) to those that have yet to return juveniles to the wild, it is possible to identify several factors that are expected to contribute to the success of headstarting efforts. Paramount among these is that headstarting programs should be conducted in-country and under local control whenever possible. Although for logistical reasons our experimental work was carried out at the zoo, on-site programs not only provide the opportunity to actively engage local community members in conservation efforts but also reduce the potential for disease transmission to wild populations and ensure that headstarted animals are acclimated to local environmental conditions (Knapp and Hudson 2004). Ideally, headstarting will have a defined endpoint, serving only as a stopgap measure to help promote population recovery while predator control programs are implemented. With further refinement, it may be possible to accelerate the headstarting process by utilizing incubation techniques that produce hatchlings capable of achieving the desired body size in a shorter time frame.

ACKNOWLEDGMENTS

We thank Jeff Lemm, Andrew Perry, Joan Price, Lori Jackintell, Tandora Grant, Rich Doyle, Mark Wharton, Lowell Nelson, Rick Hudson, Kelly Bradley, and Lisa Morici for their invaluable field assistance; Sam Karcher and Saula Ligani for assistance with the plant germination studies; Marcie Oliva, Pat Morris, Mike Worley, Sam Telford, and Don Janssen for their collaborative efforts on the health screening of released juveniles; and the U.S. military and civilian personnel at the U.S. Naval Base, Guantánamo Bay, for logistical support and continued hospitality. Blood samples were imported under U.S. Fish and Wildlife Service Permit PRT-783930. This research was supported by the National Science Foundation Conservation and Restoration Biology Program (NSF DEB-9208878 and DEB-9424471), the U.S. Department of Defense, and the Zoological Society of San Diego Conservation and Research Fund.

REFERENCES

Alberts, A. C. 2000. *West Indian Iguanas: Status Survey and Conservation Action Plan.* Gland, Switzerland: IUCN.

Alberts, A. C., Lemm, J. M., Grant, T. D., and Jackintell, L. A. 2004. Testing the utility of headstarting as a conservation strategy for West Indian iguanas. In *Iguanas: Biology and Conservation,* ed. A. C. Alberts, R. L. Carter, W. K. Hayes, and E. P. Martins, 210–19. Berkeley: University of California Press.

Alberts, A. C., Lemm, J. M., Perry, A. M., Morici, L. A., and Phillips, J. A. 2002. Temporary alteration of local social structure in a threatened population of Cuban iguanas *(Cyclura nubila). Behav. Ecol. Sociobiol.* 51:324–35.

Alberts, A. C., Oliva, M. L., Worley, M. B., Telford, S. R., Jr., Morris, P. J., and Janssen, D. L. 1998. The need for pre-release health screening in animal translocations: A case study of the Cuban iguana *(Cyclura nubila). Anim. Cons.* 1:165–72.

Alberts, A. C., Perry, A. M., Lemm, J. M., and Phillips, J. A. 1997. Effects of incubation temperature and water potential on growth and thermoregulatory behavior of hatchling rock iguanas *(Cyclura nubila). Copeia* 1997:766–76.

Auffenberg, W. 1982. Feeding strategy of the Caicos ground iguana, *Cyclura carinata.* In *Iguanas of the World: Their Behavior, Ecology, and Conservation,* ed. G. M. Burghardt and A. S. Rand, 84–116. Park Ridge, N.J.: Noyes.

Beck, B., Cooper, M., and Griffith, B. 1993. Infectious disease considerations in reintroduction programs for captive wildlife. *J. Zoo Wildl. Med.* 24:394–97.

Berovides, V. A. 1980. Notas sobre la ecologia de la iguana *(Cyclura nubila)* en Cayo Rosario. *Cien. Biol.* 5:112–15.

Brooks, R. J., Bobyn, M. L., Galbraith, D. A., Layfield, J. A., and Nancekivell, E. G. 1991. Maternal and environmental influences on growth and survival of embryonic and hatchling snapping turtles *(Chelydra serpentina). Can. J. Zool.* 69:2667–76.

Burger, J. 1989. Incubation temperature has long-term effects on behaviour of young pine snakes *(Pituophis melanoleucus). Behav. Ecol. Sociobiol.* 24:201–7.

———. 1990. Effects of incubation temperature on behavior of young black racers *(Coluber constrictor)* and kingsnakes *(Lampropeltis getulus). J. Herpetol.* 24:158–63.

Burghardt, G. M., and Rand, A. S. 1985. Group size and growth rate in hatchling green iguanas *(Iguana iguana). Behav. Ecol. Sociobiol.* 18:101–4.

Burton, F. 2000. Grand Cayman iguana: *Cyclura nubila lewisi.* In *West Indian Iguanas: Status Survey and Conservation Action Plan,* ed. A. C. Alberts, 45–47. Gland, Switzerland: IUCN.

Campbell, S. 1980. Is reintroduction a realistic goal? In *Conservation Biology: An Evolutionary-Ecological Perspective,* ed. M. E. Soulé and B. A. Wilcox, 263–69. Sunderland, Mass.: Sinauer.

Cayot, L. J., Snell, H. L., Llerena, W., and Snell, H. M. 1994. Conservation biology of Galápagos reptiles: Twenty-five years of successful research and management. In *Captive Management and Conservation of Amphibians and Reptiles*, ed. J. B. Murphy, K. Adler, and J. T. Collins, 297–305. Ithaca, N.Y.: Society for the Study of Amphibians and Reptiles.

Chiszar, D., Smith, H. M., and Radcliffe, C. W. 1993. Zoo and laboratory experiments on the behavior of snakes: Assessments of competence in captive-raised animals. *Am. Zool.* 33:109–16.

Christian, K. 1986. Aspects of the life history of Cuban iguanas on Isla Magueyes, Puerto Rico. *Carib. J. Sci.* 22:159–64.

Christian, K. A., and Lawrence, W. T. 1991. Microclimatic conditions in nests of the Cuban iguana *(Cyclura nubila). Biotropica* 23:287–93.

Christian, K. A., Lawrence, W. T., and Snell, H. L. 1991. Effect of soil moisture on yolk and fat distribution in hatchling lizards. *Comp. Biochem. Physiol.* 99A:13–19.

Collins, S. L., Knapp, A. K., Briggs, J. M., Blair, J. M., and Steinauer, E. M. 1998. Modulation of diversity by grazing and mowing in native tallgrass prairie. *Science* 280:745–47.

Congdon, J. D., Dunham, A. E., and Van Loben Sels, R. C. 1993. Delayed sexual maturity and demographics of Blanding's turtles *(Emydoidea blandingii):* Implications for conservation and management of long-lived organisms. *Cons. Biol.* 7:826–33.

Cubillas, S.O.H., and Berovides, V. A. 1991. Characteristicas de los refugios de la iguana de Cuba, *Cyclura nubila. Rev. Biol.* 5:85–87.

Dugan, B. 1982. The mating behavior of the green iguana, *Iguana iguana*. In *Iguanas of the World: Their Behavior, Ecology, and Conservation*, ed. G. M. Burghardt and A. S. Rand, 320–41. Park Ridge, N.J.: Noyes.

Duval, J. J., and Christie, W. D. 1990. Husbandry of the Cuban ground iguana, *Cyclura n. nubila*, at the Indianapolis Zoo. *Int. Zoo Yrb.* 29:65–69.

Ferguson, G. W., Brown, K. L., and DeMarco, V. G. 1982. Selective basis for the evolution of variable egg and hatchling size in some iguanid lizards. *Herpetologica* 38:178–88.

Frazer, N. B. 1992. Sea turtle conservation and halfway technology. *Cons. Biol.* 6:179–84.

Gerber, G. P., and Iverson, J. B. 2000. Turks and Caicos iguana: *Cyclura carinata carinata*. In *West Indian Iguanas: Status Survey and Conservation Action Plan*, ed. A. C. Alberts, 15–18. Gland, Switzerland: IUCN.

Germano, D. G., and Williams, G. F. 1993. Field evaluation of using passive integrated transponder (PIT) tags to permanently mark lizards. *Herpetol. Rev.* 24:54–56.

Gicca, D. 1980. The status and distribution of *Cyclura r. rileyi* (Reptilia: Iguanidae), a Bahamian rock iguana. *Carib. J. Sci.* 16:9–12.

Gorzula, S., Arocha-Piñango, C. L., and Salazar, C. 1976. A method of obtaining blood by venipuncture from large reptiles. *Copeia* 1976:838–39.

Griffith, B., Scott, J. M., Carpenter, J. W., and Reed, C. 1993. Animal translocation and potential disease transmission. *J. Zoo Wildl. Med.* 24:231–36.

Hartley, L. M., Glor, R. E., Sproston, A. L., Powell, R., and Parmerlee, J. S., Jr.

2000. Germination rates of seeds consumed by two species of rock iguanas *(Cyclura* spp.) in the Dominican Republic. *Carib. J. Sci.* 36:149–51.

Haskell, A., Graham, T. E., Griffin, C. R, and Hestbeck, J. B. 1996. Size related survival of headstarted redbelly turtles *(Pseudemys rubiventris)* in Massachusetts. *J. Herpetol.* 30:524–27.

Hayes, W. K., Carter, R. L., and Mitchell, N. C. 2000. Marking techniques. In *West Indian Iguanas: Status Survey and Conservation Action Plan,* ed. A. C. Alberts, 77–79. Gland, Switzerland: IUCN.

Hayes, W. K., Hayes, D. M., Brouhard, D., Goodge, B., and Carter, R. L. 1995. Population status and conservation of the endangered San Salvador rock iguana, *Cyclura r. rileyi. J. Int. Iguana Soc.* 4:21–30.

Henderson, R. W. 1992. Consequences of predator introductions and habitat destruction on amphibians and reptiles in the Post-Columbus West Indies. *Carib. J. Sci.* 28:1–10.

Henderson, R. W., and Sajdak, R. A. 1996. Diets of West Indian racers (Colubridae: *Alsophis)*: Composition and biogeographic implications. In *Contributions to West Indian Herpetology: A Tribute to Albert Schwartz,* ed. R. Powell and R. W. Henderson, 327–38. Ithaca, N.Y.: Society for the Study of Reptiles and Amphibians.

Heppell, S. S., Crowder, L. B., and Crouse, D. T. 1996. Models to evaluate headstarting as a management tool for long-lived turtles. *Ecol. Appl.* 6:556–65.

Hudson, R. H. 1994. Efforts to save the Jamaican iguana *(Cyclura collei). Reptiles* 1:16–17.

———. 2000. Reintroduction guidelines. In *West Indian Iguanas: Status Survey and Conservation Action Plan,* ed. A. C. Alberts, 70–75. Gland, Switzerland: IUCN.

IUCN. 2003. 2003 IUCN Red List of Threatened Species. Gland, Switzerland: IUCN.

Iverson, J. B. 1978. The impact of feral cats and dogs on populations of the West Indian rock iguana, *Cyclura carinata. Biol. Cons.* 14:63–73.

———. 1979. Behavior and ecology of the rock iguana *Cyclura carinata. Bull. Fl. St. Mus. Biol. Sci.* 24:175–358.

———. 1985. Lizards as seed dispersers? *J. Herpetol.* 19:292–93.

Jacobson, E. R. 1993. Implications of infectious diseases for captive propagation and introduction programs of threatened/endangered reptiles. *J. Zoo Wildl. Med.* 24:245–55.

Joanen, T., McNease, L., and Ferguson, M.J.W. 1987. The effects of egg incubation temperature on post-hatching growth of American alligators. In *Wildlife Management: Crocodiles and Alligators,* ed. G.J.W. Webb, S. C. Manolis, and P. J. Whitehead, 533–37. Sydney: Surrey Beatty.

King, F. W. 1990. Conservation of crocodilians: The release of captive-reared specimens. *End. Sp. Update* 8:48–51.

Kleiman, D. G. 1989. Reintroduction of captive mammals for conservation: Guidelines for reintroducing endangered species into the wild. *Bioscience* 39:152–61.

Knapp, C. R., and Hudson, R. D. 2004. Translocation strategies as a conservation tool for West Indian iguanas: Evaluations and recommendations. In *Iguanas: Biology and Conservation,* ed. A. C. Alberts, R. L. Carter, W. K. Hayes, and E. P. Martins, 199–209. Berkeley: University of California Press.

Lang, J. W. 1987. Crocodilian thermal selection. In *Wildlife Management: Crocodiles and Alligators,* ed. G.J.W. Webb, S. C. Manolis, and P. J. Whitehead, 301–17. Sydney: Surrey Beatty.

Lemm, J. M., and Alberts, A. C. 1997. Guided by nature: Conservation research and captive husbandry of the Cuban iguana. *Reptiles* 5:76–87.

McKnight, C. M., and Gutzke, W.H.N. 1993. Effects of the embryonic environment and of hatchling housing conditions on growth of young snapping turtles *(Chelydra serpentina). Copeia* 1993:475–82.

Miller, K. 1993. The improved performance of snapping turtles *(Chelydra serpentina)* hatched from eggs incubated on a wet substrate persists through the neonatal period. *J. Herpetol.* 27:228–33.

Mitchell, N. C. 1999. Effect of introduced ungulates on density, dietary preferences, home range, and physical condition of the iguana *Cyclura pinguis* on Anegada. *Herpetologica* 55:7–17.

Packard, G. C., and Packard, M. J. 1988. The physiological ecology of reptilian eggs and embryos. In *Biology of the Reptilia,* ed. C. Gans and R. B. Huey, 6:523–605. New York: Alan R. Liss.

Packard, G. C., Packard, M. J., Boardman, T. J., Morris, K. A., and Shuman, R. D. 1983. Influence of water exchanges by flexible-shelled eggs of painted turtles, *Chrysemys picta,* on metabolism and growth of embryos. *Physiol. Zool.* 56:217–30.

Packard, G. C., Packard, M. J., Miller, K., and Boardman, T. J. 1987. Influence of moisture, temperature, and substrate on snapping turtle eggs and embryos. *Ecology* 68:983–93.

Packard, G. C. and Phillips, J. A. 1994. The importance of the physical environment for incubation of reptilian eggs. In *Captive Management and Conservation of Amphibians and Reptiles,* ed. J. B. Murphy, K. Adler, and J. T. Collins, 195–208. Ithaca, N.Y.: Society for the Study of Amphibians and Reptiles.

Perera, A. 1985. Datos sobre abundancia y actividad de *Cyclura nubila* (Sauria: Iguanidae) en los alredededores de Cayo Largo del Sur, Cuba. *Poeyana* 288:1–17.

———. 2000. Cuban iguana, *Cyclura nubila nubila.* In *West Indian Iguanas: Status Survey and Conservation Action Plan,* ed. A. C. Alberts, 36–39. Gland, Switzerland: IUCN.

Phillips, J. A., and Packard, G. C. 1994. Influences of temperature and moisture on eggs and embryos of the white-throated savanna monitor *Varanus albigularis:* Implications for conservation. *Biol. Cons.* 69:1–6.

Raffaele, H., Wiley, J., Garrido, O., Keith, A., and Raffaele, J. 1998. *A Guide to the Birds of the West Indies.* Princeton, N.J.: Princeton University Press.

Rehák, I. 1994. Breeding and ethology of the Cuban ground iguanas, *Cyclura nubila,* at Prague Zoo. *Gazella* 21:61–78.

Rodda, G. H. 1992. The mating behavior of *Iguana iguana*. *Smithson. Contrib. Zool.* 534:1–40.

Rodriguez-Schettino, L. 1985. *The Iguanid Lizards of Cuba*. Gainesville: University Press of Florida.

Schall, J. J., and Pianka, E. R. 1980. The evolution of escape behavior diversity. *Am. Nat.* 115:551–66.

Schwartz, A., and Henderson, R. W. 1991. *Amphibians and Reptiles of the West Indies: Descriptions, Distributions, and Natural History*. Gainesville: University of Florida Press.

Snell, H. L., Jennings, R. D., Snell, H. M., and Harcourt, S. 1988. Intrapopulation variation in predator avoidance performance of Galápagos lava lizards: The interaction of sexual and natural selection. *Evol. Ecol.* 2:353–69.

Svahn, K. 1976. A new piroplasm *Sauroplasma boreale* sp. n. (Haemosporidia, Theileriidae) from the sand lizard *Lacerta agilis* L. *Norw. J. Zool.* 24:1–6.

Tolson, P. J. 2000. Control of introduced species. In *West Indian Iguanas: Status Survey and Conservation Action Plan*, ed. A. C. Alberts, 86–89. Gland, Switzerland: IUCN.

Tolson, P. J., and Henderson, R. W. 1993. *The Natural History of West Indian Boas*. Taunton, Somerset, England: R & A Publishing.

Van Damme, R., Bauwens, D., Braña, F., and Verheyen, R. F. 1992. Incubation temperature differentially affects hatching time, egg survival, and hatchling performance in the lizard *Podarcis muralis*. *Herpetologica* 48:220–28.

Vleck, D. 1988. Embryo water economy, egg size, and hatchling viability in the lizard *Sceloporus virgatus*. *Am. Zool.* 28:87A.

Vogel, P., Nelson, R., and Kerr, R. 1996. Conservation strategy for the Jamaican iguana, *Cyclura collei*. In *Contributions to West Indian Herpetology: A Tribute to Albert Schwartz*, ed. R. Powell and R. W. Henderson, 395–406. Ithaca, N.Y.: Society for the Study of Amphibians and Reptiles.

Wiewandt, T. A. 1977. Ecology, behavior, and management of the Mona Island ground iguana, *Cyclura stejnegeri*. Ph.D. diss., Cornell University, Ithaca, N.Y.

Wikelski, M., Carbone, C., and Trillmich, F. 1996. Lekking in marine iguanas: Female grouping and male reproductive strategies. *Anim. Behav.* 52:581–96.

Wilson, B. S., Alberts, A. C., Graham, K. S., Hudson, R. D., Kerr Bjorkland, R., Lewis, D. S., Lung, N. P., Nelson, R., Thompson, N., Kunna, J. L., and Vogel, P. 2004. Survival and reproduction of repatriated Jamaican iguanas: Head-starting as a viable conservation strategy. In *Iguanas: Biology and Conservation*, ed. A. C. Alberts, R. L. Carter, W. K. Hayes, and E. P. Martins, 220–31. Berkeley: University of California Press.

Woodford, M. H., and Kock, R. A. 1991. Veterinary considerations in re-introduction and translocation projects. *Symp. Zool. Soc. Lond.* 62:101–10.

Woodley, J. D. 1980. Survival of the Jamaican iguana, *Cyclura collei*. *J. Herpetol.* 14:45–49.

7 Endocrinology and the Conservation of New Zealand Birds

John F. Cockrem, Dominic C. Adams, Ellen J. Bennett,
E. Jane Candy, Emma J. Hawke, Sharon J. Henare,
and Murray A. Potter

SUMMARY

Some species of New Zealand birds have become extinct since the arrival of humans, and the numbers and ranges of many others have been dramatically reduced. Today there are active conservation programs for many species, the most intensive of which is the effort to save the kakapo *(Strigops habroptilus)*. In this essay, we present examples of studies in conservation endocrinology for the threatened northern brown kiwi *(Apteryx mantelli)* and for the kakapo.

Captive kiwi are held in nocturnal houses and outdoor pens for breeding and for public display. There has been debate about whether it is stressful for kiwi to be kept indoors and regularly handled for public display, but until recently there were no data available on stress in kiwi. In this study, we measured corticosterone responses to handling in kiwi living in nocturnal houses, and we compared the responses with those of kiwi in outdoor pens and free-living kiwi. We also measured responses in birds handled regularly for public display. The results suggest that kiwi in captivity do not experience chronic stress, and that regular handling of kiwi for public display is not stressful for birds that are accustomed to this procedure.

The kakapo is a large, flightless parrot which is the subject of intensive efforts to save the species from extinction. We contribute to the kakapo recovery program by participating in the management of the species, measuring fecal steroid levels in free-living kakapo, and developing a hormonal method to stimulate breeding.

The biggest challenge for avian endocrinology in conservation is to make a practical contribution to the management of rare birds, and these studies illustrate the value of direct interactions between endocrinologists and conservation managers for the application of endocrinology to conservation.

INTRODUCTION

Conservation of Birds in New Zealand

The bird fauna of New Zealand reflects the country's isolated location, with two-thirds of the species that breed in the New Zealand region being endemic and found nowhere else (Cockrem 1995). New Zealand was first settled by people from Polynesia (the Maori) approximately 1000 years ago. Although the Maori killed birds for food and brought dogs and rats to New Zealand, their impact on the native birds was relatively small (although they did exterminate the moa). Europeans arrived first in Australia and then in New Zealand in the late 1700s and began to clear extensive areas of forest. The Europeans introduced more than thirty species of birds, mainly passerines and waterfowl. They also brought more mammalian predators, including the brown and black rat, mice, stoats, weasels, ferrets, dogs, and cats.

New Zealand has no natural mammalian predators, and the combined effects of habitat reduction and mammalian predation have led to drastic reductions in the range and numbers of most native birds (McLennan et al. 1996; Basse, McLennan, and Wake 1999) and to the extinction of some species, such as the well-known huia *(Heteralocha acutirostris)*. The only habitats where native bird populations were not reduced were islands off the coast of the large North and South Islands. Some islands were completely free of introduced mammalian predators, although many islands did have some predators and hence experienced a reduction in bird numbers. These offshore islands have become crucial for the preservation of certain species of New Zealand birds and are the only places where native forest birds can be seen at densities similar to those that existed before the arrival of humans. The need to preserve bird populations on offshore islands was recognized by the government in the late 1800s, and the first translocations of birds from the mainland to the islands were made more than 100 years ago when kakapo *(Strigops habroptilus)* and kiwi *(Apteryx australis)* were moved from Fiordland to Resolution Island (Hill and Hill 1987). Predators were removed from the islands to make them safe for rare species. The translocation of birds to islands and efforts to control mammalian predators have been the focus of avian conservation efforts in New Zealand.

Conservation Endocrinology

Although translocation and predator control are important conservation strategies, reproduction is the key to survival of species, and reproduction depends on hormones. Endocrine studies can address questions in reproductive biology that are relevant to conservation, and such studies have been conducted on a range of species, including whistling frog, *Litoria ewingi* (Coddington and Cree 1995); tuatara, *Sphenodon punctatus* (Cree, Guillette, and Cockrem 1991; Cree, Cockrem, and Guillette 1992; Brown et al. 1994); green turtle, *Chelonia mydas* (Aguirre et al. 1995); American alligator, *Alligator mississippiensis* (Crain et al. 1997); tree lizard, *Urosaurus ornatus* (Moore, Thompson, and Marler 1991); northern spotted owl, *Strix occidentalis caurina* (Wasser et al. 1997); Japanese ibis, *Nipponia nippon* (Ishii et al. 1994); hare wallaby, *Lagorchestes conspicillatus,* and rock wallaby *Petrogale rothschildi* (Bradshaw 1997); African wild dog, *Lycaon pictus* (Monfort et al. 1997); clouded leopard, *Neofelis nebulosa* (Brown et al. 1995); and giant panda, *Ailuropoda melenoleuca* (Kubokawa et al. 1992). Recently we joined with Professor Susuma Ishii from Waseda University in Tokyo to define conservation endocrinology as "endocrine studies that can contribute to conservation" and to help focus endocrine work on endangered species such as the kakapo in New Zealand and the Japanese ibis in Japan (Cockrem and Ishii 1999, 413).

We have divided conservation endocrinology studies into three types: theoretical, diagnostic, and management oriented (Cockrem and Ishii 1999). Theoretical studies consider basic questions in endocrinology, such as the evolution of the structure and function of gonadotropin molecules. Diagnostic studies use measurements of variables such as plasma or fecal levels of gonadal hormones to address questions about the physiological status of individual animals. Management studies deal directly with practical questions in conservation, such as how to stimulate breeding or to improve fertility in animals of an endangered species. The three types of studies are interrelated, with diagnostic and management studies depending on a basic understanding of endocrinology and reproduction in a species. New Zealand is a small country, which means that endocrinologists can conduct studies at all three levels, interact directly with conservation managers, and contribute directly to conservation programs.

KIWI

Species, Distribution, and Conservation Status

The northern brown kiwi *(Apteryx mantelli)* is one of the four species of kiwi in New Zealand. The other species of kiwi are the southern brown kiwi *(Apteryx australis)*, the great spotted kiwi *(Apteryx haasti)*, and the little

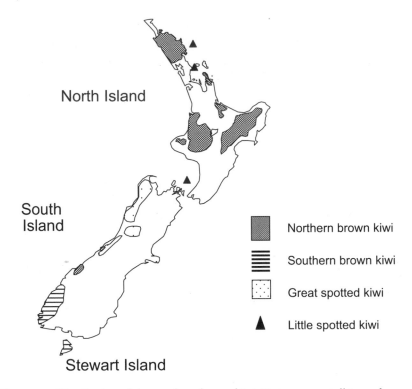

North Island

South
Island

Northern brown kiwi

Southern brown kiwi

Great spotted kiwi

▲ Little spotted kiwi

Stewart Island

Figure 7.1. Distribution of the northern brown kiwi *(Apteryx mantelli)*, southern brown kiwi *(Apteryx australis)*, great spotted kiwi *(Apteryx haasti)*, and little spotted kiwi *(Apteryx oweni)* in New Zealand today.

spotted kiwi *(Apteryx oweni)*. The northern brown kiwi, the most widely distributed kiwi (Figure 7.1), is found in the North Island and in a small area on the west coast of the South Island. It used to be found throughout the North Island, but its range is steadily decreasing. In the absence of predator control, kiwi numbers will continue to decrease on the New Zealand mainland (i.e., the North and South Islands).

Kiwi in Captivity

In New Zealand, northern brown kiwi are held in captivity for breeding and for public display. The birds are kept in outdoor pens for breeding, but since kiwi are nocturnal, they must be held in indoor nocturnal houses for the public to see them. Although some of the outdoor pens are isolated from the public, people can move close to other pens during the day and sometimes at night. In the nocturnal houses, the light–dark cycle is reversed so that people can see the birds when it is daytime outside. In the standard nocturnal houses, glass walls separate the birds from the public, although in one house, people are separated from the birds by a low wooden fence. Opinions about the value of keeping kiwi in captivity differ. It is difficult to see kiwi in the wild, and few New Zealanders ever see or hear wild kiwi. We feel that it is important for New Zealanders to have the chance to see our distinctive national bird and that captive kiwi can play an important role in education and advocacy for conservation principles. An alternative view is that kiwi should not be held indoors in nocturnal houses. In a few New Zealand institutions, kiwi are brought out of their enclosures so that the public can see the birds close-up and even touch them. In a typical display session, these birds are taken from their enclosure and placed on the ground or in an arena so that people can see the birds from less than a meter away. Some birds become accustomed to this procedure and will forage for food and appear to be undisturbed by the presence of people. Displaying birds in this way is controversial: some emphasize the value of the experience that people get when they can see and touch a kiwi, and others believe such contact between birds and people should not be allowed.

The stress on kiwi, especially birds that are brought out for people to see close-up, is often mentioned in the debates on captive kiwi, but until recently there were no data available to inform the debates. We therefore conducted a study of corticosterone in the northern brown kiwi to provide objective data on stress and stress responses in kiwi.

Stress and Corticosterone

Stress is a subjective term, and questions about the stress experienced by kiwi must be defined in physiological terms before they can be addressed by conservation endocrinologists. A surprising number of publications that discuss stress in animals do not actually define the term. Even when definitions are used, they vary: "a non-specific response of the body to any demand made on it" (Selye 1952; Von Holst 1998), "complex changes which occur in various plasma or adrenal constituents following an adverse stimulus" (Freeman 1971, 263), "an animal's defense mechanisms" (Siegel 1980, 529; Siegel 1995), "the homeostatic, physiological and behavioural responses detectable in the animal resulting from its interactions with environmental stressors" (Stephens 1980, 179), "an environmental effect that is likely to or does reduce the Darwinian fitness of an organism" (Hofer and East 1998, 412), and "the cumulative response of an animal resulting from interaction with its environment via receptors" (Fowler 1995, 57).

We use the following definitions:

Stressor—any factor that activates the hypothalamo-pituitary-adrenal axis

Stress response—the response of an animal to one or more stressors

Stress—a state in which an animal is responding to a stressor

The most intensive studies of stress in free-living mammals have been conducted on olive baboons in East Africa (Sapolsky 1983, 1984, 1985, 1991; Alberts, Sapolsky, and Altmann 1992; Brooke et al. 1994; Sapolsky, Romero, and Munck 2000). The work of Sapolsky and colleagues has shown that there are marked differences between individual baboons in their basal cortisol concentrations and in cortisol responses to acute stressors, and that these differences are related to personality. Cortisol is the major glucocorticoid in most mammals, although corticosterone is the main glucocorticoid in rodents.

Corticosterone, the major adrenal glucocorticoid in birds (Carsia and Harvey 2000), is an important metabolic hormone that increases blood glucose levels. It has a variety of other actions such as increasing locomotor activity and feeding behavior, and it can inhibit both the reproductive system and immune function. Corticosterone secretion increases when a bird perceives a stressor (Cockrem and Silverin 2002a), and the hormone promotes changes in behavior and metabolism that help birds to adjust to a stressful situation An acute rise in corticosterone levels in response to a stressor is therefore beneficial. However, if the corticosterone response to a stressor is larger than normal, if it persists for longer than usual, or if basal corticosterone levels are high, then corticosterone can have negative effects.

Basal levels of corticosterone in birds do not indicate their responsiveness to stressors, and it is important to measure corticosterone responses as well as basal levels. For example, Dunlap and Wingfield (1995) showed that corticosterone responses to handling were more informative than basal corticosterone levels for reflecting the extent to which environmental influences were affecting male fence lizards *Scleoporus occidentalis*. Figure 7.2 illustrates how three groups of birds might have similar basal corticosterone levels but markedly different corticosterone responses to a stressor.

Capture and handling are often used as standard stressors in studies of corticosterone responses in birds. The question of whether an animal is stressed can therefore be expressed as two diagnostic questions in conservation endocrinology: What are the basal plasma levels of corticosterone in this animal? and What are the magnitude and duration of the corticosterone response to handling in this animal? (Cockrem 1997).

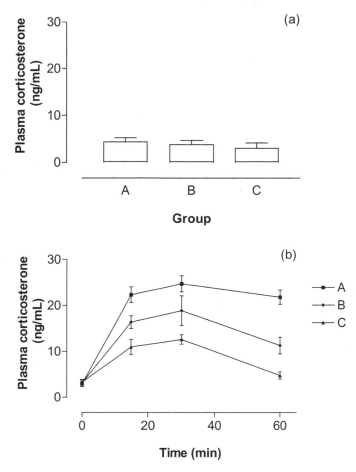

Figure 7.2. Hypothetical basal plasma corticosterone concentrations and plasma corticosterone responses to handling in three groups of birds: *(a)* basal plasma corticosterone concentrations in undisturbed birds and *(b)* plasma corticosterone responses of birds to capture and confinement for 60 min. The example shows plasma corticosterone concentrations in birds that were caught, bled immediately, and then held in a bag or box with further blood samples collected 15, 30, and 60 min after capture. The example shows how groups of birds can have similar plasma corticosterone concentrations but different corticosterone responses to a stressor.

Corticosterone Responses in Captive and Free-Living Kiwi

We used corticosterone measurements to address several questions about stress in the kiwi. We expressed the diagnostic question of whether captive kiwi are stressed as a physiological question: What are the magnitude and duration of the corticosterone response to handling in kiwi in indoor nocturnal houses, outdoor pens, and a natural habitat? We also measured corticosterone in relation to the handling of kiwi for public display. We rephrased the management question of whether kiwi should be handled for public display as a diagnostic question: Are regularly handled kiwi stressed? We then expressed it as a physiological question: What are the magnitude and duration of the corticosterone response in kiwi that are handled regularly and in those that are handled rarely?

Kiwi are flightless and nocturnal forest birds, and endocrine studies of free-living kiwi are not easy. Kiwi must first be located using muzzled dogs and then fitted with a radio transmitter. Radio transmitters allow kiwi to be located in the forest, but tracking down individual birds is still a challenge. Once a signal is detected, it is followed, and the location of the bird can be narrowed down to an area of forest perhaps 10 m². Kiwi roost in underground burrows, and the signal may be coming from beneath the ground. After the location of the burrow is precisely determined, the kiwi must then be extracted from the burrow.

Once the birds are caught, they are subjected to a standard procedure for measuring corticosterone responses. A blood sample is collected immediately. The birds are then placed in a cloth bag or box and removed at later times for additional blood samples to be collected. The birds are released when the sampling is complete. Usually we released birds after 1 hour; in one study we held them longer. We applied this procedure to captive and wild northern brown kiwi and found that they had a clear corticosterone responses to handling. Basal corticosterone levels were low. When birds were handled, corticosterone levels peaked 30 to 60 minutes after they were picked up and then declined toward basal levels over the following several hours. The shape of the corticosterone response and basal corticosterone levels were similar to those of other birds, although peak corticosterone levels were higher than normally seen in other birds.

Corticosterone responses of kiwi in nocturnal houses, outdoor pens, and the wild were compared. Because corticosterone responses may vary during the year (Astheimer, Buttemer, and Wingfield 1995; Silverin 1997), the comparison was conducted in autumn and early winter, when the birds were not breeding. Basal corticosterone levels were low in kiwi in nocturnal houses and outdoor pens, but they were elevated in wild kiwi. It is likely that the kiwi had become aware of our presence when we were locating them, and that a corticosterone response had already been initiated by the time the first sample was collected. Peak corticosterone levels in kiwi in nocturnal houses and outdoor pens were similar to or lower than peak levels in wild kiwi.

Three kiwi accustomed to regular handling for public display were picked up by their regular handlers and placed close to at least two people in the same way they were usually handled for public display. We sampled these birds 30 minutes after the start of a normal handling session, and then 1 week later at the beginning of a session. There was no corticosterone response to handling in the regularly handled birds.

This study of corticosterone in the kiwi shows that corticosterone levels and responses were not elevated in captive kiwi and that corticosterone was not increased by handling in birds that were accustomed to public display. The corticosterone measurements have provided objective data to inform the debates about stress in captive kiwi. This study shows the value of measuring corticosterone levels in birds to address questions about stress in captive birds. Corticosterone responses vary between birds (Littin and Cockrem 2001; Cockrem and Silverin 2002b), and the measurement of corticosterone responses may also prove useful for identifying birds that are less susceptible to stress and more likely to breed in captivity.

KAKAPO

Reproductive Biology

The kakapo, the world's largest parrot, is flightless, nocturnal, and long lived, and reproduction in the kakapo has some unusual features (Cockrem 1999). The kakapo is the only parrot to show lek behavior (Merton, Morris, and Atkinson 1984). Male kakapo prepare display areas called *track and bowl systems* during early summer and then spend many nights producing low-frequency calls known as *booms* to attract females for mating. The intensity and duration of the male booming displays vary greatly from year to year. The availability of food is thought to be important in determining the timing of breeding in kakapo, which naturally occurs once every 3 or 4 years in the autumn, when some forest trees have heavy crops of fruit and seed. Supplementary food has been provided to many of the free-living kakapo on different islands since 1989, but the provision of extra food has not led to yearly breeding (Powlesland and Lloyd 1994). Female kakapo come to the booming bowls to mate and then return to their home range and lay eggs in nests in natural cavities in rocks or underneath trees. The clutch size is small, and the eggs are relatively small. Eggs are incubated for about 30 days, and chicks are fed for 10 weeks, with parental care performed entirely by female kakapo.

Conservation History and Status

Efforts to save the kakapo from extinction began more than 100 years ago, and the kakapo program is now the largest single-species conservation program in New Zealand. Indeed, it may be the most intensive field conservation program for any bird in the world. The total kakapo population in late 2002 was 86 birds, consisting of 36 adult males and 26 adult females, together with 9 males and 15 females fledged from the 2002 breeding season. The great majority of the birds are currently located on Codfish Island.

Kakapo were originally found throughout New Zealand (Butler 1989). However, their numbers steadily declined after the arrival of mammalian predators such as stoats (Williams 1956), until in the early 1970s only 18 male kakapo were known to survive in remote mountain areas in Fiordland (Merton 1999). In 1977, kakapo were found on Stewart Island, and in 1980 the first female kakapo found anywhere since the late 1800s was discovered, also on Stewart Island. However, the kakapo population on Stewart Island was being steadily reduced by predation by cats (Powlesland et al. 1995). Kakapo were therefore moved to Codfish, Maud, Mana, and Little Barrier Islands (see Figure 7.3; Moorhouse and Powlesland 1991; Lloyd and Powlesland 1994). There have been extensive movements of birds between islands since then, with 75% of the birds moved in 1998 and 69% in 1999. By 1989, the number of known female kakapo had dropped to fewer than 20, and in this year the kakapo became the subject of the first single-species recovery plan for the Department of Conservation (Powlesland 1989).

The translocations of kakapo to other islands halted the dramatic decline in the population of the early 1980s, but kakapo numbers did not increase until the mid 1990s (Figure 7.4). Translocating kakapo to islands where the adult kakapo (but not necessarily kakapo eggs or chicks) were free from predation halted the decline in kakapo numbers, but on Little Barrier Island breeding did not occur until supplementary food had been provided to the free-living kakapo for 7 years (Powlesland and Lloyd 1994).

The kakapo conservation program was revised in 1995 with a new Kakapo Recovery Plan (Kakapo Management Group 1996) and the design of detailed plans to intensively manage every individual kakapo and to ensure that the chances of producing fledged young from a fertile kakapo egg were maximized. Every kakapo is now fitted with a radio transmitter so that its position can be determined regularly, and all the birds are caught and checked at

Figure 7.3. Locations where kakapo have been found or transferred to since 1980.

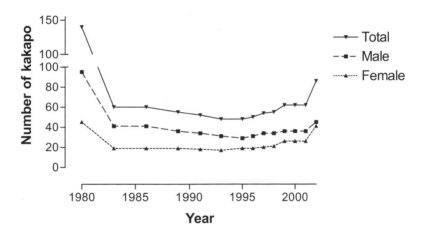

Figure 7.4. Kakapo population, 1980–2002.

approximately yearly intervals. Any health problems are treated either in the field or by taking kakapo to the mainland temporarily.

The intensive management of kakapo breeding was first implemented in 1997, when kakapo bred on Codfish Island. Six kakapo nests were located, and a video camera was placed inside each nest cavity. Activity in the nest was monitored continuously throughout the incubation period and into the chick-rearing period by an observer in a small tent nearby. A timer was started each time the female kakapo left the nest. If she did not return within a certain period, then the eggs were removed and taken for artificial incubation. A grid of rat traps and poison baits was prepared around each nest to decrease the likelihood of predation of eggs or chicks by kiore *(Rattus exulans)*. The chicks were weighed regularly, and one chick was moved to the mainland for hand rearing when its growth rate slowed markedly compared with that of other chicks. Extra food was also provided not far from nests for the female kakapo. Three kakapo chicks fledged from this breeding season; this was the largest number of chicks to fledge in one year since 1981.

Kakapo breeding in 1998 was restricted to one nest on Maud Island in the Marlborough Sounds. The birds, which had been on the island since 1990, had been provided with supplementary food and had boomed in some years, but this was the first year that a female nested on the island. The nest was situated within an area of pine forest *(Pinus radiata)* and not in native vegetation. The nest site was on a slope in a relatively open area and would have failed completely without human assistance. Therefore, a wooden shelter was constructed around and over the nest, with a small entrance tunnel for the female kakapo. All three eggs hatched, and all three young survived to fledge, including the first female chick to fledge since 1992.

In 1999, kakapo bred on Pearl and Little Barrier Islands. Kakapo had been moved from Codfish to Pearl Island in the winter of 1998 so that poison bait could be dropped on Codfish Island to eradicate kiore. The males made booming bowls in late spring even though they had been on the island for only a few months, and a full breeding season followed. Five females laid 14 eggs in 7 nests. The eggs were taken for artificial incubation because of the threat of egg predation by a flightless native rail (the weka, *Gallirallus australis*) on the island. Chicks were hand reared and later returned to Pearl Island and also to Codfish Island. On Little Barrier Island one female laid 3 eggs in a nest. The eggs all hatched, and 3 female kakapo were fledged from the nest. A total of 4 female and 2 male kakapo fledged in 1999, making it the most productive breeding season in the 20 years since the female kakapo had been rediscovered in 1980. This breeding season also saw the first complete artificial incubation of kakapo eggs, the first hand rearing of kakapo from hatch to fledging, and the first time that a female laid a second clutch of eggs after the first clutch was removed. Kakapo laid 77 eggs between 1980 and 2001, but only 19 chicks fledged in this 21-year period (Table 7.1). Four female kakapo fledged in the 18 years from 1981 to 1998, so the fledging of 4 female chicks in 1999 was a significant boost to the population.

There was no breeding in 2000 or 2001, but a spectacular breeding season on Codfish Island in 2002 led to a 39% increase in the total population.

TABLE 7.1 *Number of kakapo fledged, 1980–2002*

Year	Male	Female	Total
1981	2	1	3
1985	0	1	1
1991	2	0	2
1992	0	1	1
1997	3	0	3
1998	2	1	3
1999	2	4	6
Total 1980–2001	11	8	19
2002	9	15	24
Total 1980–2002	20	23	43

SOURCE: Based on data provided by the National Kakapo Team of the Department of Conservation, New Zealand. Years in which no chicks fledged are not represented in the list.

TABLE 7.2 *Breeding success of kakapo, 1980–2002*

Years	Eggs	Fledged Chicks	Breeding Success (%)
1980–1989	15	4	27
1990–1996	30	3	10
1997–2001	32	12	38
Total 1980–2001	77	19	25
2002	67	24	36

SOURCE: Based on data provided by the National Kakapo Team of the Department of Conservation, New Zealand.

Of the 21 adult females, 20 laid eggs and 24 chicks were fledged. The breeding season was associated with an unusually heavy fruiting of some of the forest trees. The fruiting appears to have been greater than in any year since 1980. The longevity of the kakapo (many of the adults are more than 20 years old), their high survivorship from year to year, and intensive management of individual birds and nests mean that despite the low frequency of natural breeding seasons for most birds, the population is likely to survive. The breeding of kakapo from 1980 to 2002 is summarized in Table 7.2, which shows the increase in breeding success since more-intensive management of nests was introduced in 1997 and the very productive breeding season of 2002.

Conservation Endocrinology and the Kakapo

We have contributed to the conservation program for the kakapo since 1989 and are addressing all three types of conservation endocrinology studies in our kakapo work. In collaboration with colleagues in Japan, we plan to sequence kakapo luteinizing hormone and follicle-stimulating hormone. The sequences will provide information about the evolution of gonadotropin hormones and could be used in the future to produce recombinant kakapo hormones for use in the stimulation of breeding. We are analyzing the nutrient content of natural kakapo foods in relation to the timing of breeding. Diagnostic questions about gonadal activity in individual kakapo are being addressed by the measurement of fecal steroid levels in free-living kakapo. A key management question is how to increase the frequency of breeding in kakapo. One approach to stimulating breeding is the use of exogenous hormones, and we are developing methods for hormone treatment in birds.

Regularly collecting blood samples from kakapo has not been practical, but feces can be collected from the forest floor. Non-invasive sampling is attractive, since birds do not have to be caught and handled. However, the collection of fecal samples from free-living kakapo that live in dense forest is not easy. Kakapo have home ranges of many hectares, so if samples are to be collected regularly from individual birds, their roost sites must be located and visited. Roost sites can be located using radiotelemetry, but when the site is visited the day after the bird has left, finding feces is hard, especially if the bird had been roosting in a tree.

In our first study of fecal steroids in kakapo, estradiol and testosterone were measured in samples collected from kakapo on Little Barrier Island by Department of Conservation staff (Cockrem and Rounce 1995). Samples, which were collected from tracks through the forest, could not be ascribed to individual kakapo. However, the majority of kakapo on Little Barrier Island were male, and hence it is likely that the most of the samples were from males. The samples were frozen after collection and shipped frozen to the laboratory, where they were dried. The steroids were extracted into assay buffer, and concentrations of testosterone and estradiol in the extracts were measured by radioimmunoassay. There were clear annual cycles in fecal testosterone concentrations and in the ratio of testosterone to estradiol in fecal samples (Figure 7.5). Testosterone concentrations and the testosterone/

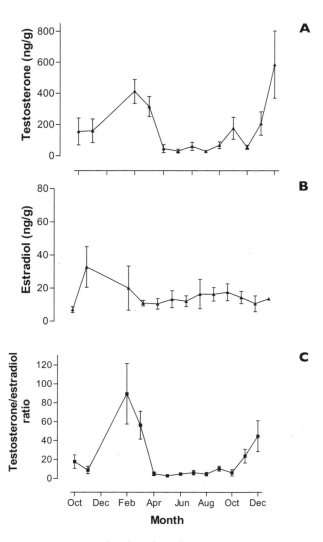

Figure 7.5. Annual cycles of *(A)* fecal testosterone concentration, *(B)* fecal estradiol concentration, and *(C)* the ratio of fecal testosterone to fecal estradiol concentrations in kakapo feces collected from free-living kakapo on Little Barrier Island from October 1989 to December 1990 (data from Cockrem and Rounce 1995). Data are plotted as mean ± standard error. Sample sizes were three to six fecal samples per month except in October 1989 (nine samples) and March 1990 (twenty-four samples).

estradiol ratio peaked in February, at the time of copulation and maximum sexual behavior, declined rapidly in April, and then rose again the following spring, when male sexual behavior was starting again. Annual cycles of plasma testosterone levels in birds generally reflect similar cycles in testis size (Wingfield and Farner 1978; Follett and Robinson 1980; Dawson 1983; Wingfield 1984). The kakapo data suggest that male kakapo are likely to undergo an annual cycle of plasma levels of testosterone and hence of gonad size, with peak hormone levels and maximum gonad size in late summer. We are now analyzing a larger set of samples from Codfish Island in which the samples can be assigned to individual kakapo.

Most of the female kakapo are at least 25 years old, and some have not bred in the last 10 years. We are using Japanese quail *(Coturnix coturnix japonica)* to develop a method for the hormonal stimulation of breeding that could be applied to the kakapo in the future. The aim of this work is to use exogenous hormones to stimulate ovarian growth and ovulation leading to oviposition and the hatching of viable young. There are many challenges associated with this work, including the choice of hormones and hormone-delivery methods, variation between birds in their response to treatment, and the need for practical methods to assess ovarian size and function.

We have used pregnant mare serum gonadotropin (PMSG) and gonadotropin-releasing hormone (GnRH). Some birds respond to a given dose of PMSG with full ovarian development leading to egg laying, whereas others have little or no response. Variation between birds in their response to hormone treatment has commonly been found in studies using a variety of hormones and treatment methods (Mitchell 1967; Mitchell 1970; Imai 1972; Wakabayashi et al. 1992; Wakabayashi, Kikuchi, and Ishii 1996). It will be necessary to develop methods to test the responsiveness of individuals to hormone treatment before initiating a full treatment schedule, and a considerable amount of work is still needed to develop a practical method that can be applied to endangered birds such as the kakapo.

CHALLENGES IN CONSERVATION ENDOCRINOLOGY

There are many challenges in the application of conservation endocrinology to the management of endangered species. The non-invasive measurement of hormones is attractive, but collecting samples from free-living birds is difficult. We do have good endocrine tools to use in the measurement of stress, whereas the development of practical methods for the hormonal stimulation of breeding will require considerable research. In all of these studies, we deal with small sample sizes and individual variation. Applying conservation endocrinology becomes more challenging as we move from theoretical questions to diagnostic questions to management questions. The biggest challenge for avian endocrinology in conservation is to make a practical contribution to the management of rare birds.

ACKNOWLEDGMENTS

The authors are pleased to acknowledge help from the National Kakapo Team of the Department of Conservation (especially Graeme Elliott, Paul Jansen, and Don Merton); from Comalco New Zealand, the primary sponsor of the Kakapo Recovery Program; and from the Royal Forest and Bird Protection Society of New Zealand, a member of the Threatened Species Trust Programme.

REFERENCES

Aguirre, A. A., Balazs, G. H., Spraker, T. R., and Gross, T. S. 1995. Adrenal and hematological responses to stress in juvenile green turtles *(Chelonia mydas)* with and without fibropapillomas. *Physiol. Zool.* 68:831–54.

Alberts, S. C., Sapolsky, R. M., and Altmann, J. 1992. Behavioral, endocrine, and immunological correlates of immigration by an aggressive male into a natural primate group. *Horm. Behav.* 26:167–78.

Astheimer, L. B., Buttemer, W. A., and Wingfield, J. C. 1995. Seasonal and acute changes in adrenocortical responsiveness in an Arctic-breeding bird. *Horm. Behav.* 29:442–57.

Basse, B., McLennan, J. A., and Wake, G. C. 1999. Analysis of the impact of stoats, *Mustela erminea,* on northern brown kiwi, *Apteryx mantelli,* in New Zealand. *Wildl. Res.* 26:227–37.

Bradshaw, S. D. 1997. Water metabolism of endangered marsupial species. In *Advances in Comparative Endocrinology: Proceedings of the XIIIth International Congress of Comparative Endocrinology, Yokohama, Japan, November 16–21, 1997,* ed. S. Kawashima and S. Kikuyama, 1701–5. Bologna: Monduzzi Editore.

Brooke, S. M., de Haas-Johnson, A. M., Kaplan, J. R., Manuck, S. B, and Sapolsky, R. M. 1994. Dexamethasone resistance among nonhuman primates associated with a selective decrease of glucocorticoid receptors in the hippocampus and a history of social instability. *Neuroendocrinol.* 60:134–40.

Brown, J. L., Wildt, D. E., Graham, L. H., Byers, A. P., Collins, L., Barrett, S., and Howard, J. 1995. Natural versus chorionic gonadotropin-induced ovarian responses in the clouded leopard *(Neofelis nebulosa)* assessed by fecal steroid analysis. *Biol. Reprod.* 53:93–102.

Brown, M. A., Cree, A., Daugherty, C. H., Dawkins, B. P., and Chambers, G. K.

1994. Plasma concentrations of vitellogenin and sex steroids in female tuatara *(Sphenodon punctatus punctatus)* from northern New Zealand. *Gen. Comp. Endocrinol.* 95:201–12.

Butler, D. 1989. *Quest for the Kakapo.* Auckland: Heinemann Reed.

Carsia, R. V., and Harvey, S. 2000. Adrenals. In *Sturkie's Avian Physiology,* ed. G. C. Whittow, 489–537. 5th ed. San Diego, Calif.: Academic Press.

Cockrem, J. F. 1995. The timing of seasonal breeding in birds, with particular reference to New Zealand birds. *Reprod. Fert. Devel.* 7:1–19.

———. 1997. Contributions of comparative endocrinology to conservation biology. In *Advances in Comparative Endocrinology: Proceedings of the XIIIth International Congress of Comparative Endocrinology, Yokohama, Japan, November 16–21, 1997,* ed. S. Kawashima and S. Kikuyama, 1695–700. Bologna: Monduzzi Editore.

———. 1999. Saving the kakapo. In *The Use of Wildlife in Research,* eds. D. Mellor and V. Monomay, 60–66. Glen Osmond, Australia: Australia and New Zealand Council for the Care of Animals in Research and Teaching.

Cockrem, J. F., and Ishii, S. 1999. Conservation endocrinology—A new field of comparative endocrinology. In *Recent Progress in Molecular and Comparative Endocrinology,* ed. H. B. Kwon, J.M.P. Joss, and S. Ishii, 413–18. Kwangju, Korea: Hormone Research Center.

Cockrem, J. F., and Rounce, J. R. 1995. Non-invasive assessment of the annual gonadal cycle in free-living kakapo *(Strigops habroptilus)* using fecal steroid measurements. *Auk* 112:253–57.

Cockrem, J. F., and Silverin, B. 2002a. Sight of a predator can stimulate a corticosterone response in the great tit *(Parus major). Gen. Comp. Endocrinol.* 125:248–55.

———. 2002b. Variation within and between birds in corticosterone responses of great tits *(Parus major). Gen. Comp. Endocrinol.* 125:197–206.

Coddington, E. J., and Cree, A. 1995. Effect of acute captivity stress on plasma concentrations of corticosterone and sex steroids in female whistling frogs, *Litoria ewingi. Gen. Comp. Endocrinol.* 100:33–38.

Crain, D. A., Guillette, L. J., Jr., Rooney, A. A., and Pickford, D. B. 1997. Alteration in steroidogenesis in alligators *(Alligator mississippiensis)* exposed naturally and experimentally to environmental contaminants. *Environ. Health Perspect.* 105:528–33.

Cree, A., Cockrem, J. F., and Guillette, L. J. 1992. Reproductive cycles of male and female tuatara *(Sphenodon punctatus)* on Stephens Island, New Zealand. *J. Zool., Lond.* 226:199–217.

Cree, A., Guillette, L. J., and Cockrem, J. F. 1991. Identification of female tuatara in ovulatory condition using plasma sex steroid concentrations. *N.Z. J. Zool.* 18:421–26.

Dawson, A. 1983. Plasma gonadal steroid levels in wild starlings *(Sturnus vulgaris)* during the annual cycle and in relation to the stages of breeding. *Gen. Comp. Endocrinol.* 49:286–94.

Dunlap, K. D., and Wingfield, J. C. 1995. External and internal influences on indices of physiological stress. I. Seasonal and population variation in adreno-cortical secretion of free-living lizards, *Scleoporus occidentalis. J. Exp. Zool.* 271:36–46.

Follett, B. K., and Robinson, J. E. 1980. Photoperiod and gonadotropin secretion in birds. *Prog. Reprod. Biol.* 5:39–61.

Fowler, M. E. 1995. Stress. In *Restraint and Handling of Wild and Domestic Animals,* 57–66. 2d ed. Ames: Iowa State University Press.

Freeman, B. M. 1971. Stress and the domestic fowl: A physiological appraisal. *World's Poult. Sci. J.* 27:263–75.

Hill, S., and Hill, J. 1987. *Richard Henry of Resolution Island—a Biography.* Dunedin, N.Z.: John McIndoe.

Hofer, H., and East, M. L. 1998. Biological conservation and stress. *Adv. Stud. Behav.* 27:405–525.

Imai, K. 1972. Effects of avian and mammalian pituitary preparations on follicular growth in hens treated with methallibure or fasting. *J. Reprod. Fert.* 31:387–97.

Ishii, S., Wada, M., Wakabayashi, S., Sakai, H., Kubodera, Y., Yamaguchi, N., and Kikuchi, M. 1994. Endocrinological studies for artificial breeding of the Japanese ibis, *Nipponia nippon,* an endangered avian species in Asia. *J. Biosci.* 19:491–502.

Kakapo Management Group. 1996. *Kakapo Recovery Plan, 1996–2005.* Threatened species recovery plan no. 21, Department of Conservation, Wellington, N.Z.

Kubokawa, K., Ishii, S., Tajima, H., Saitou, K., and Tanabe, K. 1992. Analysis of sex steroids in feces of giant pandas. *Zool. Sci.* 9:1017–23.

Littin, K. E., and Cockrem, J. F. 2001. Individual variation in corticosterone secretion in laying hens. *Brit. Poult. Sci.* 42:536–46.

Lloyd, B. D., and Powlesland, R. G. 1994. The decline of kakapo *Strigops habroptilus* and attempts at conservation by translocation. *Biol. Cons.* 69:75–85.

McLennan, J. A., Potter, M. A., Robertson, H. A., Wake, G. C., Colbourne, R., Dew, L., Joyce, L., McCann, A. J., Miles, J., Miller, P. J., and Reid, J. 1996. Role of predation in the decline of kiwi, *Apteryx spp.,* in New Zealand. *N.Z. J. Ecol.* 20:27–35.

Merton, D. V. 1999. Kakapo. In *Handbook of Australian, New Zealand, and Antarctic Birds,* ed. P. J. Higgins, 633–46. Melbourne, Aust.: Oxford University Press.

Merton, D. V., Morris, R. B., and Atkinson, I.A.E. 1984. Lek behaviour in a parrot: The kakapo *Strigops habroptilus* of New Zealand. *Ibis* 126:277–83.

Mitchell, M. E. 1967. Stimulation of the ovary in hypophysectomized hens by an avian pituitary preparation. *J. Reprod. Fert.* 14:249–56.

———. 1970. Treatment of hypophysectomized hens with partially purified avian FSH. *J. Reprod. Fert.* 22:233–41.

Monfort, S. L., Wasser, S. K., Mashburn, K. L., Burke, M., Brewer, B. A., and

Creel, S. R. 1997. Steroid metabolism and validation of noninvasive endocrine monitoring in the African wild dog (Lycaon pictus). Zoo Biol. 16:533–48.

Moore, M. C., Thompson, C. W., and Marler, C. A. 1991. Reciprocal changes in corticosterone and testosterone levels following acute and chronic handling stress in the tree lizard, Urosaurus ornatus. Gen. Comp. Endocrinol. 81:217–26.

Moorhouse, R. J., and Powlesland, R. G. 1991. Aspects of the ecology of kakapo Strigops habroptilus liberated on Little Barrier Island (Hauturu). N.Z. Biol. Cons. 56:349–65.

Powlesland, R. G., comp. 1989. Kakapo Recovery Plan, 1989–1994. Wellington: Department of Conservation.

Powlesland, R. G., and Lloyd, B. D. 1994. Use of supplementary feeding to induce breeding in free-living kakapo Strigops habroptilus in New Zealand. Biol. Cons. 69:97–106.

Powlesland, R. G., Roberts, A., Lloyd, B. D., and Merton, D. V. 1995. Number, fate, and distribution of kakapo (Strigops habroptilus) found on Stewart Island, New Zealand, 1979–92. N.Z. J. Zool. 22:239–48.

Sapolsky, R. M. 1983. Individual differences in cortisol secretory patterns in the wild baboon: Role of negative feedback sensitivity. Endocrinol. 113:2263–67.

———. 1984. Stress-induced suppression of testosterone titers in the freely-living olive baboon—Role of glucocorticoids. Int. J. Primatol. 5:376.

———. 1985. Stress-induced suppression of testicular function in the wild baboon: Role of glucocorticoids. Endocrinol. 116:2273–78.

———. 1991. Testicular function, social rank, and personality among wild baboons. Psychoneuroendocrinol. 16:281–93.

Sapolsky, R. M., Romero, L. M., and Munck, A. U. 2000. How do glucocorticoids influence stress responses? Integrating permissive, suppressive, stimulatory, and preparative actions. Endocrinol. Rev. 21:55–89.

Selye, H. 1952. The Story of the Adaptation Syndrome. Montreal, Canada: Acta.

Siegel, H. S. 1980. Physiological stress in birds. Biosci. 30:529–34.

———. 1995. Stress, strains, and resistance. Brit. Poult. Sci. 36:3–22.

Silverin, B. 1997. The stress response and autumn dispersal behaviour in willow tits. Anim. Behav. 53:451–59.

Stephens, D. B. 1980. Stress and its measurement in domestic animals: A review of behavioural and physiological studies under field and laboratory situations. Adv. Vet. Sci. Comp. Med. 24:179–210.

Von Holst, D. 1998. The concept of stress and its relevance for animal behavior. Adv. Stud. Behav. 27:1–131.

Wakabayashi, S., Kikuchi, M., and Ishii, S. 1996. Hormonal induction of ovulation and oviposition in Japanese quail kept under a short-day regimen. Poult. Avian Biol. Rev. 7:183–92.

Wakabayashi, S., Kikuchi, M., Wada, M., Sakai, H., and Ishii, S. 1992. Induction of ovarian growth and ovulation by administration of a chicken gonadotropin preparation to Japanese quail kept under a short-day regimen. *Brit. Poult. Sci.* 33:847–58.

Wasser, S. K., Bevis, K., King, G., and Hanson, E. 1997. Non-invasive physiological measures of disturbance in the northern spotted owl. *Cons. Biol.* 11:1019–22.

Williams, G. R. 1956. The kakapo *(Strigops habroptilus,* Gray). *Notornis* 7:29–57.

Wingfield, J. C. 1984. Environmental and endocrine control or reproduction in the song sparrow, *Melospiza melodia.* I. Temporal organization of the breeding cycle. *Gen. Comp. Endocrinol.* 56:406–16.

Wingfield, J. C., and Farner, D. S. 1978. The endocrinology of a natural breeding population of the white-crowned sparrow *(Zonotrichia leucophrys pugetensis). Physiol. Zool.* 51:188–205.

8 Conservation of Australian Arid-Zone Marsupials

Making Use of Knowledge of Their Energy and Water Requirements

Ian D. Hume, Lesley A. Gibson, and Steven J. Lapidge

SUMMARY

Measurement of the energy and water requirements of free-living bilbies *(Macrotis lagotis)* and yellow-footed rock-wallabies *(Petrogale xanthopus)* in the arid zone of western Queensland has provided information that confirms the suspicion that both marsupials have exceptionally low water requirements and are probably independent of a source of free water. Although these low requirements provide these threatened marsupials with some natural protection against introduced competitors (feral goats and rabbits) and exotic predators (foxes and feral cats), more could and should be done to ensure the long-term conservation of these arid-zone marsupials by progressively removing artificial sources of water wherever possible throughout their remaining ranges.

INTRODUCTION

The arid and semi-arid zones of Australia cover 70% of the island continent (Perry 1967) and are important ecosystems in terms of natural resources. These zones have been greatly modified by the introduction of exotic herbivores such as domestic sheep, cattle, goats, and horses. In addition, feral herbivores including the domestic species listed above that have escaped or have been released from captivity, as well as camels, donkeys, pigs (which in the feral state are largely herbivorous), and rabbits have impacted markedly on these environments (Hume 1987). The changes wrought by these herbivores include the removal of much of the original tussock grassland and replacement with subclimax shorter grasses and herbs (Newsome 1975). This change has removed important refuges for small to medium-sized marsupials, leaving them much more exposed to predation by exotic carni-vores such as the European red fox *(Vulpes vulpes)*, which was deliberately introduced in the 1860s and 1870s (Coman 1995); feral cats *(Felis catus)*, whose introduction to Australia probably predates European settlement in 1788 (Newsome 1995); and dingoes *(Canis lupis dingo)*, which were intro-duced much earlier, 3500–4000 years ago (Corbett 1995). As a result, several small marsupial species have become extinct (Figure 8.1), and others survive only on one or a few offshore islands that are free of cats and foxes (Burbidge et al. 1988; Burbidge and McKenzie 1989) or in small pockets of their former range on the mainland (Figure 8.2).

This chapter describes recent studies on two medium-sized arid-zone marsupials, both of which are listed as endangered or threatened (Maxwell, Burbidge, and Morris 1996). The bilby, or rabbit-eared bandicoot *(Macrotis lagotis)*, and the yellow-footed rock-wallaby *(Petrogale xanthopus)* suffered dramatic declines in their numbers *(P. xanthopus)* or in both their range and their numbers *(M. lagotis)* (see Figure 8.2d) in the late 1800s and early 1900s. The declines have continued, albeit at a lower rate. The need for actions to ensure the long-term survival of both species is urgent. Results from recent studies of their energy and water requirements can suggest successful actions. However, the way forward is rendered complex by a number of ecological and socioeconomic factors.

THE BILBY

The bilby was once widespread throughout arid and semi-arid regions and even some temperate regions of Australia, but it is now found in only a small fraction of its former range, in the deserts of central Australia as well as in isolated pockets of grassy and stony plains in far western Queensland (Figure 8.2d). The bilby is one of eight extant species of the marsupial fam-ily Peramelidae in Australia. All peramelid species are omnivores that dig for most of their food, which consists of invertebrates and their larvae, plant roots and seeds, and fungi (Hume 1999). The bilby is the only semifossorial member of the family, living in deep burrows by day and foraging on the surface at night. The reasons for its decline are complex and hotly debated, but Lunney (2001) recently concluded that it can be primarily attributed to the impact of sheep on the fragile environment after they were introduced into the semi-arid zone in the mid 1800s. By the 1880s, the impact of mil-lions of sheep resulted in the removal of much of the low vegetation that provided refuge for small native mammals, including bilbies. The encroach-ment of European rabbits *(Oryctolagus cuniculus)* during the 1880s exacer-

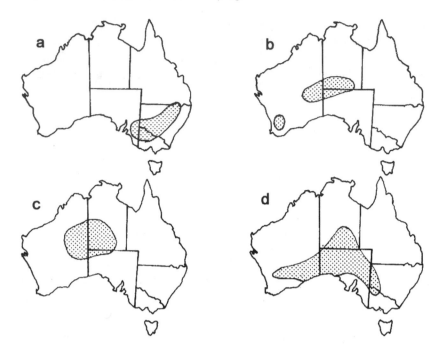

Figure 8.1. Past distributions of four extinct small to medium-sized marsupials in Australia (from Strahan 1983): *(a)* eastern hare-wallaby *(Lagorchestes leporides)*, *(b)* crescent nailtail wallaby *(Onychogalea lunata)*, *(c)* desert bandicoot *(Perameles eremiana), (d)* pig-footed bandicoot *(Chaeropus ecaudatus)*.

bated the problem of removal of cover, and predation by the fox in the last few years of the nineteenth century expedited the demise of many local populations of bilbies, especially in the temperate zone. Predation by feral cats presumably also had an effect, but this has not been measured (Newsome 1995). Cats were probably first introduced into Australia even before European settlement, since many Dutch ships carrying cats for rodent control were blown ashore on the west coast during the seventeenth century while sailing to the Dutch East Indies (now Indonesia).

The work described here was based on an isolated population of bilbies in an area of far western Queensland that has recently been declared a national park (Astrebla Downs National Park) in order to protect the species. The semifossorial habits of the bilby provide some protection against the extremes of climate in its refuge of largely inhospitable arid regions, but we wanted to know what features of the bilby's physiology were also likely to be helpful in an arid environment. Energy and water requirements were measured in free-living bilbies with the doubly-labeled-water method

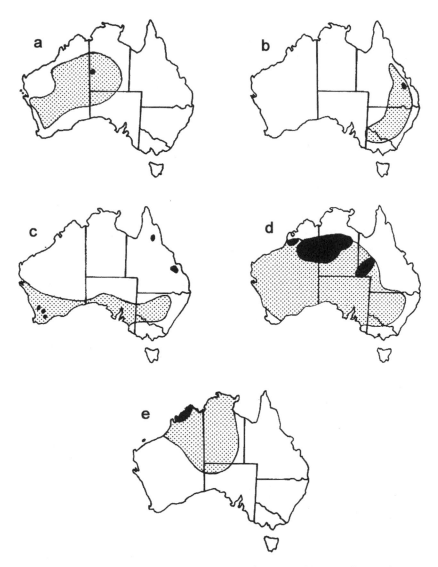

Figure 8.2. Present *(solid)* and past distributions *(dotted)* of five small to medium-size marsupials in Australia (from Strahan 1983): *(a)* rufous hare-wallaby *(Lagorchestes hirsutus)*, *(b)* bridled nailtail wallaby *(Onychogalea fraenata)*, *(c)* brushtailed bettong *(Bettongia penicillata)*, *(d)* bilby *(Macrotis lagotis)*, and *(e)* golden bandicoot *(Isoodon auratus)*.

(Lifson and McClintock 1966). This method provides estimates of the total body water content of the animal (usually expressed as a percentage of body mass), the rate of water turnover (WTR; assumed to be equivalent to water requirement), and the rate of CO_2 production (a measure of metabolic rate). The metabolic rate of a free-living animal is its field metabolic rate (FMR), or total energy cost of free existence (Hume 1999).

We measured a mean total body water of 79.7% of body mass over two summers and two winters (Gibson and Hume 2000). This value is high compared to the range of 57–62% measured by Hulbert and Dawson (1974) in captive bilbies, and it suggests that free-living bilbies store little fat. Total body water as a proportion of body mass varies inversely with body fat (Holleman and Dieterich 1973; Degen 1997).

There were no significant differences in WTR between the seasons. The overall mean of 65.5 mL $\cdot$ kg$^{-0.71}$ $\cdot$ day^{-1} (n = 39) is well below the means recorded for most marsupials (Hume 1999) and is similar to those of three small marsupials from maritime arid environments, the golden bandicoot *(Isoodon auratus)* and spectacled hare-wallaby *(Lagorchestes conspicillatus)* from Barrow Island, and the quokka *(Setonix brachyurus)* from Rottnest Island in Western Australia. The low water requirements of the bilby may reflect its semifossorial lifestyle and nocturnal activity pattern but is interesting in that the bilby's former range included temperate as well as arid zones.

In contrast to the WTR, the overall FMR of the bilby (595 kJ $\cdot$ kg$^{-0.58}$ $\cdot$ day^{-1} [n = 39]) did not differ from that predicted for a marsupial of its body mass, regardless of habitat. Generally, desert-dwelling eutherians have significantly lower FMRs than predicted (Nagy 1994). Not enough arid-zone marsupials have been studied in this regard to allow similar predictions among this group of mammals. However, the relatively high energetic costs of burrowing, and of digging for subterranean food (Gibson and Hume 2000), may explain the unexpectedly high FMR of bilbies.

THE YELLOW-FOOTED ROCK-WALLABY

The yellow-footed rock-wallaby *(Petrogale xanthopus)*, a medium-sized wallaby, has always been confined to rocky hills throughout a range that originally extended from the Flinders Ranges in South Australia through the Barrier Range in western New South Wales to the Grey Range in southwestern Queensland. There is now a disjunction between the Queensland populations and those in NSW and South Australia (Figure 8.3). The time of separation of 180,000 years ago has been sufficient for the two popula-

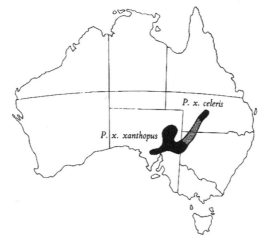

Figure 8.3. Present *(solid)* and past *(dotted)*
distribution of two subspecies of the yellow-footed
rock-wallaby *(Petrogale xanthopus)* (from Strahan
1983). The two subspecies have been separate for
approximately 180,000 years.

tions to be regarded as distinct subspecies, *P. x. xanthopus* in South Australia
and NSW and *P. x. celeris* in Queensland. Although the introduction of
domestic stock (sheep and cattle) into the semi-arid areas inhabited by *P.
xanthopus* had little direct impact because of the refuges afforded by rocky
hills, removal of low plant cover on the plains between isolated hills (mesas)
exposed dispersing rock-wallabies to increased levels of predation. More
recently, this effect has been exacerbated by the clearing of native vegetation
(mainly trees) on the plains, making successful dispersal more difficult and
leading to local extinctions of the species. In the late nineteenth century,
there were also heavy losses through shooting of *P. x. xanthopus* for their
attractively marked skins, which were exported to London from Adelaide. In
addition to these pressures, there has been direct dietary competition from
feral goats. Both yellow-footed rock-wallabies (Sanson 1989) and goats are
mixed feeders. Allen (2001) has recently demonstrated that the two herbivores compete directly; both prefer grasses and forbs (nonwoody dicots)
when available, but in dry periods both switch to a diet of browse. In high
numbers, feral goats can remove virtually all green vegetation to a height of
1.8 m (Henzell 1995). These factors, together with predation by foxes on
young animals, have kept rock-wallaby numbers low throughout their
range and, in NSW, have reduced numbers to dangerously low levels.

In parallel with Allen's (2001) work on their diet, Lapidge's 2001 study measured WTR and FMR in free-living *P. x. celeris* in southwestern Queensland using the methods described above for the bilby. Measurements were made in one (wet) summer and one (dry) winter. As with the bilby, the FMR of the yellow-footed rock-wallaby was not different from the marsupial "mean," at 576 ± 248 kJ $\cdot$ kg$^{-0.58}$ $\cdot$ day^{-1} (n = 8) in summer and 607 ± 194 kJ $\cdot$ kg$^{-0.58}$ $\cdot$ day^{-1} (n = 8) in winter. WTR in summer, which was relatively wet, was 106 ± 31 mL $\cdot$ kg$^{-0.71}$ $\cdot$ day^{-1} (n = 8), a value that is similar the values measured so far for most other macropodid marsupials (Hume 1999). However, in the dry winter, WTR was only 49 ± 14 mL $\cdot$ kg$^{-0.71}$ $\cdot$ day^{-1} (n = 8), which is as low as any measured in a marsupial and is similar to that of the bilby.

Thus for both arid-zone marsupials, energy requirements did not differ significantly from those of many other marsupials and therefore did not suggest an energetically conservative lifestyle. However, in both the bilby and the yellow-footed rock-wallaby (at least the Queensland subspecies), water requirements were among the lowest recorded in free-living marsupials. Whether this holds for the southern subspecies of *P. xanthopus* is not clear. WTRs reported for *P. x. xanthopus* range from 276 mL $\cdot$ kg$^{-0.71}$ $\cdot$ day^{-1} in a wet winter to 101 mL $\cdot$ kg$^{-0.71}$ $\cdot$ day^{-1} in a dry summer (Green 1989). This latter value is double that recorded for our *P. x. celeris* in a dry winter. This high WTR may have been because the *P. x. xanthopus* had access to free water close to their habitat at all times. In contrast, our *P. x. celeris* were radio collared, and at no time were they recorded traveling to free water, which was available 1–3 km from their hills (Lapidge 2001).

CONSERVATION IMPLICATIONS

These data firmly establish the low water requirements often suspected for both bilbies and yellow-footed rock-wallabies, and lead to the conclusion that it is likely that neither marsupial needs access to free water. By contrast, feral goats, significant competitors with rock-wallabies, are well adapted to the Australian arid zone but need to drink in dry seasons (Henzell 1995). Newsome (1995) felt that feral cats are probably independent of free water if feeding on fresh prey, but this is unlikely given the high nitrogen excretory loads, limited kidney concentrating abilities, and high radiant heat loads, which collectively serve to increase water flux (Rochelle Buffenstein, letter to author, November 10, 2001). In any case, feral cats probably have higher water requirements than bilbies or yellow-footed rock-wallabies and tend to retreat from dry areas in drought (Newsome 1995). Foxes may be more dependent on free water.

These differences in water requirements must provide the two marsupials with a measure of natural protection from exotic predators and competitors. This knowledge of their frugal use of water should be utilized; a lack of free water is no ecological barrier to either the bilby or the yellow-footed rock-wallaby. The level of natural protection could be increased by removing artificial watering points from locations in which there are extant populations of bilbies or yellow-footed rock-wallabies.

Artificial watering points made possible the rapid and massive invasion of the Australian semi-arid and arid zones by both exotic herbivores and exotic predators. The Great Artesian Basin, which underlies 22% of the Australian landmass, provided the water (Bentley and Pegler 2000); the first bore was drilled in 1878, and by 1938 there were almost 2000 artesian bores in Queensland. Over the same period, the number of sheep in Queensland doubled to 20 million, and the number of cattle increased to 6 million. Rabbits, foxes, and feral cats followed, as did feral goats a little later. It is the presence of artesian bores that helps to maintain levels of predation on the bilby and yellow-footed rock-wallaby at artificially high levels. Clearly one way to reduce this pressure might be to close off the artesian bores.

There are two obstacles to what might seem a simple solution. First, without this source of water, sheep and cattle ranching would all but collapse in large parts of the semi-arid zone. However, closing the artesian bores is an option in nonranching areas, such as Astrebla Downs National Park (created from a cattle ranch in the hope of ensuring the survival of the bilby), and in Idalia National Park (established partly to protect the habitat of the Queensland yellow-footed rock-wallaby). In these areas, cattle and sheep are no longer present, and in Idalia National Park, open earth dams traditionally filled from natural rainfall are being progressively decommissioned. Also, as Noble et al. (1998) proposed, in some ranching areas a better balance between the provision of water for livestock and the conservation of biological diversity could be achieved by maintaining a patchwork of areas remote from artificial water.

In Astrebla Downs National Park, established in an area in which ranching is largely dependent on artesian bores rather than earth dams, ecological arguments are raised by the prospect of completely closing off existing artesian bores (Noble et al. 1998). Open bore drains, built to reticulate the artesian water around the property, support a variety of plant and animal species in distinctive ecosystems. The drain from the main bore on Astrebla

Downs National Park supports dense reed beds that are home to a rare avian species, the yellow chat *(Epthianura crocea)*. Denying cats and foxes access to the drains by fencing in the bore head and drains would be virtually impossible.

Despite this conflict, the progressive elimination of artificial watering points in conservation areas should remain a priority in order to assist in the long-term survival of bilbies, yellow-footed rock-wallabies, and other small marsupials in Australia's arid zone. Of the two larger arid-zone marsupials, neither of which is threatened, the red kangaroo *(Macropus rufus)* is highly mobile and thus behaviorally adapted for survival in the arid zone (Hume 1999). In contrast, the sedentary euro, or hill kangaroo *(M. robustus erubescens)*, has a thrifty water economy (Freudenberger and Hume 1993) and is thus physiologically adapted to life in the arid zone. Neither kangaroo should be affected significantly by a reduction in the density of artificial watering points.

REFERENCES

Allen, C. B. 2001. Analysis of the diets of three sympatric herbivores from arid west Queensland: The competition of goats past. Ph. D. diss., University of Sydney.

Bentley, D. C., and Pegler, L. K. 2000. The environmental legacy of bore drains. In *Symposium Papers: Australian Rangeland Society Centenary Symposium*, 29–32. Perth: Australian Rangeland Society.

Burbidge, A. A., Johnson, K. A., Fuller, P. J., and Southgate, R. I. 1988. Aboriginal knowledge of the mammals of the central deserts of Australia. *Aust. Wildl. Res.* 15:9–39.

Burbidge, A. A., and McKenzie, N. L. 1989. Patterns in the modern decline of Western Australia's vertebrate fauna: Causes and conservation implications. *Biol. Cons.* 50:143–98.

Coman, B. J. 1995. Fox *Vulpes vulpes*. In *Mammals of Australia*, ed. R. Strahan, 698–99. Chatswood: Reed Books.

Corbett, L. K. 1995. Dingo *Canis lupis dingo*. In *Mammals of Australia*, ed. R. Strahan, 696–98. Chatswood: Reed Books.

Degen, A. A. 1997. *Ecophysiology of Small Desert Mammals*. Berlin: Springer.

Freudenberger, D. O., and Hume, I. D. 1993. Effects of water restriction on digestive function in two macropodid marsupials from divergent habitats and the feral goat. *J. Comp. Physiol. B* 163:247–57.

Gibson, L. A., and Hume, I. D. 2000. Seasonal field energetics and water influx rates of the greater bilby *(Macrotis lagotis)*. *Aust. J. Zool.* 48:225–39.

Green, B. 1989. Water and energy turnover in free-living macropodoids. In *Kangaroos, Wallabies, and Rat-Kangaroos*, ed. G. Grigg, P. Jarman, and I. Hume, 223–29. Sydney: Surrey Beatty.

Henzell, R. 1995. Goat *Capra hircus* Linnaeus 1758. In *Mammals of Australia,* ed. R. Strahan, 728–30. Chatswood: Reed Books.

Holleman, D. F., and Dieterich, R. A. 1973. Body water content and turnover in several species of rodents as evaluated by the tritiated water method. *J. Mammal.* 54:456–65.

Hulbert, A. J., and Dawson, T. J. 1974. Water metabolism in perameloid marsupials from different environments. *Comp. Biochem. Physiol.* 47A:617–33.

Hume, I. D. 1987. Native and introduced herbivores in Australia. In *The Nutrition of Herbivores,* ed. J. B. Hacker and J. H. Ternouth, 1–22. Sydney: Academic Press.

———. 1999. *Marsupial Nutrition.* Cambridge: Cambridge University Press.

Lapidge, S. J. 2001. Re-introduction biology of yellow-footed rock-wallabies *(Petrogale xanthopus celeris* and *P. x. xanthopus).* Ph. D. diss., University of Sydney.

Lifson, N., and McClintock, R. 1966. Theory of the use of the turnover rates of body water for measuring energy and material balance. *J. Theor. Biol.* 12:46–74.

Lunney, D. 2001. Causes of the extinction of native mammals of the western division of New South Wales: An ecological interpretation of the nineteenth century historical record. *Rangel. J.* 23:44–70.

Maxwell, S., Burbidge, A. A., and Morris, K. D. 1996. *The 1996 Action Plan for Australian Marsupials and Monotremes.* Canberra: Wildlife Australia.

Nagy, K. A. 1994. Field bioenergetics of mammals: What determines field metabolic rates? *Aust. J. Zool.* 42:43–53.

Newsome, A. E. 1975. An ecological comparison of the two arid zone kangaroos of Australia, and their anomalous prosperity since the introduction of ruminant stock to their environment. *Quart. Rev. Biol.* 50:389–424.

———. 1995. Cat *Felis catus.* In *Mammals of Australia,* ed. R. Strahan, 700–702. Chatswood: Reed Books.

Noble, J. C., Habermehl, M. A., James, C. D., Landsberg, J., Langton, A. C., and Morton, S. R. 1998. Biodiversity implications of water management in the Great Artesian Basin. *Rangel. J.* 20:275–300.

Perry, R. A. 1967. The need for rangeland research in Australia. *Proc. Ecol. Soc. Aust.* 2:1–14.

Sanson, G. D. 1989. Morphological adaptations of teeth to diets and feeding in the Macropodoidea. In *Kangaroos, Wallabies, and Rat-kangaroos,* ed. G. Grigg, P. Jarman, and I. Hume, 151–68. Sydney: Surrey Beatty.

Strahan, R. 1983, ed. *Complete Book of Australian Mammals.* Sydney: Australian Museum.

9 The Population Decline of Steller Sea Lions

Testing the Nutritional Stress Hypothesis

Russel D. Andrews

INTRODUCTION

Steller sea lions *(Eumetopias jubatus)* are the largest of the otariids, or eared seal subfamily, and certainly one of the most difficult to study. This difficulty has been an important determinate of our unfortunate inability, despite over a decade of intensive research, to satisfactorily explain the precipitous population decline of Steller sea lions (a decline that continues, albeit at a more moderate rate than at the beginning of our population monitoring). The decline was so severe that in November 1990 the United States National Marine Fisheries Service (NMFS) listed Steller sea lions as threatened under the Endangered Species Act. In June 1997, because of continued population declines, NMFS declared Steller sea lions as endangered throughout much of their range. As soon as Steller sea lions became a listed species, the amount of research devoted to determining the cause or causes of the decline increased dramatically. Although field experiments have not been conducted, a combination of laboratory experimentation and field observations has resulted in many insights and, one hopes, will eventually lead to an understanding of the factors that are currently limiting the recovery of Steller sea lions.

Adult male Steller sea lions are an impressive sight, weighing up to 1100 kg, but in this sexually dimorphic species, the females average only 250 kg. Pupping occurs from late May through early July, on rookeries that span the Pacific Rim from northern Japan to central California (Pitcher and Calkins 1981). Females give birth to a single pup, and although the exact timing of weaning is unknown, most mothers are thought to wean their pup just prior to giving birth in the subsequent breeding season. Interbirth intervals have not been determined for individual sea lions. Some females have been observed to continue nursing their young for as long as 3 years. After giving birth, females remain on the rookery to nurse their pups for approximately 9 days. After this perinatal period, females alternate periods of foraging at sea with periods of suckling on shore. Until pups are about 3 months old, they usually remain on the rookery during their mothers' foraging trips.

Although Steller sea lions occur throughout the Pacific Rim, the center of their abundance and distribution was, in the 1970s, from the western end of the Aleutian Islands to the center of the Gulf of Alaska. Evidence from mitochondrial DNA and morphology suggests that there are at least two distinct stocks, a western and an eastern stock, with the dividing line at 144° west longitude, on the edge of Prince William Sound, in the eastern Gulf of Alaska (Bickham, Patton, and Loughlin 1996; Loughlin 1997). In the mid-1960s, there were probably between 200,000 and 300,000 Steller sea lions in the western stock, but by the time that stock was declared endangered in 1997, it had declined by over 75%, to approximately 40,000 individuals (Loughlin 1998). In contrast, over the same period the eastern stock remained stable or increased slightly, with a population size of around 20,000 (Calkins, McAllister, and Pitcher 1999). Therefore, when the NMFS modified the listing status of the Steller sea lion in 1997, the western stock was declared endangered and the eastern stock remained in the threatened category.

While there are a large number of possible causes of the population decline of the western stock, early concerns were raised about the potentially negative impact of the developing groundfish fishery in the Gulf of Alaska and the Bering Sea (Braham, Everitt, and Rugh 1980; Loughlin and Merrick 1989). The take of this fishery was increasing during the period that the sea lion population was dwindling, and the fishery is now valued at nearly U.S.$1 billion per year, making it one of the largest fisheries in the world. Because of the potential interaction between this fishery and Steller sea lions, management regulations have been imposed upon the fishery, with some parties stating that the management actions have been too severe and others claiming they do not go far enough to protect the sea lions. This disagreement has led to a great deal of interest, and controversy, regarding the cause or causes of the decline in the Steller sea lion population.

POTENTIAL CAUSES OF THE POPULATION DECLINE

There are undoubtedly a large number of possible proximate causes of the Steller sea lion population decline, but there are really only four ultimate

TABLE 9.1 *Potential causes of population decline of the western stock of Steller sea lions*

Ultimate Causes	Proximate Causes
Decreased adult survival	Disease
Decreased juvenile survival	Pollution
Decreased fecundity	Direct human killing
Increased emigration	Increased predation
	Environmental change
	Nutritional stress: changes (either natural or anthropogenic) in prey availability or quality

mechanisms to consider: a decline in adult survival, a decline in juvenile survival, a decline in fecundity, or an increase in emigration out of the western stock (Table 9.1). Emigration is an unlikely explanation because rangewide surveys conducted since 1989 have failed to find large numbers of "missing" sea lions (Loughlin 1998). Decreased fecundity, on the other hand, may play an important role. The pregnancy rate of adult females collected in the mid-1970s was 67%, whereas the pregnancy rate was only 55% in females collected in the mid-1980s (Pitcher et al. 1998). This difference, however, was not statistically significant. A modeling study of the population suggested that a much larger decrease in fecundity would have been necessary to explain the observed population decrease (York 1994).

Direct measurements of survival, such as from mark–recapture studies, have not been systematically made on Steller sea lions. There is, however, some evidence that juvenile survival has decreased. The mean age of the adult females collected in the 1980s (Calkins and Goodwin 1988) was about 1.5 years less than the mean age of females collected in the 1970s, and modeling suggested that this reflects a decrease in juvenile survival (York 1994). The model also showed that a 10–20% decrease in juvenile survival was the most likely cause of the population decline in the western stock. This model, however, was based on samples of sea lions that were not randomly collected, and therefore these conclusions must be viewed with caution. Further support for a decrease in juvenile survival comes from the observations of a low number of juveniles compared with adults during aerial surveys of haulouts (Loughlin 1998), although the low number of juveniles could also be the result of low fecundity.

Proximate Causes

A few of the proximate causes for a decrease in either fecundity or juvenile survival are listed in Table 9.1. Although much more work is under way, there is no evidence yet that either disease or pollution played a large role in the population decline (Loughlin 1998). Some Steller sea lions are likely to have died as incidental catch in fishing nets or as a result of having become entangled in marine debris, but this loss was unlikely to have been a major determinant of the decline. Direct killing of sea lions, whether for subsistence or to prevent damage to gear, may have removed a large number of sea lions, especially during the 1970s, but it probably accounted for only a small part of the decline (Trites and Larkin 1992). Currently, subsistence take is only a few hundred sea lions per year, and illegal killing is thought to have been nearly eliminated. The impact of predation by killer whales *(Orcinus orca)* and sharks is difficult to estimate because there have been few observations of such attacks, and these predators are difficult to census. Estimates of the population size of marine mammal–eating killer whales in western Alaska and their energy intake needs have led to the conclusion that although killer whale predation was probably not the primary cause of the Steller sea lion population decline during the 1970s and 1980s, killer whales may currently be responsible for a large fraction of sea lion mortality, now that the population is so much smaller (Barrett-Lennard et al. 1995).

Environmental change, such as increases in sea surface temperature, changes in ocean current patterns, or increases in the frequency of big storms, could impact the survival of Steller sea lions in many ways. For one thing, these changes are likely to affect the sea lions' prey, and it has been frequently proposed that the population decline of Steller sea lions was caused by food limitation, a proposition that has been termed the "nutritional stress hypothesis" (e.g., Alaska Sea Grant 1993; DeMaster and Atkinson 2002). The main evidence that Steller sea lions in the western stock were nutritionally stressed during the period when their population was declining comes from a comparison of female sea lions collected from 1975 to 1978 with females collected from 1985 to 1986 (Calkins and Goodwin 1988). The earlier collection coincided with the approximate start of the decline, and the later collection occurred during a time when the population was declining sharply. The comparison showed that the 1980s females were shorter, thinner, and less massive, so that overall body growth was reduced compared with that of the 1970s females (Calkins, Becker, and Pitcher 1998). Additional support for the nutritional stress hypothesis was that late gestation pregnancy rates fell from 67% in the 1970s to only 55% in the 1980s (Pitcher et al. 1998). Although this difference was not statistically significant, the population decline may have already been under way at the time the first collections were made (Calkins, Becker, and Pitcher 1998), so perhaps the surprisingly low pregnancy rate from the 1970s was depressed below a previously unstressed level. Although these results have been taken to indicate that nutritional stress may have contributed to the decline of Steller sea lions throughout the range of the western stock, it should be noted that the collections were made primarily in the central Gulf of Alaska, with the majority in the Kodiak Island area. In the mid-1970s, the total population in the central Gulf of Alaska region was only about 30% of the entire western stock (Trites and Larkin 1996), so these samples may not necessarily be representative of the entire declining population.

It has been suggested that additional support for the nutritional stress hypothesis comes from the concurrence of the timing of the start of the population decline and a "regime shift" of climatic and oceanographic conditions in the North Pacific that is postulated to have occurred around 1977 (Alaska Sea Grant 1993; Francis et al. 1998; Anderson and Piatt 1999). This regime shift consisted primarily of an intensification and a shift in the location of the wintertime Aleutian low-pressure cell, which led to stronger westerly winds and warmer surface waters in the Gulf of Alaska. These physical changes are thought to be linked to biological responses through their effect on ocean circulation and mixing and air–sea heat exchange (Francis et al. 1998). Numerous authors have concluded that the 1977 regime shift reduced the carrying capacity for many populations of marine mammals and birds in the North Pacific because of a transition from an ecosystem dominated by shrimp and forage fish to one dominated by pollock and large flatfish. This switch in the marine community might have been important to Steller sea lions because pollock usually have a much lower energy density than fatty forage fishes such as herring. Furthermore, there is some indication that the diet of Steller sea lions in at least part of the western range did change around this time, with pollock becoming more common in the diet after the mid-1970s (Alaska Sea Grant 1993; Merrick and Calkins 1996).

Some caveats, however, should be noted concerning the regime shift's proposed effect on prey availability and the resulting possible exposure to nutritional stress. Much of the evidence for the regime shift's effect on the ecosystem is based upon trawl surveys that were intended to assess the abundance of shrimp and that were conducted in limited inshore areas of the Gulf of Alaska (Anderson and Piatt 1999; Mueter and Norcross 2000).

Although the results of those studies might lead one to conclude that pollock were not very plentiful in the North Pacific prior to the mid-1970s, more wide-scale assessments that specifically targeted pollock show that in fact they were abundant throughout the Gulf of Alaska and the Eastern Bering Sea in the 1960s and early 1970s, although there clearly was a strong year class in 1978 that might have benefited from the 1977 regime shift (Bakkala 1993). Herring have been cited as an example of fatty forage fish that once were prevalent in sea lion diets but became less common because of their decline after the 1977 regime shift. Herring in the eastern Bering Sea, however, appear to have been increasing throughout the 1970s (Wespestad 1991). Furthermore, during the 1960s and 1970s, eastern Bering Sea pollock abundance was nearly an order of magnitude higher than herring abundance.

The data on Steller sea lion food habits prior to the mid-1970s are so limited in sample size and geographic coverage that participants in a 1991 workshop entitled Is It Food? concluded that no firm conclusions on the composition of the Steller sea lion diet before 1975 could be drawn. Subsequent diet data from the stomachs of the sea lions collected from 1975 to 1978 and from 1985 to 1986 did indicate that the proportion of sea lions that consumed pollock increased in the Kodiak Island area, but the limited geographic range of those results should prevent one from drawing conclusions that apply to the entire western stock. Additionally, because the mean size of the pollock found in the stomachs from the mid-1980s collections was smaller than that of the pollock from the mid-1970s stomachs, the total amount of pollock consumed by individual sea lions may actually have decreased, not increased (Merrick and Calkins 1996). Nonetheless, the cumulative evidence in support of the nutritional stress hypothesis was so strong that for most of the 1990s it was the leading hypothesis to explain the decline of Steller sea lions and therefore the hypothesis that received the greatest amount of testing. The following review is not meant to be comprehensive, but I will attempt to highlight some of the tests of the nutritional stress hypothesis that I am most familiar with.

LABORATORY EXPERIMENTS AND THE NUTRITIONAL STRESS HYPOTHESIS

Laboratory projects that have been designed to contribute to understanding the role of nutritional stress in the decline of Steller sea lions can be divided into at least three categories: studies that provide basic information on the life history, behavior, or physiology of sea lions; studies to develop or validate techniques that will be subsequently used in the field; and studies that directly test elements of the nutritional stress hypothesis. All three types of studies have been conducted by investigators at the University of British Columbia and the Vancouver Aquarium Marine Science Center, where research with captive Steller sea lions has been pioneered over the last 9 years. Such work has also recently been accomplished at the Mystic Aquarium, in Connecticut, and at the Alaska SeaLife Center, a new facility designed from the ground up to facilitate research on captive marine mammals and other organisms from the North Pacific.

Studies designed to supply basic biological information include those that are providing measurements of the bioenergetic parameters required to estimate the daily energy expenditures of wild sea lions and to estimate how energy balance might be affected by changes in the environment. For example, a recent study on juvenile Steller sea lions swimming in a swim flume has determined the cost of transport at different speeds (Rosen and Trites 2002). These results can be used to estimate the additional costs that might be incurred by sea lions if the distribution of their prey source shifted further offshore, which would require greater swimming distances (Rosen and Trites 2002). One must be mindful, however, of the potential artifacts or biases associated with data collected in a laboratory experiment. In this case, for example, the costs of transport are likely to be overestimated because the swim flume requires the sea lions to swim at a depth less than three times their body diameter. In the wild, sea lions are likely to spend a considerable portion of their time swimming at deeper depths, where the absence of wave drag reduces the cost of swimming.

Although bioenergetic modeling based on data collected in the lab can be quite useful, it would be very valuable if energetic measurements of free-ranging Steller sea lions could be made routinely. The doubly-labeled-water method has been frequently used on pinnipeds in the past, but this method is less than ideal. The method requires that subjects be recaptured for a blood sample, the measurement period is severely limited (5–10 days), and the result obtained is simply the average metabolic rate over the measurement period, with no information on a finer timescale or for individual behaviors. Butler (1993) has suggested that because there appears to be a strong relationship between metabolic rate and heart rate in many animals, the recording of heart rate in seabirds and marine mammals might overcome the limitations imposed by the doubly-labeled-water method. A relationship between heart rate and oxygen consumption for fasted juvenile Steller sea lions was determined by monitoring both variables from captive sea lions at the Vancouver Aquarium over a range of workloads, from resting in air to swimming at up to 1.5 m s^{-1} in a swim flume (McPhee et al. 2003). However, the validity of applying these equations to free-ranging sea lions was called into question when we tested the relationship on a captive male sea lion performing trials that involved the ingestion of a large meal just prior to a 3-hour swim period. In those trials, the heart rate–metabolic rate relationship from fasted animals significantly underestimated oxygen consumption. In addition to the effects of feeding, we are also concerned about the effects of diving and its accompanying cardiovascular adjustments on the relationship. Therefore, we will now have to consider conducting further experiments that more closely match the conditions expected in free-ranging sea lions before attempting to apply this method to animals in the wild.

To test some of the predictions of the nutritional stress hypothesis, we need tools to remotely monitor the foraging behavior of Steller sea lions. Satellite tracking devices and time–depth recorders (TDRs) allow investigators to monitor beyond visual range as animals migrate out into the open ocean and dive beneath the surface, but these techniques do not provide direct evidence of foraging. Most investigators have usually inferred foraging behavior from the shape of time–depth profiles and movements at sea. A direct method of measuring prey ingestion was suggested decades ago (Mackay 1964), but it was not applied to studies of marine animal foraging until recently (Wilson, Cooper, and Plotz 1992). This method relies on the drop in stomach temperature that occurs when a warm endotherm ingests much cooler, ectothermic prey. Captive juvenile Steller sea lions at the Vancouver Aquarium were used to successfully develop and validate the use of stomach temperature monitoring to determine the timing and quantity of prey ingestion in Steller sea lions (Andrews 1998). Estimation of the quantity of ingested prey was complicated by many factors (e.g., body temperature and stomach heat flux changes, movement of the stomach temperature transmitter within the stomach, diverse prey size and shape, potentially concomitant water ingestion, and insulation of the transmitter by previously swallowed prey) and suffered a large margin of error. Determination of the timing of ingestion, however, was much more accurate, at least for the first few ingestion events in a bout of feeding, and was subsequently used in some field studies that are described below.

Laboratory experiments have also been designed to directly test the nutritional stress hypothesis, and they have been particularly useful for examining the potential consequences of reductions in prey quantity or quality. One prediction of the nutritional stress hypothesis is that the health or body condition of Steller sea lions, and therefore eventually their survival or fecundity, will be reduced in response to switching to a diet that has a relatively low energy density. However, most wild animals are known to be adept at coping with fluctuations in energy supply, and they can compensate through various mechanisms, such as by increasing food intake or by reducing activity or even maintenance metabolism (King and Murphy 1985). Rosen and Trites (2000) examined the effect of switching the diet of six captive juvenile Steller sea lions from their normal diet of 100% herring (approximate energy density of 7 kJ g^{-1}) to a diet of 100% pollock (approximate energy density of 4.6 kJ g^{-1}) for an 11–23-day period. The sea lions did not increase their daily food intake and were unable to maintain body mass on the pollock diet. Despite a 15% reduction in resting metabolic rate, the sea lions lost an average of 6.5% of their body mass by the end of the 2-week-long pollock trials. Ingested food mass did not change compared with the previous or subsequent herring feeding periods, but the authors calculated that gross energy intake was substantially reduced during pollock feeding because of the lower energy density, lower digestive efficiency, and higher percentage of gross energy intake devoted to the heat increment of feeding (Rosen and Trites 2000).

Taken at face value, this experiment would seem to strongly support the hypothesis that Steller sea lion populations declined at least in part because of the poor nutritional quality of pollock (assuming one accepts the premise that the consumption of pollock increased after the mid-1970s). However, this conclusion might be premature, for a number of reasons. Limitations of this study included the constraint that sea lions were allowed access to food only during the two to three times per day that their trainers offered it by hand, that the experimental period might have been too short to allow a compensatory adaptation to the new prey source, and that there was no control for seasonal effects, such as programmed body mass or food intake changes. Researchers at the Vancouver Aquarium are hoping to repeat these experiments utilizing a protocol that allows the sea lions free access to their prey, a truly *ad libitum* feeding trial, and a longer time period (D. Rosen, personal communication). A similar feeding experiment was conducted on harbor seals that were switched from a 100% herring diet to a 100% pollock diet. In this study, however, each experimental diet period lasted for 4 months and was repeated at three different times of the year. The results showed that seals can compensate so as to maintain body mass and fat stores even when feeding exclusively on pollock (Castellini et al. 2001).

An experiment that more closely approximates the diet shift that may have occurred among some Steller sea lions in the western stock is currently under way at the Alaska SeaLife Center. This experiment involves three different mixed diets, which are based on the predecline and postdecline diets of the Kodiak Island area and the current diet in southeast Alaska, where sea lion numbers appear to be increasing. Sea lions are maintained on each diet for 4 months, and the trials are repeated at different times of the year to control for seasonal effects. Preliminary results suggest that, like harbor seals, Steller sea lions are indeed capable of adjusting to diets of lower energy density and that it is important to consider seasonal changes in body mass and metabolism (Castellini et al. 2001).

FIELD STUDIES

A decrease in juvenile survival could be due to the inability of mothers to adequately nourish their pups during lactation or to the inability of weaned juveniles to successfully forage on their own. Other pinniped species have been observed to respond to apparent nutritional stress by increasing female foraging trip durations during lactation, increasing energy expenditure during foraging, or both. Studying the foraging behavior and energetics of pups and lactating females should reveal whether Steller sea lions are food stressed in the areas where their numbers continue to decline. Predictions from the nutritional stress hypothesis include the following: lactating female Steller sea lions will increase their foraging effort in the area of population decline, and this increase may be reflected in increased energy expenditure or a change in diving strategy, such as a reduction in the time spent resting; foraging trip durations will be longer in the area of decline; and sea lions in the area of decline will travel for a longer period, cover greater distances before successfully finding and ingesting prey, or both.

In June 1997, a test of the nutritional stress hypothesis was conducted (Andrews et al. 2002). Steller sea lions were studied at two of the central Aleutian Islands, Seguam and Yunaska, and at the Forrester Island rookery complex in southeast Alaska. At the time of the study, the population at Seguam Island was declining by about 5% per year, whereas sea lion numbers at Forrester Island were stable. In the central Aleutians, five lactating Steller sea lions were captured and instrumented with devices to remotely monitor their foraging behavior (Andrews 1998). Four of the sea lions were recaptured, but because one had lost her instrument package, only three foraging records were recovered from this area. Near Forrester Island, ten lactating Steller sea lions were captured and instrumented. Five of these were recaptured and produced successful data records. During the research cruise near Forrester Island, real-time satellite tracking data on the at-sea locations of sea lions were relayed to a vessel conducting a fish assessment around Forrester Island, and a similar fish assessment occurred around Seguam Island that summer.

Although a great deal of variability in foraging behavior was observed (both at the individual and group level), some basic differences between Steller sea lions from the different regions were identified. Trip durations and the percentage of time spent at sea were much shorter for Steller sea lions from Seguam Island compared with those from the Forrester Island rookery. The short trips at Seguam Island generally consisted of a single bout of uninterrupted dive cycles, whereas at Forrester Island the trips were broken into dive bouts of varying length separated by periods spent traveling or resting at the surface. On average, however, the proportion of a trip spent submerged was not different. Another measure of foraging effort, the vertical travel distance per unit time at sea, was about 1.5 times as great for Steller sea lions at Forrester Island. The at-sea field metabolic rates, however, were similar for both groups. Data on the time and distance elapsed from departure on a foraging trip until commencement of foraging dives showed that at both rookeries Steller sea lions appeared to begin searching for prey soon after entering the water. If sea lions swam in a straight line away from the rookery at 1.5 to 2.5 m sec^{-1}, they would have traveled between 1 and 5 km away from the rookery before commencing foraging dives at both Seguam and Forrester. However, the mean time from departure until the first prey-ingestion event identified on the stomach temperature record was about 5 times as long for Steller sea lions at Forrester Island compared to those at Seguam Island. The rough estimation of prey intake rate at Seguam Island was about 2 times that at Forrester Island. Therefore, it would appear that in 1997, adult female Steller sea lions at Seguam Island found suitable prey much quicker, and once they found it, they were able to ingest it at a much higher rate than Steller sea lions at Forrester Island.

The higher prey-capture rate of Steller sea lions at Seguam apparently allowed them to spend shorter periods away from their pups and thereby spend a greater proportion of total time suckling their pups. This increase may account for the nearly doubled pup-growth rates measured in the central Aleutians compared with Forrester Island (Brandon 2000). Surprisingly, all these results were opposite the predictions of our original hypothesis. Our hypothesis was based on the premise that prey availability around the central Aleutian Islands, where the sea lion population continued to decline, was lower than that around Forrester Island. The results of the fish assessments, however, did not support this premise. Catch per unit effort for the fishing vessel at Seguam and another central Aleutians rookery was much higher than that at Forrester Island.

Several factors restrict our ability to infer causes of either the past or current Steller sea lion population decline from this limited comparison of the foraging ecology of Steller sea lions from the declining and stable populations: extremely small sample sizes, the possibility of adverse effects of the instruments on foraging behavior and energetics, the difference between the current rate of decline and the larger rate from 1979 to 1990, density-dependent effects on individual foraging success (reduced population size implies reduced intraspecific competition), and the potential interannual variations in many environmental parameters (e.g., the 1997 El Niño and the anomalous conditions in the Bering Sea that year). Nonetheless, the direct comparison between two similarly handled groups should allow at least a tentative conclusion to be drawn. There was no evidence that lactating females at Seguam Island were nutritionally stressed in 1997.

Other studies of maternal attendance, pup birth weights, and pup growth seem to support the conclusion that throughout the 1990s adult females did not appear to be suffering from nutritional stress (Merrick et al. 1995; Davis et al. 2002). These results, of course, tell us little about the role of nutritional stress before the 1990s, when the population decline was most severe. It is also worth noting that even during the 1990s, studies of lactating females were conducted only at a small subset of rookeries in the range of the western stock. There is at least one other reason that nutritional stress cannot be ruled out as a factor in the population declines of the 1990s. Adult females without pups have not been studied, and if birth rates continue to be as low as 55%, this omission may be important. It is possible that females may have to reach a threshold body condition to implant a fetus or carry it to term. If this is the case, then the healthy females with pups that have been studied to date may not be truly representative of the population. A large number of new studies are currently in progress, so perhaps a more complete understanding of the problem will soon be attained.

Given the theme of this book, it would have been useful if I could have reviewed the results of an experiment that tested the efficacy of the some of the management actions that have been taken over the last decade to protect Steller sea lions. Such experiments, however, have not been conducted. Many experts seem to agree that the existing data are inadequate to determine whether fisheries in the Gulf of Alaska and the Bering Sea are adversely affecting Steller sea lions. Therefore, it might be quite worthwhile to conduct large-scale experiments to determine the effects of fishing on Steller sea lion prey and on the sea lions themselves. Such a proposal raises the ethical question of whether experiments of this type should be conducted on endangered species. There should be little disagreement, however, that the laboratory experiments conducted on Steller sea lions have provided valuable information. As long as they are scientifically justifiable, these experiments should continue because they are essential to our understanding of Steller sea lions and the factors that might be contributing to their population decline.

ACKNOWLEDGMENTS

This manuscript benefited from the constructive comments of Una Swain and an anonymous reviewer. Funding was provided by the Alaska SeaLife Center and the Natural Sciences and Engineering Research Council of Canada. Steller sea lion research conducted by the author and collaborators was funded by the North Pacific Marine Science Foundation and the National Oceanic and Atmospheric Administration (NOAA) through grants to the North Pacific Universities Marine Mammal Research Consortium, the National Marine Mammal Laboratory, and the Alaska Department of Fish and Game, and was permitted by the National Marine Fisheries Service Office of Protected Resources. Critical support was received from individuals at the National Marine Mammal Laboratory, the National Marine Fisheries Service, NOAA, the Alaska Department of Fish and Game, the University of British Columbia, the University of Alaska Fairbanks, and the University of California at Santa Cruz.

REFERENCES

Alaska Sea Grant. 1993. *Is It Food? Addressing Marine Mammal and Seabird Declines.* Publ. no. AK-SG-93–01, Alaska Sea Grant, Fairbanks.

Anderson, P. J., and Piatt, J. F. 1999. Community reorganization in the Gulf of Alaska following ocean climate regime shift. *Marine Ecology Progress Series* 189:117–23.

Andrews, R. D. 1998. Remotely releasable instruments for monitoring the foraging behaviour of pinnipeds. *Marine Ecology Progress Series* 175:289–94.

Andrews, R. D., Calkins, D. G., Davis, R. W., Norcross, B. L., Peijnenberg, K., and Trites, A. W. 2002. Foraging behavior and energetics of adult female Steller sea lions. In *Steller Sea Lion Decline: Is It Food II?* ed. D. DeMaster and S. Atkinson, 19–22. Publ. no. AK-SG-02–02, Alaska Sea Grant, Fairbanks.

Bakkala, R. G. 1993. *Structure and Historical Changes in the Groundfish Complex of the Eastern Bering Sea.* NOAA Technical Report, NMFS 114. Seattle, Wash.

Barrett-Lennard, L. G., Heise, K., Saulitas, E., Ellis, G., and Matkin, C. 1995. The impact of killer whale predation on Steller sea lion populations in British Columbia and Alaska. Unpublished report, North Pacific Universities Marine Mammal Research Consortium, Vancouver, B.C.

Bickham, J. W., Patton, J. C., and Loughlin, T. R. 1996. High variability for control-region sequences in a marine mammal: Implications for conservation and biogeography of Steller sea lions (*Eumetopias jubatus*). *Journal of Mammalogy* 7:95–108.

Braham, H. W., Everitt, R. D., and Rugh, D. J. 1980. Northern sea lion decline in the eastern Aleutian Islands. *Journal of Wildlife Management* 44:25–33.

Brandon, E. A. 2000. Maternal investment in Steller sea lions in Alaska. Ph.D. diss., Texas A&M University, Galveston.

Butler, P. J. 1993. To what extent can heart rate be used as an indicator of metabolic rate in free-living marine mammals? *Symposia of the Zoological Society of London* 66:317–32.

Calkins, D. G., Becker, E. F., and Pitcher, K. W. 1998. Reduced body size of female Steller sea lions from a declining population in the Gulf of Alaska. *Marine Mammal Science* 14:232–44.

Calkins, D. G., and Goodwin, E. 1988. Investigation of the declining sea lion population in the Gulf of Alaska. Unpublished report, State of Alaska, Department of Fish and Game, Anchorage Regional Office.

Calkins, D. G., McAllister, D. C., and Pitcher, K. W. 1999. Steller sea lion status and trend in Southeast Alaska: 1979–1997. *Marine Mammal Science* 15:462–77.

Castellini, M., Calkins, D., Burkanov, V., Castellini, J. M., Trumble, S., and Mau, T. 2001. Long-term feeding trials with harbor seals and Steller sea lions: Impact of different diets on health and nutrition. In *Abstracts of the 14th Biennial Conference on the Biology of Marine Mammals*, 40. Lawrence, Kans.: Society for Marine Mammalogy.

Davis, R. W., Adams, T. C., Brandon, E. A., Calkins, D. G., and Laughlin, T. R. 2002. Female attendance, lactation, and pup growth in Steller sea lions. In *Steller Sea Lion Decline: Is It Food II?* eds. D. DeMaster and S. Atkinson, 23–27. Publ. no. AK-SG-02-02, Alaska Sea Grant, Fairbanks.

DeMaster, D., and Atkinson, S., eds. 2002. *Steller Sea Lion Decline: Is It Food II?* Publ. no. AK-SG-02–02, Alaska Sea Grant, Fairbanks.

Francis, R. C., Hare, S. R., Hollowed, A. B., and Wooster, W. S. 1998. Effects of interdecadal climate variability on the oceanic ecosystems of the NE Pacific. *Fisheries Oceanography* 7:1–21.

King, J. R., and Murphy, M. E. 1985. Periods of nutritional stress in the annual cycles of endotherms: Fact or fiction? *American Zoologist* 25:955–64.

Loughlin, T. R. 1997. Using the phylogeographic method to identify Steller sea lion stocks. In *Molecular Genetics of Marine Mammals*, ed. A. Dizon, S. J. Chivers, and W. F. Perrin, 159–71. Lawrence, Kans.: Society for Marine Mammalogy.

———. 1998. The Steller sea lion: A declining species. *Biosphere Conservation* 1:91–98.

Loughlin, T. R., and Merrick, R. L. 1989. Comparison of commercial harvest of walleye pollock and northern sea lion abundance in the Bering Sea and Gulf of Alaska. In *Proceedings of the International Symposium on the Biology and Management of Walleye Pollock*, 679–700. Publ. no. AK-SG-89–01, Alaska Sea Grant, Fairbanks.

Mackay, R. S. 1964. Deep body temperature of untethered dolphin recorded by ingested radio transmitter. *Science* 144:864–66.

McPhee, J. M., Rosen, D.A.S., Andrews, R. D., and Trites, A. W. 2003. Predicting metabolic rate from heart rate in juvenile Steller sea lions, *Eumetopias jubatus*. *Journal of Experimental Biology* 206:1941–51.

Merrick, R. L., Brown, R., Calkins, D. G., and Loughlin, T. R. 1995. A comparison of Steller sea lion, *Eumetopias jubatus*, pup masses between rookeries with increasing and decreasing populations. *Fishery Bulletin* 93:753–58.

Merrick, R. L., and Calkins, D. G. 1996. Importance of juvenile walleye pollock, *Theragra chalcogramma*, in the diet of Gulf of Alaska Steller sea lions, *Eumetopias jubatus*. In NOAA Technical Report, NMFS 126, pp. 153–66.

Mueter, F. J., and Norcross, B. L. 2000. Changes in species composition of the demersal fish community in nearshore waters of Kodiak Island, Alaska. *Canadian Journal of Fisheries and Aquatic Sciences* 57:1169–80.

Pitcher, K. W., and Calkins, D. G. 1981. Reproductive biology of Steller sea lions in the Gulf of Alaska. *Journal of Mammalogy* 62:599–605.

Pitcher, K. W., Calkins, D. G., and Pendleton, G. W. 1998. Reproductive performance of female Steller sea lions: an energetics-based reproductive strategy? *Canadian Journal of Zoology* 76:2075–83.

Rosen, D.A.S., and Trites, A. W. 2000. Pollock and the decline of Steller sea lions: Testing the junk-food hypothesis. *Canadian Journal of Zoology* 78:1243–50.

———. 2002. Cost of transport in Steller sea lions, *Eumetopias jubatus*. *Marine Mammal Science* 18:513–24.

Trites, A. W., and Larkin, P. A. 1992. *The Status of Steller Sea Lions Populations and the Development of Fisheries in the Gulf of Alaska and Aleutian Islands*. Report to Pacific States Marine Fisheries Commission, Fisheries Centre, University of British Columbia, Vancouver.

———. 1996. Changes in the abundance of Steller sea lions *(Eumetopias jubatus)* in Alaska from 1956 to 1992: How many were there? *Aquatic Mammals* 22:153–66.

Wespestad, V. G. 1991. Pacific herring population dynamics, early life history, and recruitment variation relative to eastern Bering Sea oceanographic factors. Ph.D. diss., University of Washington, Seattle.

Wilson, R. P., Cooper, J., and Plotz, J. 1992. Can we determine when marine endotherms feed? A case study with birds. *Journal of Experimental Biology* 167:267–75.

York, A. 1994. The population of northern sea lions 1975–1985. *Marine Mammal Science* 10:38–51.

Control or Elimination
of Exotic and Invasive Species

10 Overview

Pamela J. Mueller

Conservation biologists summarize the multitude of factors that act together to threaten biodiversity under the acronym HIPPO: habitat destruction, invasive species, pollution; (over)population by humans, and overharvesting (Wilson 2002). Invasive species are considered second only to habitat destruction as a force driving species toward extinction and thus depressing biodiversity worldwide.

Exotic and invasive species are organisms freely living beyond their natural or historical range. Also called foreign, alien, non-indigenous, or nonnative, exotics have reached their new homes through transportation, intentional or otherwise, by humans, and subsequent release, either intentional, accidental, or incidental. With the increasing globalization of the past century, the pace of exotic introductions has increased dramatically. Introduced species, be they plant or animal, vertebrate or insect, microbe, disease or parasite, frequently reproduce and spread exceptionally well in their new environment, where they may lack disease, competition, parasites, or predators. This attribute of invasives has led biologist E. O. Wilson to label exotics as "stealth destroyers": insidious, creeping cancers that can alter entire ecosystems through their establishment and subsequent proliferation.

Not all exotic species are necessarily undesirable. Humans have introduced their favorite plants and animals around the world for agricultural purposes. Wheat, for example, grows in much of the world as an introduced and yet highly desirable exotic plant. The same is true of the majority of important agricultural crops and domestic animals. In North America, a large proportion of ornamental plants are exotic, having been brought from other continents or locales for decorative purposes. It is the ability of certain exotics to escape from confinement and rapidly overwhelm or displace existing organisms that renders exotics dangerous. Hence *invasive* is the most appropriate descriptor for these unwelcome foreigners, and the term has now officially been adopted by U.S. federal agencies to denote exotics whose capacity for conquest has rendered them dangerous to native ecosystems.

Most exotic organisms that arrive on foreign shores do not develop populations large enough to become problematic invaders. But those that do can cause enormous damage to crops, to native species, and to biodiversity. Introduced species threaten biodiversity by preying upon and outcompeting native organisms, particularly island-dwelling mammals and birds (where native species have evolved in isolation from predation pressures) and Australian marsupial fauna (Atkinson 1989; Groombridge 1992). Invasive organisms can cause massive economic and agricultural losses, as well as losses of the biological resources and identity of the invaded region (Pimentel 2002). The spread of invasives reduces available food and habitat for native wildlife, causes declines in populations of endangered species, and introduces novel parasites, pathogens, and toxic compounds. Unwelcome invaders change fire regimes and soil moisture and chemistry; hybridize with native plants; and alter processes of community succession. Currently the U.S. Senate Office of Technology Assessment estimates that 4000 plant and 500 animal species in the United States are introduced exotics, although others put the estimate at 50,000 (Wilson 2002).

Animals introduced for recreational hunting (e.g., rabbits) or as predators—often in misguided attempts to deal with previously introduced invaders (e.g., fox and mongoose to catch rats and rabbits, especially on islands)—were frequent invaders in decades past. Today many of these failed attempts at biological control continue to decimate local small mammal and bird fauna, leaving us with a new set of exotics to contemplate. From starlings to brown tree snakes, purple loosestrife to zebra mussels, and kudzu to gypsy moths, invasive exotic species are a worldwide problem. Less obvious but no less important invaders include parasites, diseases, and microbes, against which native species have evolved no protection (Dobson and May 1986; Pimentel 2002). Corbin, D'Antonio, and Bainbridge (this volume) describe a battle against exotic grasses, a type of invader that frequently escapes notice by the untutored eye.

The scope of the problem is vast. Much attention is focused on dealing with these stealth destroyers; many methods of both prevention and eradication have been, and continue to be, tried in problem sites around the world, with varying levels of success. What then, does this text contribute to the discussion? In keeping with the theme of the conference upon which this book is based, this section highlights the work of scientists who are attacking the invasion problem by applying the techniques of experimental biology. To the extent that it is possible to do so, these scientists apply accepted traditional techniques of experimental design while calling upon modern methodologies to expand their repertoire of tools, in their quest to gather experimentally based information regarding the biology of both invading and indigenous species. Such information forms the backbone of solutions to the invasion problem.

The chapters in this section discuss two cases—grasses (11) and trees (12)—of invasions in two California locations as well as in South Africa, combining discussion of the authors' experiments with their ideas for curbing the invaders. We conclude this section with Mark Hoddle's (13) overview of the history and utility of biological control, one of the more promising strategies for addressing invasion.

Biological control, or the introduction of natural enemies of targeted invasive pests, is the cornerstone of the strategy of integrated pest management, or IPM. Used for decades in agriculture, IPM relies upon a multipronged approach to dealing with pests. Multiple modes of attack—frequently combinations of chemical, mechanical and biological control agents—are coordinated in an attempt to manage a pest species at an acceptable level, without necessarily eliminating it altogether (which has been deemed both impossible and unnecessary). Although IPM has not traditionally been used in the service of conservation goals, its widespread and successful use in agricultural systems suggests its utility as a tool in managing invasive species. The authors in this section consider different facets of IPM in relation to the particular cases they have studied.

In addition to the integration and application of multiple, often simultaneous, strategies, principles of IPM include sampling the population to determine the level of infestation before proceeding with control, accepting that complete eradication of the pest is usually impossible, and acquiring a thorough understanding of the biology of both the target pest to be eradicated and of the natural enemies chosen for biological control programs. Invasive organisms gain footholds in part because they often have no enemies; therefore, introducing a natural enemy should fill an empty niche. Ideally, the enemy reduces the invasive pest population and is itself reduced through time as its food supply—the pest—diminishes. Hoddle (13) amply details how failure to select a natural enemy with tight host specificity for the pest can lead to disastrous consequences.

Chapters in this section demonstrate several aspects of IPM approaches. David Richardson (12) shows how even with long-lived organisms such as trees, judicious use of available demographic and geographic information system (GIS) information, combined with insightful science, can yield data for development of predictive models. Such models can be used by the IPM practitioner to choose the most appropriate strategy for attacking invasive tree populations on the basis of known and identified characteristics of the targeted problem organism. Richardson uses "natural experiments" (often the only kind of experiment possible with long-lived organisms such as trees) to identify which life-history traits make certain tree species, such as those of the genus *Pinus*, good at invading particular environments. Biologists lament their inability to predict which of the many exotic species washing up yearly on foreign shores are likely to become successful invaders. Richardson's approach offers hope that, by taking an experimental approach and integrating knowledge and techniques from diverse disciplines, we *can* create models to allow for such prediction. Richardson's work exemplifies an important IPM tenet: a thorough understanding of a target's biology, achieved through basic investigation and experimentation, is necessary for the development of the most efficient, cost effective, and appropriate strategies for control.

Since 9.2 million hectares in California, nearly 25% of the state, have been swamped by Mediterranean grasses and forbs, the battle to retake California grassland is important and extensive. Jeffrey Corbin and his colleagues (11) not only demonstrate the value of multifactorial field experimentation for understanding the dynamics of an affected ecosystem but also show that a variety of approaches may be needed to solve the problem of the invasive takeover of California native grassland by exotic grasses. Although many strategies have been tried for restoration of California grassland—including fire, grazing, mechanical clearing, and alteration of soil pH—no single strategy has consistently been effective. A combination of judiciously chosen methods—based upon known factors of each vegetation area and incorporating the additional, frequently overlooked, component of adding in seeds, or seedlings, to increase the propagule pool of native grasses—can improve the success rate. Both Corbin et al. and other research groups (Kaiser 2002) have employed testing that uses classical experimental designs from agronomy to investigate and assess factors affecting plant growth.

All the papers in this section demonstrate that the use of experimental biology (I define *experimental* classically here while still allowing for the occasional modern modification) can yield data providing significant assistance in the fight against invasive species. These chapters bring new concepts—collection of experimental data, integration of multiple control methods, principles of IPM, correct application of biological control—into the conversation.

REFERENCES

Atkinson, I. 1989. Introduced animals and extinction. In *Conservation for the Twenty-First Century,* ed. D. Western and M. C. Pearl, 54–75. New York: Oxford University Press.

Dobson, A. P., and May, R. M. 1986. Disease and conservation. In *Conservation Biology: The Science of Scarcity and Diversity,* ed. M. Soulé and M. S. Sinauer, 345–65. Sunderland, Mass.: Sinauer.

Groombridge, B., ed. 1992. *Global Biodiversity: Status of the Earth's Living Resources.* London: Chapman and Hall.

Kaiser, J. 2002. An ecological oasis in the desert. *Science* 297:1635–37.

Pimentel, D., ed. 2002. *Biological Invasions: Economic and Environmental Costs of Alien Plant, Animal, and Microbe Species.* Boca Raton, Fla.: CRC Press.

Wilson, E. O. 2002. *The Future of Life.* New York: Knopf.

11   Tipping the Balance in the
      Restoration of Native Plants

Experimental Approaches to Changing the
Exotic:Native Ratio in California Grassland

Jeffrey D. Corbin, Carla M. D'Antonio,
and Susan J. Bainbridge

SUMMARY

As exotic species increasingly threaten native biodiversity, habitat managers have turned to a variety of tools designed to increase the efficiency of plant-restoration projects. These efforts include eliminating exotic competitors through mechanical removal, herbicide application, or fire, and increasing native species' competitiveness relative to that of exotic species through reduction of soil nitrogen availability, grazing, prescribed burning, or biological control. In this chapter, we evaluate the ability of experimental tests of these techniques to favor native species in California grassland ecosystems. We found no evidence that any of the strategies consistently favored native species relative to exotic species. Outcomes were highly case specific and likely varied with biotic and abiotic conditions in the experimental systems. Several studies suggest that these techniques are more successful in reducing specific invasive plant species in California grasslands rather than in increasing the success of native revegetation.

Limited availability of native propagules in the experimental systems likely limited the extent to which restoration techniques actually promoted native species. The most promising strategy for increasing native components in invaded ecosystems is likely to be the coordination of multiple strategies that address exotic-species abundance, native-seed or -seedling availability, and the postestablishment competitiveness of the native species. Such an application of an integrated "pest" management approach to the restoration of degraded habitats holds greater promise for the successful reestablishment of native biodiversity than simply targeting exotic species for removal.

154

INTRODUCTION

Exotic species increasingly threaten native biodiversity in natural habitats worldwide. Habitat managers trying to restore native-species richness and abundance face the daunting challenge that exotic species frequently are superior competitors in sites where the natives and exotics co-occur. For example, invasive *Spartina* spp. (cordgrass) in western North American estuaries are capable of excluding such native salt-marsh species as *Spartina foliosa* (California cordgrass) and *Salicornia virginica* (pickleweed) (Daehler and Strong 1996). Reintroduction of native species into these invaded ecosystems is unlikely to succeed as long as the exotics are competitively superior.

Habitat managers and restoration ecologists must utilize a variety of tools to tip the competitive balance toward native species and away from exotic species while increasing the efficiency of plant-restoration projects. Current tools include eliminating exotic competitors through mechanical removal, herbicide application, or fire, and increasing native species' competitiveness relative to exotics' through reduction of soil nitrogen (N) availability, grazing, or prescribed burning or through the introduction of biological control agents. These tools are often applied repeatedly or in combination, but to provide a lasting increase in the native component of degraded ecosystems they frequently need to be used in conjunction with reintroduction of native-plant species either as seeds or seedlings. The application of these tools may also be constrained by practical considerations such as safety or toxicity concerns (e.g., fire or herbicide application) or limited habitat area (e.g., fire or grazing).

The need to enhance the success of native-plant restoration and to increase native species' competitiveness is particularly urgent in California grassland ecosystems. Grasslands are a major component of the state's natural vegetation, comprising nearly 10 million ha, or 25% of the state's surface area (Heady et al. 1991). The state's grasslands are used extensively for livestock production (Wagner 1989) and recreation and are habitat for many of California's state-listed threatened and endangered plants. During the last two centuries, invasion by European annual grasses and forbs into California grasslands, modifications of land use, and, possibly, changes in the region's climate have resulted in a dramatic, large-scale conversion from dominance by perennial bunchgrasses, forbs, or both to dominance by Eurasian annual species (Burcham 1970; Crampton 1974; Bartolome, Klukkert, and Barry 1986; Baker 1989; Hamilton 1997). Whereas perennial species such as *Nassella pulchra* (purple needlegrass), *Bromus carinatus* (California brome), *Elymus glaucus* (blue wildrye), *Danthonia californica* (California oatgrass),

Poa secunda (pine bluegrass), and *Festuca* spp. (fescue) were thought to dominate some of the region's grasslands prior to European settlement, introduced grasses such as *Bromus diandrus* (ripgut brome), *Bromus hordeaceus* (soft chess), *Avena* spp. (wild oat), and *Vulpia* spp. (annual fescue) are dominant today, even in stands where some native bunchgrasses have persisted.

Life-history and growth characteristics of exotic annual species offer substantial advantages over those of native perennial species in disturbed habitats that are frequently the targets of restoration efforts. Seedbank composition in California grasslands is highly skewed toward exotic annual species (Champness and Morris 1948; Major and Pyott 1966; Dyer, Fossum, and Menke 1996; Holl et al. 2000; Alexander 2001). Seed production by annual species substantially exceeds the number of seeds necessary to replace the population (Young and Evans 1989), whereas the establishment of perennial species has frequently been shown to be limited by seed availability (e.g., Peart 1989a; Kotanen 1996; Hamilton, Holzapfel, and Mahall 1999). Furthermore, annual seeds in growth chambers have been shown to germinate earlier and under a wider range of temperatures than native perennial seeds (Reynolds, Corbin, and D'Antonio 2001). The more abundant and earlier-germinating annual grass species can form dense stands and monopolize resources, thereby restricting the growth and survival of native seedlings (Bartolome and Gemmill 1981; Dyer, Fossum, and Menke 1996; Dyer and Rice 1997; Hamilton, Holzapfel, and Mahall 1999; Brown and Rice 2000). As a result, competitive interactions between native and exotic grasses in California have usually been shown to strongly favor the exotic species, especially in recently established native populations (Dyer, Fossum, and Menke 1996; Dyer and Rice 1997; Hamilton, Holzapfel, and Mahall 1999; Brown and Rice 2000).

The large competitive advantages that some exotic species enjoy over natives suggest that efforts to restore native-plant biodiversity in exotic-dominated grasslands in California must improve the competitiveness of native species relative to that of exotic species. In this chapter, we review the successes and failures of techniques either being proposed or employed over a large scale to improve conditions for native species during grassland restoration. These techniques include the reduction of N in N-enriched habitats through sawdust addition or repeated biomass removal, grazing, prescribed burning, herbicide application, and biological control. We focus on efforts to alter composition in sites that are currently grassland rather than on the *de novo* creation of grassland from recently plowed or otherwise heavily disturbed sites.

REDUCTION OF PLANT-AVAILABLE NITROGEN

Ecosystem nitrogen enrichment is a common barrier to native-plant restoration. Past fertilization (Vitousek et al. 1997), atmospheric nitrogen deposition (Bobbink 1991; Jefferies and Maron 1997), fire (Wan, Hui, and Luo 2001), habitat disturbance (Hobbs and Mooney 1985), and invasion by nitrogen-fixing shrubs (Vitousek et al. 1987; Maron and Connors 1996) can all increase soil nitrogen availability. While general characteristics of non-native invading plant species have proven elusive (Mack et al. 2000), enhancement of N availability has been shown to favor fast-growing invasive species in a variety of habitats (e.g., Huenneke et al. 1990; Vinton and Burke 1995; Maron and Connors 1996). Restoration in N-enriched habitats must, therefore, deal with the question of how to promote slower-growing native species in competition with faster-growing exotic species.

Invasion of northern coastal prairie grasslands in California by a variety of N-fixing shrubs commonly known as brooms—for example, *Genista monspessulana* (French broom) and *Cytisus scoparius* (Scotch broom)— and *Ulex europeus* (gorse) has been shown to have significant impacts on soil N availability and plant community composition (Randall, Rejmánek, and Hunter 1998; Haubensak 2001). Haubensak (2001) found that N availability was three times as high in a broom-invaded grassland than in an adjacent uninvaded grassland. The colonization of coastal prairie grasslands by the native shrub *Lupinus arboreus* (bush lupine) has had similar effects on N cycling and community composition as broom invasion (Maron and Jefferies 1999). Individual shrubs grow rapidly, producing a dense canopy that shades out native grassland species. In northern California coastal prairies, repeated cycles of lupine colonization and death lead to a doubling of total soil N, greatly increased N availability, and thus increased vegetative production (Maron and Jefferies 1999). Maron and Connors (1996) documented that these cycles cause a large-scale shift in grassland composition from native perennial to exotic annual species. Increased N levels even after broom removal or lupine dieback may continue to favor exotic species and hamper efforts to reintroduce native species.

Two promising methods to reduce plant-available N and increase the competitiveness of slower-growing natives in such N-enriched habitats are (1) the addition of a labile carbon source such as sucrose or sawdust (e.g., Morgan 1994; Alpert and Maron 2000; Paschke, McLendon, and Redente 2000) and (2) repeated mowing followed by biomass removal (Collins et al. 1998; Maron and Jefferies 2001). The addition of a carbon source is assumed to increase microbial N immobilization and decrease plant-available N (Morgan 1994; Alpert and Maron 2000; Paschke, McLendon, and Redente 2000). Repeated mowing and biomass removal are assumed to remove N in plant biomass that would otherwise be remineralized as plant litter is produced and decomposes. Under lower N conditions, growth of all vegetation would be expected to decrease, but if faster-growing exotic species are disproportionately affected by lower soil N concentrations, slower-growing native species may benefit indirectly owing to reduced competition.

Carbon Addition

Carbon addition has successfully reduced the abundance of exotic species in California grassland (Alpert and Maron 2000), shrubland (Zink and Allen 1998), and sagebrush–bunchgrass (Young et al. 1998) communities, as well as in shortgrass steppe ecosystems in Colorado (Reever Morghan and Seastedt 1999; Paschke, McLendon, and Redente 2000). Alpert and Maron (2000) tilled 1.5 kg m^{-2} of sawdust into bare N-rich patches left after the death of bush lupine individuals in a coastal prairie site. The patches were not experimentally seeded with natives, and recruitment into them depended on seed rain or the seedbank. Native biomass comprised only 8–12% of the total biomass in all treatments, reflecting the highly invaded nature of these ecosystems. Sawdust addition significantly reduced the aboveground biomass of exotic grasses, although the biomass of exotic forbs was unaffected. Sawdust addition also showed no significant benefit for native-species richness or biomass. Thus, while sawdust addition successfully reduced exotic grass abundance, there was no evidence that it increased the occurrence of native species.

Two studies in our lab employed similar experimental approaches to assess the ability of sawdust addition to benefit native species in ecosystems invaded by two different types of N-fixing shrubs (Corbin and D'Antonio 2004a; Haubensak 2001). Following removal of shrubs and understory vegetation in a broom-invaded coastal scrub ecosystem and another lupine-invaded coastal prairie, seedlings of three species of native perennial grasses were transplanted into experimental plots, half of which were seeded with exotic annual grasses. In the postlupine site, we added a third treatment consisting of three species of exotic perennial grass competitors. Sawdust was added to half the plots (600 g m^{-2} yr^{-1}) for 2 years.

Native species did not benefit from the addition of sawdust following removal of French and Scotch broom in either growing season (Haubensak 2001). Instead, exotic annual grasses significantly reduced the growth and survival of all three native species, whether sawdust had been added or not.

Apparently, the effect of competition with annual species was so strong that it overwhelmed any potential effect of sawdust addition.

Sawdust addition showed greater promise in reducing the competitive advantage of exotic annual grasses in the lupine-invaded coastal prairie (Corbin and D'Antonio 2004a). In the first growing season, sawdust decreased the competitive suppression of seedlings of two native grass species by exotic annual grasses, but there was no benefit for native species competing with exotic perennial grasses. In the second year, sawdust addition did not affect the competitive interactions between natives and either exotic annual or exotic perennial grasses. In fact, the native perennial grasses that survived the first year of competition with annual grasses significantly reduced the aboveground productivity of annual grasses, even without sawdust addition. We concluded that sawdust addition provided no significant benefit to native plants in this system, where target individuals were planted as seedlings, and survival was high in all treatments. Competition between native species and exotic annual grasses was most asymmetric (in favor of the exotics) in the first growing season, after which native species were capable of significantly reducing the productivity of annual grasses (Corbin and D'Antonio 2004b). The possibility remains that sawdust addition may provide greater benefit to restoration projects in which seedling survival in the first year is less certain or in which native species are introduced as seeds.

Mowing and Biomass Removal

Maron and Jefferies (2001) examined the effectiveness of mowing and removing aboveground biomass in reducing soil N and favoring native species in a coastal prairie grassland that had experienced lupine invasion and dieback. The mowing and removal of plant biomass for five growing seasons reduced exotic grass biomass and doubled the number of forb species present as compared to unmanipulated control plots. However, mowing had no effect on the number of perennial grass species, most likely because of a lack of native propagules. The 5-year experiment removed approximately 9% of the total soil N as plant biomass but was not sufficient to induce N limitation of vegetation. In fact, unmowed plots experienced a significant reduction in soil N in the form of nitrate leaching losses in the fall and early winter, a reduction that was nearly equivalent to the biomass removal in mowed plots. The authors concluded that while mowing was effective in reducing exotic biomass and increasing the species richness of forbs (although many were exotic), mowing was unable to reduce soil N levels enough to favor the reestablishment of native grasses. The study also suggested that reduction of soil N content in ecosystems that have become suitably enriched may require long-term treatment, owing to the slow turnover of soil organic N pools.

Livestock Grazing

Over the past decade there has been increased interest in the use of livestock grazing to reduce the biomass of introduced species and increase the diversity and abundance of native species in California grassland settings (e.g., Menke 1982; Edwards 1995, 1996; Reeves and Morris 2000). The California Cattlemen's Association, for example, suggested that carefully controlling the timing and intensity of livestock grazing can promote native diversity in California grasslands (Reeves and Morris 2000). Grazing may benefit native vegetation by disproportionately targeting exotic biomass, thereby reducing the exotics' competitive advantages; by reducing exotic seed production; or both. By contrast, some conservationists believe that livestock grazing has contributed to the degradation of many California grasslands and that its persistence is inimical to restoration of native-species richness (Fleischner 1994; Painter 1995).

D'Antonio et al. (2001) reviewed livestock-grazing studies from throughout California in an attempt to quantitatively evaluate the use of grazing as a tool to reduce exotic-species cover and promote native biological diversity (see Table 11.1 for a complete listing). Their initial goal was to conduct a meta-analysis of the size and direction of grazing's effects on native and exotic plants using all the available published and unpublished data sets from California. Meta-analysis is a statistical way of synthesizing results from different studies on a common topic (Gurevitch and Hedges 1993). They calculated an "effect size" in each study for each response variable (e.g., native forb cover) based on the ratio of the variable in the treated area (grazed) compared to the control (ungrazed). They assessed the effect of grazing on the measured response variables across studies using the mean of the pooled effect sizes. They found that most studies lacked adequate controls, lacked replication, or had no available measurement of among-plot variability and hence were not useable for meta-analysis. A summary of the six studies that fit the meta-analysis criteria demonstrated that livestock grazing was associated with an increase in the cover of native perennial grasses for those sites (Figure 11.1). Contrary to the claims of others (Thomsen et al. 1993; Kephart 2001), these studies showed a slight negative effect of livestock grazing on native-forb abundance and a positive effect of grazing on the abundance of exotic forbs. However, the results should be interpreted with caution because this small number of studies is inadequate for a true meta-analysis, and the

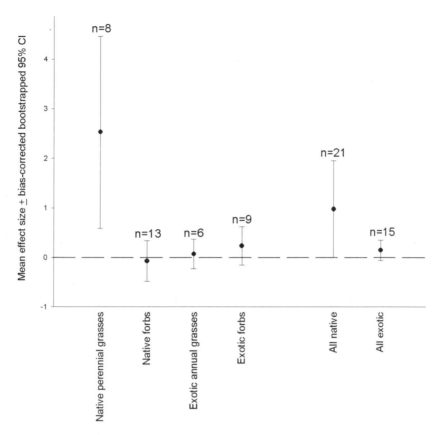

Figure 11.1. Effect of grazing on grassland plant life-form groups (based on studies reviewed by D'Antonio et al. [2001]). Values are the cumulative effect sizes (mean natural log of the response ratio [X_{grazed} / $X_{control}$] weighted by study variances ± 95% C.I.; n = number of effect sizes). Number of effect sizes may be greater than the number of published studies, owing to multiple comparisons within the same study.

addition of just a few studies could reverse the overall outcome. In addition, these studies represent a small subset of the California grassland and may not be representative of the state as a whole. Sadly, the often heated debates over the impact of livestock grazing and the role, if any, that grazing should play in grassland restoration are likely to continue until many additional careful quantitative studies are conducted across the full range of grassland habitats in the state.

Many grazing studies in California focus on the responses of particular native species, such as the native perennial bunchgrass *Nassella pulchra*, to

TABLE 11.1 *Studies of the impact of fire, grazing, or both on the species composition of California grasslands*

Reference	Study Type	Grassland Type
Ahmed 1983	Fire	Valley and foothill
Arguello 1994	Fire	Bald hills
Bartolome, Stroud, and Heady 1980; Jackson and Bartolome 2002	Grazing	Annual, valley and foothill
Bartolome et al. in press	Fire and grazing	Valley and foothill
Bartolome and Gemmill 1981	Grazing	Various
Betts 2003	Fire	Valley and foothill
Cooper 1960	Grazing	Coastal prairie
Cox and Austin 1990	Fire	Vernal pool
Delmas 1999	Fire	Wildflower field
DiTomaso, Kyser, and Hastings 1999	Fire	Valley and foothill
Dyer and Rice 1997	Fire and grazing	Vernal pool
Dyer, Fossum, and Menke 1996; Fossum 1990	Fire	Vernal pool
Eller 1994	Grazing	Annual
Elliot and Wehausen 1974	Grazing	Coastal prairie
Foin and Hektner 1986	Grazing	Coastal prairie
Garcia-Crespo 1983	Fire	Valley and foothill
Graham 1956	Fire	Annual grassland–savannah
Hansen 1986	Fire	Alkali grassland–vernal pool
Harrison 1999	Grazing	Serpentine-forb
Hatch, Bartolome, and Hillyard 1991	Fire and grazing	Valley and foothill

the cessation of grazing or to particular grazing regimes. This species is, arguably, the most commonly used species in grassland restoration projects and the best studied of the native grasses. Yet a review of the relevant literature readily demonstrates the difficulties inherent to generalizing about the effects of livestock grazing. Huntsinger et al. (1996) and Dennis (1989) found substantial variation in the response of *N. pulchra* individuals to simulated grazing (clipping) among different populations. Genetic differences among the populations may explain the differential population response, although this variable was not explicitly examined. Likewise, exclosure

TABLE 11.1 *(continued)*

Reference	Study Type	Grassland Type
Hatch et al. 1999	Fire and grazing	Coastal prairie
Heady 1956	Grazing	Valley and foothill
Keeley, Lubin, and Fotheringham 2003	Grazing	Oak woodland
Kephart 2001	Fire and grazing	Valley and foothill
Kneitel 1997	Fire	Annual
Langstroth 1991	Fire and grazing	Vernal pool
Larson and Duncan 1982	Fire	Annual
Marty 2001, 2002	Fire and grazing	Vernal pool
Merenlender et al. 2001	Grazing	Valley and foothill
Meyer and Schiffman 1999	Fire	Annual
Micallef 1998	Grazing	Annual, valley, and foothill
Parsons and Stolhgren 1989	Fire	Annual
Pollack and Kan 1998	Fire	Vernal pool
Porter and Redak 1996	Fire	Valley and foothill
Reeves and Morris 2000	Grazing	Various
Saenz and Sawyer 1986	Grazing	Bald hills, woodland
Stromberg and Griffin 1996	Grazing	Valley and foothill
Thomsen et al. 1993	Grazing	Annual
TNC 2000	Fire and grazing	Wildflower field
White 1967	Grazing	Valley and foothill
York 1997	Fire	Wildflower field
Zavon 1977	Fire and grazing	Annual

studies from several parts of central and northern California have demonstrated increases (Hatch et al. 1991), decreases (Hatch et al. 1999), and no change (White 1967; Stromberg and Griffin 1996) in the abundance of *N. pulchra* in response to protection from grazing. Some studies have observed fluctuations of *Nassella* abundance at the same site, but because of a lack of appropriate controls, the changes could not clearly be attributed to cessation of grazing (Bartolome and Gemmill 1981; Merenlender et al. 2001).

Several investigators have attempted to use livestock grazing to control particular exotic species, such as *Centaurea solstitialis* (yellow starthistle), in California grasslands. For example, Thomsen et al. (1993) found that the timing of grazing was critical to the outcome of their experimental grazing treatments: late-spring and early-summer grazing greatly reduced yellow starthistle abundance relative to controls, although grazing did not eliminate the starthistle populations. At the same time, though native-plant-species richness was not recorded, the investigators observed an increase in populations of three disturbance-responsive native forbs with this late-spring grazing. Such targeted grazing may be useful in grassland restoration projects if the goal is simply to reduce a noxious weed.

Careful timing of grazing to coincide with the period of exotic seed production has the potential to benefit native species by reducing exotic germination and productivity in subsequent growing seasons. However, few studies have quantified the effect of grazing on the seed production of exotic species. Stromberg and Kephart (1996) argued that mowing or grazing for 2–3 years following native-plant restoration is likely to reduce exotic annual biomass and exotic seed production. Maron and Jefferies (2001) found that mowing reduced annual propagules and induced seed limitation of one of the most abundant exotic grass species, *Bromus diandrus*. We do not know, however, whether livestock grazing is capable of producing the same effect.

Overall, the existing data are insufficient to conclusively discern a relationship between livestock grazing and California's native grassland plants, or to evaluate the potential of grazing to enhance native-species richness and cover. Grazing has been shown to benefit native species in some individual studies, but its effects do not appear to be generalizable among studies or among years. Studies such as those of Stromberg and Griffin (1996) and Safford and Harrison (2001) suggest that grazing does not have as strong an effect on native species as has previously been suspected, but more research is needed to explore the generality of such conclusions. Many managers are now using controlled burning in combination with grazing to reduce exotic species and promote natives, and grazing may have a more predictable impact in combination with other techniques such as fire. Specific grazing regimes, with modest levels of grazing carefully timed to coincide with critical periods of exotic vegetation growth and seed production, have the greatest potential to be useful in a restoration context. Further research that employs both extensive quantitative surveying of properties with different grazing histories over a range of carefully recorded environmental conditions plus properly designed experiments is necessary to clarify the utility of grazing in increasing native competitiveness.

Prescribed Fire

Land managers are increasingly turning to prescribed fire in an attempt to reduce the dominance of exotic plant species such as N-fixing shrubs (e.g., brooms and gorse), herbaceous forbs (e.g., yellow starthistle), and exotic grasses (e.g., *Taeniatherum caput-medusae* [medusahead]) (Pollack and Kan 1998; DiTomaso, Kyser, and Hastings 1999; Bossard, Randall, and Hoshovsky 2000; Alexander 2001). Fire has the potential to instantaneously reduce exotic vegetation biomass, including standing biomass and residual litter, and can be applied to a relatively large landscape. Fire can also influence the seed crop and germination of native and nonnative species in subsequent growing seasons. Although fire can directly kill seeds on adult plants or fire-intolerant seeds in the soil, transient increases in light availability, soil surface temperatures, and soil nitrogen availability frequently associated with fire can also increase seed germination and seedling survival of fire-tolerant species. Frequent fires, however, may reduce available soil N and grassland productivity because they cause repeated volatilization of N and increased root death (Seastedt, Briggs, and Gibson 1991).

Efforts to generalize as to the impacts of fire on native-plant abundance and species diversity have proven difficult, in part because of the differential responses of various life-form groups to fire. Whereas some studies have shown dramatic increases in native-forb abundance in the first year following fire (Pollack and Kan 1998; Meyer and Schiffman 1999; DiTomaso, Kyser, and Hastings 1999), other studies have demonstrated minimal or negative effects of fire on native perennial grasses (Dyer, Fossum, and Menke 1996; Hatch et al. 1999). The effects of fire on the grassland vegetation also vary with time because the reductions in annual productivity that are frequently observed following fire tend to be temporary (Keeley 1981).

D'Antonio et al. (2001) reviewed the role of fire in structuring California grassland vegetation and the abundances of four life-form groups: native perennial grasses, native forbs, exotic annual grasses, and exotic forbs (see Table 11.1 for a complete listing). They conducted a meta-analysis on nineteen studies of prescribed or natural fires in California grassland. The investigators did not augment native propagules or seed availability, so the effect of seed limitation on the grassland response was not a controlled factor. They found that fire tended to shift grassland composition toward native forb species in the first year, but native perennial grasses were generally negatively affected by fire. The abundance of exotic species was, for the most part, unaffected by fire in the first growing season, apparently because the composition of exotic vegetation shifted from annual grasses, which decreased, to annual forbs, which increased, after fire. In subsequent seasons following a single burn, total plant biomass increased to an average of 13% more in burned areas than in unburned areas. Where native perennial grasses were studied, *Nassella pulchra* abundance generally rebounded during the second postfire year, whereas *Danthonia californica* was slower to recover. Although germination of native grasses increased following fire, there was no detectable increase in native-grass abundance in subsequent years. Similarly, Dyer, Fossum, and Menke (1996) found that the establishment of native perennial grass seedlings was about the same in burned and unburned areas, whereas seedling mortality in burned areas was high. By the third year, the cover of native species relative to that of exotic species was not significantly different in burned areas and unburned areas, probably because of the rapid recovery of exotic annual grasses.

The observation that repeated burning reduces soil N availability suggests that prescribed burning could be used to tip the balance of competition in favor of native species if nitrogen is an important resource for both native and nonnative grassland species. For example, Seastedt, Briggs, and Gibson (1991) found that productivity of frequently burned tallgrass prairie grasslands was limited by N availability. The influence of fire frequency on soil N levels and N limitation of productivity in California grasslands is not known, though 2 or 3 consecutive years of burning have been shown to either decrease (Parsons and Stohlgren 1989; Delmas 1999) or have no significant effect on (Hansen 1986; DiTomaso, Kyser, and Hastings 1999) aboveground productivity beyond the effect of a single burn. D'Antonio et al. (2001) concluded that annual burning (after two or three burns) in ungrazed grassland resulted in higher native-forb and exotic-forb abundance than a single burn, but exotic annual grasses apparently did not respond further to the additional fires. Unfortunately, data were insufficient to conduct a meta-analysis on the effects of repeated burning on native perennial grasses. A single study of the effects of repeated fire reported a dramatic increase in native perennial grasses, particularly *Nassella pulchra* (DiTomaso, Kyser, and Hastings 1999), but more work is needed at other sites to evaluate the generality of this finding. Further investigation of the importance of fire frequency to soil N levels, N limitation, grassland productivity, and native species' competitiveness in California grasslands would help to determine whether repeated burning provides benefits to native biodiversity.

The meta-analysis by D'Antonio et al. (2001) determined that a combination of a single burn and cattle grazing likely did not improve the magnitude of the benefits of a single burn for native forbs, but grazing did sustain the benefits of a single fire for native forbs into the third postfire growing season. Likewise, grazing sustained the decrease in exotic annual grasses observed in the first year after a single fire into the third year. Grazing also lessened the negative impact of fire on native perennial grasses in the first growing season. Exotic forb suppression was successful only when a site was burned annually for several consecutive years and also grazed. However, this suppression did not appear to benefit native forbs: there was no increase in the abundance of native forbs by the third year in repeatedly burned and grazed sites. So, as with other techniques described in this chapter, combining fire and grazing reduced exotic species but did not increase the diversity and abundance of native forbs.

The time of year in which controlled burns are performed may have a significant effect on the impact on grassland species composition. D'Antonio et al. (2001) found that the month in which grasslands were burned significantly influenced native perennial grasses, with growing-season burns (e.g., November–June) having significantly more detrimental impacts on native cover than summer or fall burns. Burn season did not have a strong effect on native forbs or exotic annual grasses. Burns can also be specifically timed to limit invasive species' seed dispersal. Fires targeting medusahead and yellow starthistle before mature plants dispersed their seeds effectively suppressed these species and, apparently, their soil seedbanks (Pollack and Kan 1998; DiTomaso, Kyser, and Hastings 1999).

In addition to trying to manipulate the abundance of native and exotic grasses and forbs using fire, many managers use fire to control woody invaders in California grasslands. Alexander (2001) surveyed species composition and broom seedbank density and aboveground cover in numerous managed grassland sites in northern California to determine whether controlled burning was capable of controlling invasive shrubs (primarily *Genista monspessulana* and *Cytisus scoparius*) and opening suitable habitat for native species. She found that although fire reduced aboveground biomass of adult broom plants, it stimulated germination of broom seeds from the soil seedbank, which resulted in very dense broom stands in the first few years. The germination occurring after the first fire significantly reduced the number of live broom seeds in the soil seedbank, but because of the resulting increase in number of new broom plants, the seedbank has a great potential to become large again if the new crop of seedlings is not controlled. Hence, later fires must occur before the new crop of broom seedlings becomes reproductive (within 3–4 years). Alexander (2001) also found that the postbroom grasslands created by controlled burning were dominated largely by non-indigenous grasses and forbs. The only places where the post-broom grassland had substantial native cover were a few sites where broom had been pulled repeatedly by hand and no burning had been conducted. In these few sites, native perennial grasses and forbs codominated with exotic annual grasses. Overall, Alexander's study suggests that repeated fire effectively reduces the broom seedbank and the cover of adult plants. However, the study points out that the reduction of broom dominance through prescribed burning alone is unlikely to increase native-species richness and cover in landscapes in which exotic forbs and grasses are so abundant.

Herbicide Application

Herbicide application has been suggested as a way to reduce established exotic vegetation in heavily invaded ecosystems and to control the flush of exotic annual species from the soil seedbank prior to planting of native species (e.g., Wilson and Gerry 1995; Stromberg and Kephart 1996; Rice and Toney 1998). Herbicides such as glyphosate, picloram, and clopyralid have been shown to substantially reduce exotic biomass and increase native-seedling establishment in a variety of grassland systems (e.g., Wilson and Gerry 1995; Rice et al. 1997; Rice and Toney 1998). Stromberg and Kephart (1996) recommended repeated herbicide application to reduce the exotic annual seedbanks prior to native-plant establishment in coastal California old fields. Our own experience in a northern coastal prairie grassland dominated by a mixture of exotic annual and perennial grasses and biennial forbs supports the utility of herbicide application in favoring native-grass establishment (Corbin and D'Antonio 2004a). Though we are not aware of controlled experiments comparing the efficacy of herbicides in restoration of California grasslands, herbicide application is likely an effective tool to facilitate site preparation and reduce exotic reestablishment (Anderson and Anderson 1996; Stromberg and Kephart 1996).

Biological Control

The introduction of biological control agents holds great promise in reducing the competitiveness of invasive plants in cases where insects, pathogens, or vertebrates specifically target undesirable species (Hoddle, this volume). However, in spite of the advantages of biological control—which include relatively low costs and long-term, self-sustaining management of invasive species—cases of successful control of nonnative grassland species in California are rare. DeLoach (1991) found that of twenty-three native and exotic weed species in western rangelands (including the northwestern United States and western Canada) where biological control had been attempted, only seven were successfully controlled. In California several insects have been introduced to control seed production of the pernicious yellow starthistle, and although the insects have been established successfully in many areas, starthistle remains a widespread, abundant weed (Turner, Johnson, and McCaffrey 1995; Villegas 1998; Roché et al. 2001). The introduction of three biological control agents in coastal Oregon grasslands to control *Senecio jacobaea* (tansy ragwort) has been far more successful, reducing the exotic to 1% of its former abundance (McEvoy, Cox, and Coomes 1991; McEvoy and Coomes 1999). One of the few investigations of biological control of exotic grasses (Carsten et al. 2000) found evidence that augmentation of a natural crown rust of *Avena* spp. (wild oat) on San Clemente Island reduced seed production of these annual grasses. In the cases of both yellow starthistle and wild oats, the biological control agents have been more effective at reducing seed production than at reducing exotic population sizes, but the agents could reduce exotic competitiveness in combination with other control techniques (DiTomaso 2000).

DISCUSSION

A review of attempts to increase the efficiency of California grassland restoration did not yield a strategy that consistently favored native species over exotic species. Although some individual techniques showed promise for increasing native-plant growth or seedling survival, no technique consistently increased native-grass or native-forb diversity or biomass. Rather, the outcomes were highly case specific and likely varied with such factors as initial vegetation composition, nutrient availability, past land-use history, and climatic conditions. Further exploration of the restoration tools under a wider range of habitat conditions is required before habitat managers can predictably apply them to revegetate degraded ecosystems.

Some techniques showed promise in reducing the exotic components of degraded ecosystems even though they were unable to increase the native component. Reduction of plant-available N (Alpert and Maron 2000; Maron and Jefferies 2001), mowing or grazing (Thomsen et al. 1993; Stromberg and Kephart 1996), prescribed fire (Pollack and Kan 1998; DiTomaso et al. 1999; Alexander 2001), and herbicide application (Stromberg and Kephart 1996) were able to reduce specific invasive plant species in California grasslands. While these techniques would be of even greater use in a restoration context if they were capable of consistently increasing the competitiveness or abundance of native species, the control of exotic species is frequently a primary goal of habitat management (Ehrenfeld 2000).

The absence of native plant species, either as individuals or as seeds from nearby populations, frequently limits the success of efforts to restore degraded habitats. Many natural areas that are chosen for restoration are chosen precisely because their native component has been lost. For example, invasion by exotic species can be a major factor in the elimination of native-plant populations in natural habitats (Baker 1989; Bossard, Randall, and Hoshovsky 2000). Agricultural activities, especially plowing and other intense farming activities, are also capable of directly removing native individuals and likely exhausting the seedbank of the former dominants (Milberg 1992). The effects of agricultural activities on native-grass abundances have been shown to persist for decades after cessation of the agricultural activity (Stromberg and Griffin 1996). In cases where exotic species invasion or past land-use history have degraded native abundance, native species are unable to take advantage of even the most diligent efforts to remove exotic species or modify the competitive environment.

Habitat restoration strategies must, therefore, take into account not only the vulnerability of native individuals to competition with exotic species but also the limited source of native propagules in degraded ecosystems. There is strong evidence that the colonization of appropriate habitats by native-plant populations is often limited by seed availability. The soil seedbanks of native grasses and some forb species have been shown to be negligible in both disturbed (Kotanen 1996) and undisturbed (Peart 1989a) habitats in California. Seed rain of native grasses and forbs is generally substantially lower than that of exotic species when all groups are growing in the same environment (Hobbs and Mooney 1985; Peart 1989a; Kotanen 1996). Native species also have limited ability to repopulate degraded sites from nearby remnant populations (if such populations even exist), owing to low seed production (Hobbs and Mooney 1985; Peart 1989a; Kotanen 1996) and limited seed dispersal relative to exotic species (Hobbs and Mooney 1985; Peart 1989a–c; Kotanen 1996, 1997). We believe that restoration projects that augment the pool of native propagules via seed addition or seedling out-planting are much more likely to succeed than projects that rely on natural seed dispersal and recolonization. Some efforts to increase the competitiveness of native species, including sawdust addition (Alpert and Maron 2000), mowing and biomass removal (Maron and Jefferies 2001), grazing (Hatch et al. 1999) and prescribed burning (e.g., Alexander 2001 and others), may have had more success if more native propagules had been available to take advantage of the modified competitive environment.

Seedling establishment and persistence exert a major influence on plant population dynamics (Harper 1977), particularly in a restoration context in which native-plant species must revegetate habitats from which they have been extirpated. A variety of studies in California grasslands have demonstrated that exotic grasses restrict the establishment of native perennial grass (e.g., Peart 1989a; Dyer and Rice 1997; Hamilton et al. 1999; Brown and Rice 2000). However, there are indications that mature native perennial grasses are capable of competing with exotic species and reducing future species invasion (Peart 1989b; Corbin and D'Antonio 2004b; but see Hamilton, Holzapfel, and Mahall 1999, in *N. pulchra*–dominated grassland). We have found that mixed communities of native perennial bunchgrasses are able to reduce the aboveground productivity of exotic annual grasses and resist invasion by exotic forb species within 2 years after native-seedling establishment (Corbin and D'Antonio 2004b). These results indicate that the period of seedling establishment in the first year after seed or seedling planting is a critical phase for native-grass restoration. Habitat managers should, therefore, concentrate on increasing native competitiveness during this window of establishment, after which mature native individuals may be better able to compete with exotic species.

We believe that a strategy that uses multiple tools to both reduce the competitiveness of exotic plant species and increase the establishment of native species holds promise for more successful restoration of native biomass. The coordination of multiple approaches to deal with undesirable species, a hallmark of integrated pest management (IPM) theory (Buhler, Liebman, and Obrycki 2000; Hoddle, this volume), has been well-developed in agricultural systems and to a lesser extent in rangeland management (reviewed in DiTomaso 2000). The most common application of IPM in the control of exotic species in natural systems is the introduction of biological control agents, but IPM can also include such strategies as prescribed burning and mechanical removal. Stromberg and Kephart (1996) argued that the establishment of native grasses in abandoned agricultural fields in central California is facilitated by a three-step program that includes site preparation by either plowing or applying herbicides to exotic grasses, seeding or planting seedlings of the desired native species, and instituting postestablishment management practices designed to increase the competitiveness of the native species. Such a program, though likely more expensive and labor intensive than other restoration techniques, should be considered if it is more likely to accomplish the goal of increasing population sizes of native-plant species and creating communities with greater resistance to further invasion.

ACKNOWLEDGMENTS

The authors would like to acknowledge Coleman Kennedy, Meredith Thomsen, Karen Haubensak, Sally Reynolds, and other members of the

D'Antonio lab group and James Bartolome for valuable discussions that contributed to this chapter. Cynthia Brown, John Maron, and an anonymous reviewer made valuable comments to an earlier version. Jeffrey Corbin would also like to thank Malcolm Gordon and Soraya Bartol for organizing the IoE Conference under difficult circumstances. The National Science Foundation (DEB 9910008) and the David and Lucille Packard Foundation supported the preparation of this chapter.

REFERENCES

Ahmed, E. O. 1983. Fire ecology of *Stipa pulchra* in California annual grassland. Ph.D. diss., University of California, Davis.

Alexander, J. M. 2001. The effects of prescribed burns on post-broom invasion grassland community composition. Master's thesis, University of California, Berkeley.

Alpert, P., and Maron, J. L. 2000. Carbon addition as a countermeasure against biological invasion by plants. *Biol. Invasions* 2:33–40.

Anderson, J. H., and Anderson, J. L. 1996. Establishing permanent grassland habitat with California native perennial grasses. *Valley Hab.* 14:1–12.

Arguello, L. A. 1994. Effects of prescribed burning on two perennial bunchgrasses in the Bald Hills of Redwood National Park. Master's thesis, Humboldt State University, Arcata, Calif.

Baker, H. G. 1989. Sources of the naturalized grasses and herbs in California grasslands. In *Grassland Structure and Function: California Annual Grassland,* ed. L. F. Huenneke and H. Mooney, 29–38. Dordrecht, Netherlands: Kluwer Academic Publishers.

Bartolome, J. W., Fehmi, J. S., Jackson, R. D., and Allen-Diaz, B. In press. Response of a native perennial grass stand to disturbance in California's Coast Range Grassland. *Restor. Ecol.*

Bartolome, J. W., and Gemmill, B. 1981. The ecological status of *Stipa pulchra* (Poaceae) in California. *Madroño* 28:172–84.

Bartolome, J. W., Klukkert, S. E., and Barry, W. J. 1986. Opal phytolyths as evidence for displacement of native California grassland. *Madroño* 33:217–22.

Bartolome, J. W., Stroud M. C., and Heady, H. F. 1980. Influence of natural mulch on forage production on differing California annual range sites. *J. Rangel. Manage.* 33:4–8.

Betts, A. 2003. Ecology and control of goatgrass *(Aegilops triuncialis)* and medusaehead *(Taeniatherum caput-medusae)* in California annual grasslands. Ph.D. diss., University of California, Berkeley.

Bobbink, R. 1991. Effects of nutrient enrichment in Dutch chalk grasslands. *J. Appl. Ecol.* 28:28–41.

Bossard, C. C., Randall, J. M., and Hoshovsky, M. C. 2000. *Invasive Plants of California's Wildlands.* Berkeley: University of California Press.

Brown, C. S., and Rice, K. J. 2000. The mark of Zorro: Effects of the exotic annual grass *Vulpia myuros* on California native perennial grasses. *Restor. Ecol.* 8:10–17.

Buhler, D. D., Liebman, M., and Obrycki, M. M. 2000. Theoretical and practical challenges to an IPM approach to weed management. *Weed Sci.* 48:274–80.

Burcham, L. T. 1970. Ecological significance of alien plants in California grasslands. *Proc. Ass. Am. Geog.* 2:36–39.

Carsten, L. D., Johnston, M. R., Douglas, L. I., and Sands, D. C. 2000. A field trial of crown rust (*Puccinia coronata* f. sp. avenae) as a biocontrol agent of wild oats on San Clemente Island. *Biol. Control* 19:175–81.

Champness, S. S., and Morris K. 1948. Populations of buried viable seeds in relation to contrasting pasture and soil types. *J. Ecol.* 36:149–73.

Collins, S. L., Knapp, A. K., Briggs, J. M., Blair, J. M., and Steinauer E. M. 1998. Modulation of diversity by grazing and mowing in native tallgrass prairie. *Science* 280:745–47.

Cooper, D. W. 1960. Fort Baker ranges returned to champagne grasses. *J. Rangel. Manage.* 13:203–5.

Corbin, J. D., and D'Antonio, C. M. 2004a. Can carbon addition increase competitiveness of native grasses: A case study from California. *Restor. Ecol.* 12.

———. 2004b. Competition between native perennial and exotic annual grasses: Implications for a historic species invasion. *Ecology* 85.

Cox, G. W., and Austin, J. 1990. Impacts of a prescribed burn on vernal pool vegetation at Miramar Naval Air Station, San Diego, California. *Bull. South. Calif. Acad. Sci.* 89:67–85.

Crampton, B. 1974. *Grasses of California.* Berkeley: University of California Press.

Daehler, C. C., and Strong, D. R. 1996. Status, prediction, and prevention of introduced cordgrass *Spartina* spp. invasions in Pacific estuaries, USA. *Biol. Cons.* 78:51–58.

D'Antonio, C. M., Bainbridge, S., Kennedy, C., Bartolome, J., and Reynolds, S. 2001. Ecology and restoration of California grasslands with special emphasis on the influence of fire and grazing on native grassland species. Unpublished report to David and Lucile Packard Foundation, Los Altos, Calif.

Delmas, A. 1999. The effect of fire on California's native grasslands in the absence of grazing at the Nature Conservancy's Vina Plains Preserve, in southern Tehama County. Master's thesis, California State University, Chico.

DeLoach, C. J. 1991. Past successes and current prospects in biological control of weeds in the United States and Canada. *Nat. Areas J.* 11:129–42.

Dennis, A. 1989. Effects of defoliation on three native perennial grasses in the California annual grassland. Ph.D. diss., University of California, Berkeley.

DiTomaso, J. M. 2000. Invasive weeds in rangelands: Species, impacts, and management. *Weed Sci.* 48:255–65.

DiTomaso, J. M., Kyser, G. B., and Hastings, M. S. 1999. Prescribed burning for control of yellow starthistle *(Centaurea solstitialis)* and enhanced native plant diversity. *Weed Sci.* 47:233–42.

Dyer, A. R., Fossum, H. C., and Menke, J. W. 1996. Emergence and survival of *Nassella pulchra* in a California grassland. *Madroño* 43:316–33.

Dyer, A. R., and Rice, K. J. 1997. Intraspecific and diffuse competition: The response of *Nassella pulchra* in a California grassland. *Ecol. Appl.* 7:484–92.

Edwards, S. W. 1995. Notes on grazing and native plants in central California. *Four Seasons* 10:61–65.

———. 1996. A Rancholabrean-age, latest Pleistocene bestiary for California botany. *Four Seasons* 10:5–34.

Ehrenfeld, J. G. 2000. Defining the limits of restoration: The need for realistic goals. *Restor. Ecol.* 8:2–9.

Eller, K. G. 1994. The potential value of fire for managing Stephen's kangaroo rat habitat at Lake Perris. Master's thesis, University of California, Riverside.

Elliot, H. W., and Wehausen, J. D. 1974. Vegetation succession on coastal rangeland of Point Reyes Peninsula. *Madroño* 22:231–38.

Fleischner, T. L. 1994. Ecological costs of livestock grazing in western North America. *Cons. Biol.* 8:629–44.

Foin, T. C., and Hektner, M. M. 1986. Secondary succession and the fate of native species in a California coastal prairie community. *Madroño* 33:189–206.

Fossum, H. C. 1990. Effects of prescribed burning and grazing on *Stipa pulchra* (Hitch.) seedling emergence and survival. Master's thesis, University of California, Davis.

Garcia-Crespo, D. 1983. Multiple treatments to renovate depleted bunchgrass *(Muhlenbergia rigens* and *Stipa pulchra)* range sites in southern California. Ph.D. diss., Loma Linda University, Loma Linda, Calif.

Graham, C. A. 1956. Some reactions of annual vegetation to fire on Sierra Nevada Foothill range land. Master's thesis, University of California, Berkeley.

Gurevitch, J., and Hedges, L. V. 1993. Mixed-effects models in meta-analysis. In *Design and Analysis of Ecological Experiments,* ed. S. M. Scheiner and J. Gurevitch, 378–98. New York: Chapman and Hall.

Hamilton, J. G. 1997. Changing perceptions of pre-European grasslands in California. *Madroño* 44:311–33.

Hamilton, J. G., Holzapfel, C., and Mahall, B. E. 1999. Coexistence and interference between a native perennial grass and non-native annual grasses in California. *Oecologia* 121:518–26.

Hansen, R. B. 1986. The effect of fire and fire frequency on grassland species composition in California's Tulare Basin. Master' thesis, California State University, Fresno.

Harper, J. L. 1977. *Population Biology of Plants.* London: Academic Press.

Harrison, S. 1999. Native and alien species diversity at the local and regional scales in a grazed California grassland. *Oecologia* 121:99–106.

Hatch, D. A., Bartolome, J. W., Fehmi, J. S., and Hillyard, D. S. 1999. Effects of burning and grazing on a coastal California grassland. *Restor. Ecol.* 7:376–81.

Hatch, D. A., Bartolome, J. W., and Hillyard, D. S. 1991. Testing a management strategy for restoration of California's native grasslands. In *Yosemite Centennial Symposium Proceedings: Natural Areas and Yosemite, Prospects for the Future, a Global Issues Symposium Joining the 17th Annual Natural Areas Conference with the Yosemite Centennial Celebration*, 343–49. Denver: National Park Service, Branch of Publications and Graphic Design, Denver Service Center.

Haubensak, K. A. 2001. Controls over invasion and impact of broom species *(Genista monspessulana* and *Cytisus scoparius)* in California coastal prairie ecosystems. Ph.D. diss., University of California, Berkeley.

Heady, H. F. 1956. Changes in a California annual plant community induced by manipulation of natural mulch. *Ecology* 37:798–812.

Heady, H. F., Bartolome, J. W., Pitt, M. D., Stroud, M. G., and Savelle, G. D. 1991. California Prairie. In *Natural Grasslands: Ecosystems of the World*, ed. R. T. Coupland, 8A:313–25. Amsterdam: Elsevier.

Hobbs, R. J., and Mooney H. A. 1985. Community and population dynamics of serpentine grassland annuals in relation to gopher disturbance. *Oecologia* 67:342–51.

Holl, K. D., Steele, H. N., Fusari, M. H., and Fox, L. R. 2000. Seedbanks of maritime chaparral and abandoned roads: Potential for vegetation recovery. *J. Torrey Bot. Soc.* 127:207–20.

Huenneke, L. F., Hamburg, S. P., Koide, R., Mooney, H. A., and Vitousek, P. M. 1990. Effects of soil resources on plant invasion and community structure in California serpentine grassland. *Ecology* 71:478–91.

Huntsinger, L., McClaran, M. P., Dennis, A., and Bartolome, J. 1996. Defoliation response and growth of *Nassella pulchra* (A. Hitchc.) Barkworth from serpentine and non-serpentine grasslands. *Madroño* 43:46–57.

Jackson, R. D., and Bartolome, J. W. 2002. A state-transition approach to understanding nonequilibrium plant community dynamics of California grasslands. *Plant Ecol.* 162:49–65.

Jefferies, R. L., and Maron, J. L. 1997. The embarrassment of richness: Atmospheric deposition of nitrogen and community and ecosystem processes. *Trends Ecol. Evol.* 12:74–78.

Keeley, J. E. 1981. Reproductive cycles and fire regimes. In *Proceedings of the Conference on Fire Regimes and Ecosystem Properties*, ed. H. A. Mooney, T. M. Bonnicksen, N. L. Christensen, J. E. Lotan, and W. A. Reiners, 231–77. General Technical Report WO-26, U.S. Department of Agriculture, Forest Service.

Keeley, J. E., Lubin, D., and Fotheringham, C. J. 2003. Fire and grazing impacts on plant diversity and alien plant invasions in the southern Sierra Nevada. *Ecol. Appl.* 13:1355–74.

Kephart, P. 2001. Resource management demonstration at Russian Ridge Preserve. *Grasslands* 11:8–11.

Kneitel, J. M. 1997. The effects of fire and pocket gopher *(Thomomys bottae)* disturbances in a California valley grassland. Master's thesis, California State University, Northridge.

Kotanen, P. M. 1996. Revegetation following soil disturbance in a California meadow: The role of propagule supply. *Oecologia* 108:652–62.

———. 1997. Effects of experimental soil disturbance on revegetation by natives and exotics in coastal Californian meadows. *J. Appl. Ecol.* 34:631–44.

Langstroth, R. P. 1991. Fire and grazing ecology of *Stipa pulchra* grassland: A field study at Jepson Prairie. Master's thesis, University of California, Davis.

Larson, J. R., and Duncan, D. A. 1982. Annual grassland response to fire retardant and wildfire. *J. Rangel. Manage.* 35:700–703.

Mack, R. M., Simberloff, D., Lonsdale, W. M., Evans, H., Clout, M., and Bazzazz, F. A. 2000. Biotic invasions: Causes, epidemiology, global consequences, and control. *Ecol. Appl.* 10:689–710.

Major, J., and Pyott, W. T. 1966. Buried, viable seeds in two California bunchgrass sites and their bearing on the definition of a flora. *Vegetatio* 13:253–82.

Maron, J. L., and Connors, P. G. 1996. A native nitrogen-fixing shrub facilitates weed invasion. *Oecologia* 105:302–12.

Maron, J. L., and Jefferies, R. L. 1999. Bush lupine mortality, altered resource availability, and alternative vegetation states. *Ecology* 80:443–54.

———. 2001. Restoring enriched grasslands: Effects of mowing on species richness, productivity, and nitrogen retention. *Ecol. Appl.* 11:1088–100.

Marty, J. T. 2001. Fire effects in and around vernal pools at the Valensin Ranch, Cosumnes River Preserve. Unpublished data. Nature Conservancy, Galt, Calif.

———. 2002. Managing and restoring California annual grassland species: An experimental field study. Ph.D. diss., Agronomy and Range Science, University of California, Davis.

McEvoy, P., and Coomes, E. M. 1999. Biological control of plant invaders: Regional patterns, field experiments, and structured population models. *Ecol. Appl.* 9:387–401.

McEvoy, P., Cox, C., and Coomes, E. M. 1991. Successful biological control of ragwort, *Senecio jacobaea,* by introduced insects in Oregon. *Ecol. Appl.* 1:430–42.

Menke, J. W. 1982. Grazing and fire management for native perennial grass restoration in California grasslands. *Fremontia* 20:22–25.

Merenlender, A., Heise, K., Bartolome, J. W., and Allen-Diaz, B. H. 2001. Monitoring long-term vegetation change at the Hopland Research and Extension Center. *Calif. Agric.* 55:42–65.

Meyer, M. D., and Schiffman, P. M. 1999. Fire season and mulch reduction in a California grassland: A comparison of restoration strategies. *Madroño* 46:25–37.

Micallef, S. B. 1998. Grazing effects on the grassland vegetation of Mount Diablo State Park, California. Master's thesis, San Francisco State University, San Francisco, Calif.

Milberg, P. 1992. Seedbank in a 35-year-old experiment with different treatments of a semi-natural grassland. *Acta Oecol.* 13:743–52.

Morgan, J. P. 1994. Soil impoverishment: A little known technique holds potential for establishing prairie. *Restor. Manage. Notes* 12:55–56.

Painter, E. L. 1995. Threats to the California flora: Ungulate grazers and browsers. *Madroño* 42:180–88.

Parsons, D. J., and Stohlgren, T. J. 1989. Effects of varying fire regimes on annual grasslands in the southern Sierra Nevada of California. *Madroño* 36:154–68.

Paschke, M. W., McLendon, T., and Redente, E. F. 2000. Nitrogen availability and old-field succession in a short-grass steppe. *Ecosystems* 3:144–58.

Peart, D. R. 1989a. Species interactions in a successional grassland. I. Seed rain and seedling recruitment. *J. Ecol.* 77:236–51.

———. 1989b. Species interactions in a successional grassland. II. Colonization of vegetated sites. *J. Ecol.* 77:252–66.

———. 1989c. Species interactions in a successional grassland. III. Effects of canopy gaps, gopher mounds, and grazing on colonization. *J. Ecol.* 77:267–89.

Pollack, O., and Kan, T. 1998. The use of prescribed fire to control invasive exotic weeds at Jepson Prairie Preserve. In *Ecology, Conservation, and Management of Vernal Pool Ecosystems—Proceedings from a 1996 Conference*, ed. C. W. Witham, E. T. Bauder, D. Belk, W. R. Ferren, Jr., and R. Ornduff, 241–49, Sacramento, Calif.: California Native Plant Society.

Porter, E. E., and Redak R. A. 1996. Short-term recovery of the grasshopper communities (Orthoptera: Acrididae) of a California native grassland after prescribed burning. *Env. Entomol.* 26:234–40.

Randall, J. M., Rejmánek, M., and Hunter, J. C. 1998. Characteristics of the exotic flora of California. *Fremontia* 26:3–12.

Reever Morghan, K. J., and Seastedt, T. R. 1999. Effects of soil nitrogen reduction on nonnative plants in restored grasslands. *Restor. Ecol.* 7:51–55.

Reeves, K., and Morris, J. 2000. Hollister Hills State Vehicular Recreation Area 2000 Biological Monitoring Report. Unpublished report, T. O. Cattle Company, San Juan Batista, Calif.

Reynolds, S. A., Corbin, J. D., and D'Antonio, C. M. 2001. The effects of litter and temperature on the germination of native and exotic grasses in a coastal California grassland. *Madroño* 48:230–35.

Rice, P. M., and Toney, J. C. 1998. Exotic weed control treatments for conservation of fescue grassland in Montana. *Biol. Cons.* 85:83–95.

Rice, P. M., Toney, J. C., Bedunah, D. J., and Carlson, C. E. 1997. Plant community diversity and growth form responses to herbicide application for control of *Centaurea maculosa*. *J. Appl. Ecol.* 34:1397–412.

Roché, C. T., Harmon, B. L., Wilson, L. M., and McCaffrey, J. P. 2001. *Eustenopus villosus* (Coleoptera: Curculionidae) feeding of herbicide-resistant yellow starthistle (*Centaurea solstitialis* L.). *Biol. Control* 20:279–86.

Saenz, L., and Sawyer, J. O. 1986. Grasslands as compared to adjacent *Quercus garryana* woodland understories exposed to different grazing regimes. *Madroño* 33:40–46.

Safford, H. D., and Harrison, S. P. 2001. Grazing and substrate interact to affect native vs. exotic diversity in roadside grasslands. *Ecol. Appl.* 11:1112–22.

Seastedt, T. R., Briggs, J. M., and Gibson, D. J. 1991. Controls on nitrogen limitation in tallgrass prairie. *Oecologia* 87:72–79.

Stromberg, M. R., and Griffin, J. R. 1996. Long-term patterns in coastal California grasslands in relation to cultivation, gophers, and grazing. *Ecol. Appl.* 6:1189–211.

Stromberg, M. R., and Kephart, P. 1996. Restoring native grasses in California old fields. *Restor. Manage. Notes* 14:102–11.

Thomsen, C. D., Williams, W. A., Vayssieres, M., Bell, F. L., and George M. R. 1993. Controlled grazing on annual grassland decreases yellow starthistle. *Calif. Agric.* 47:36–40.

TNC (The Nature Conservancy). 2000. Effects of prescribed fire and cattle grazing on a vernal pool grassland landscape: Recommendations for monitoring. Unpublished report prepared for the Environmental Protection Agency, San Francisco, Calif.

Turner, C. E., Johnson, J. B., and McCaffrey, J. P. 1995. Yellow starthistle. In *Biological Control in the Western United States,* ed. J. R. Nechols, L. A. Andres, J. W. Beardsley, R. D. Goeden, and C. G. Jackson, 405–10. Natural Resources Publication 3361, University of California, Division of Agriculture, Oakland, Calif.

Villegas, B. 1998. Distribution of yellow starthistle seedhead flies, *Chaetorellia succinea* and *Chaetorellia australis,* in California. In *Biological Control Program Annual Summary, 1997,* ed. D. M. Woods, 55–56. Sacramento, Calif.: California Department of Food and Agriculture, Plant Health and Pest Prevention Services.

Vinton, M. A., and Burke, I. C. 1995. Interactions between individual plant species and soil nutrient status in shortgrass steppe. *Ecology* 76 1116–33.

Vitousek, P. M., Mooney, H. A., Lubchenco, J., and Melillo, J. M. 1997. Human domination of Earth's ecosystems. *Science* 277:494–99.

Vitousek, P. M., Walker, L. R., Whitaker, L. D., Mueller-Dombois, D., and Matson, P. M. 1987. Biological invasion by *Myrica faya* alters ecosystem development in Hawai'i. *Science* 238:802–4.

Wagner, F. H. 1989. Grazers, past and present. In *Grassland Structure and Function: California Annual Grassland,* ed. L. F. Huenneke and H. Mooney, 151–62. Dordrecht, Netherlands: Kluwer Academic Publishers.

Wan, S., Hui, D., and Luo Y. (2001). Fire effects on nitrogen pools and dynamics in terrestrial ecosystems: A meta-analysis. *Ecol. Appl.* 11:1349–56.

White, K. L. 1967. Native bunchgrass *(Stipa pulchra)* on Hastings Reservation, California. *Ecology* 48:949–55.

Wilson, S. D., and Gerry, A. K. 1995. Strategies for mixed-grass prairie restoration: Herbicide, tilling, and nitrogen manipulation. *Restor. Ecol.* 3:290–98.

York, D. 1997. A fire ecology study of a Sierra Nevada foothill basaltic mesa grassland. *Madroño* 44:374–83.

Young, J. A., and Evans, R. A. 1989. Seed production and germination dynamics in California annual grasslands. In *Grassland Structure and Function: California Annual Grassland,* ed. L. F. Huenneke and H. Mooney, 39–45 Dordrecht, Netherlands: Kluwer Academic Publishers.

Young, J. A., Trent, J. D., Blank, R. R., and Palmquist D. E. 1998. Nitrogen interactions with medusahead (*Taeniatherum caput-medusae* spp. asperum) seed-banks. *Weed Sci.* 46:191–95.

Zavon, J. A. 1977. Grazing and fire effects on annual grassland composition and sheep diet selectivity. Master's thesis, University of California, Davis.

Zink, T. A., and Allen, M. F. 1998. The effects of organic amendments on the restoration of a disturbed coastal sage scrub habitat. *Restor. Ecol.* 6:52–58.

12 Using Natural Experiments in the Study of Alien Tree Invasions

Opportunities and Limitations

David M. Richardson, Mathieu Rouget,
and Marcel Rejmánek

> The only rules of scientific method are honest observations and
> accurate logic.
>
> ROBERT MACARTHUR,
Geographical Ecology

SUMMARY

For a variety of reasons, hundreds of tree species have been planted well outside their natural ranges. In numerous parts of the world, alien trees are the
foundation of commercial forestry and agroforestry enterprises, and many
are planted for a wide range of other uses. A small sample of tree species that
are widely cultivated as aliens have spread from planting sites, and some
taxa are among the most damaging of invasive alien plants.

The configuration of alien tree plantings—many species planted in a
wide range of habitats, subject to many types, intensities, and combinations
of natural and human-mediated disturbances across continents—provides
us with a series of "natural" (i.e., nonmanipulative) experiments, with
opportunities for gaining useful insights on central issues in invasion ecology. For long-lived organisms like trees, such natural experiments are usually the only available source of information for management, especially in
the face of rapid global change.

This chapter reviews the value of natural experiments for studying
aspects of alien tree invasions, with special reference to South Africa. Case
studies, involving mainly species of *Acacia, Hakea, Pinus, Eucalyptus*, and
Metrosideros, are reviewed to explore the extent to which natural experiments have improved (or could improve) our ability to understand and
manage these invasions.

Suggestions are made on further insights that could be gained through the careful evaluation of natural experiments. Innovations in remote sensing and statistics for spatial data are providing us with new tools.

INTRODUCTION

Biological invasions—that is, the spread of organisms that have been introduced to areas outside their natural ranges by human activity—are among the most important agents (in some areas *the* most important agent) of habitat transformation in many parts of the world. Invasions have numerous effects, including local species extinctions and the disruption of ecosystem processes and services.

Charles Elton's 1958 book *The Ecology of Invasions by Animals and Plants* was a milestone in what is now known as *invasion ecology*. In the last two decades, research interest in all facets of invasions has exploded. Invasion ecologists grapple with three fundamental questions. What makes some organisms better invaders than others? What features of an ecosystem or community affect its susceptibility to invasion by particular organisms? What can be done to reduce or mitigate the harmful effects of invasions?

Invasions are notoriously difficult to predict, which has led many to express pessimism regarding the prospects for making predictions at scales that are useful to managers (Crawley 1987; Whitmore 1991; Williamson 1999). Nevertheless, plant-invasion ecology has made substantial strides in the last decade toward developing a set of generalizations that have important applications in management (see Rejmánek et al. forthcoming for a recent summary).

For a variety of reasons, scientists in many disciplines use *natural experiments* (i.e., experiments in which the researcher does not directly manipulate the site but rather selects sites or other entities where manipulation has already taken place; see Diamond 1986). Natural experiments have been applied in the fields of ecology, economics, human ecology and health (heredity, immunology, psychology, sociology, virology, etc.), legal studies, marketing research, parasitology, and many other scientific disciplines. Such experiments clearly offer insights that could not be obtained from more formal laboratory or field experiments for ethical or practical reasons.

Natural experiments are an extremely rich source of knowledge in ecology. Many authors have reviewed the value of such experiments, though not always calling them that (Cook and Campbell 1979; Diamond 1986; Davis 1989; Pickett 1989; Pickett et al. 1994; Gotelli and Graves 1996; Underwood 1997). Many of the most important generalizations in plant-invasion ecology have been derived from natural experiments (Table 12.1),

1. Which taxa invade?

  1.1. Stochastic approach

    1.1.1. The probability of invasion success increases with initial population size and the number of introduction attempts ("propagule pressure").

    1.1.2. Long residence time improves the chances of invasion.

    1.1.3. Many/most plant invasions are preceded by a lag phase that may last many decades.

  1.2. Empirical, taxon-specific approach

    1.2.1. If a species is invasive anywhere in the world, there is a good chance that it will invade similar habitats in other parts of the world.

    1.2.2. Among invasive alien plants, some families (Amaranthaceae, Brassicaceae, Fabaceae, Hydrocharitaceae, Papaveraceae, Pinaceae, Poaceae, and Polygonaceae) are significantly overrepresented.

  1.3. The role of biological characters

    1.3.1. Fitness homeostasis is an important determinant of invasiveness.

    1.3.2. Genetic change can facilitate invasions, but many species have sufficient phenotypic plasticity to exploit new environments.

    1.3.3. Small genome size has value as an indicator of invasiveness within closely related taxa.

    1.3.4. Several characters linked to reproduction and dispersal are key indicators of invasiveness.

    1.3.5. Seed dispersal by vertebrates is implicated in many plant invasions.

    1.3.6. Low relative growth rate of seedlings and low specific leaf area are good indicators of low plant invasiveness in many environments.

    1.3.7. Large native range is an indicator of potential invasiveness.

    1.3.8. Vegetative reproduction is responsible for many plant invasions.

    1.3.9. Alien taxa are more likely to invade a given area if native members of the same genera (and family) are absent, partly because many herbivores and pathogens cannot switch to phylogenetically distant taxa.

    1.3.10. The ability to utilize generalist mutualists greatly improves an alien taxon's chances of becoming invasive.

    1.3.11. Efficient competitors for limited resources are likely to be the best invaders in natural and seminatural ecosystems.

    1.3.12. Characters favoring passive dispersal by humans greatly improve an alien plant taxon's chance of becoming invasive.

TABLE 12.1 *(continued)*

1.4. Environmental compatibility

　1.4.1.　Climate matching is a useful first step in screening alien species for invasiveness.

　1.4.2.　Resource enrichment or release, often just intermittent, supports many invasions.

　1.4.3.　Propagule pressure (see 1.1.1) can override biotic and abiotic resistance of a community to invasion.

　1.4.4.　Determinants of invasibility (macroscale climate factors, microclimatic factors, soils, various community/ecosystem properties) interact in complicated ways; evaluation of invasibility must therefore be context specific.

1.5. Relationship between species richness and invasibility

　1.5.1.　At the landscape scale, invasibility seems to be positively correlated with native-plant-species richness; at smaller (neighborhood) scales the correlation seems to be negative.

2. How fast do alien plants spread?

　2.1.　Spread is determined primarily by reproduction and dispersal, but various extrinsic factors interact with these factors to mediate spread rates.

　2.2.　Spread rates based on local dispersal syndromes greatly underestimate spread potential.

　2.3.　Rare, long-distance dispersal is hugely important for explaining population growth and spread over medium and long time scales.

3. What is the impact of alien plant invasions?

　3.1.　Predicting the impact of invasive alien plants is much more difficult than predicting invasiveness.

　3.2.　Alien species that add a new function to an invaded ecosystem are much more likely to have big impacts than those that alter only existing resource use levels.

4. How can invasions be controlled, contained, or eradicated?

　4.1.　Early detection and initiation of management can make the difference between being able to employ feasible offensive strategies (eradication) and the necessity of retreating to a more expensive defensive strategy.

SOURCE:　Based on information reviewed in Rejmánek, Richardson, Higgins, Pitcairn, and Grotkopp, forthcoming.

and only recently have manipulative experiments begun to be used to address certain aspects of invasions (reviewed in Rejmánek et al. forthcoming). The intentional or accidental movement of thousands of species to areas outside their natural range at different times, in different numbers and various mixtures, accompanied by a radical array of changes to every conceivable feature of the environments in which the alien species find themselves provides a natural experiment at a grand scale (Holland writes of "cultural landscapes as biogeographical experiments" with reference to invasions in New Zealand [2000, 39]).

Plant species of all growth forms have become *naturalized* or *invasive* (terminology in this chapter follows Richardson, Pyšek, et al. 2000). Trees, although sometimes not considered typical weedy plants, are invasive aliens in many parts of the world. Among the most invasive and best-studied genera are *Acacia* (Fabaceae), *Acer* (Aceraceae), *Ailanthus* (Simaroubaceae), *Elaeagnus* (Oleaceae), *Fraxinus* (Oleaceae), *Maesopsis* (Rhamnaceae), *Pinus* (Pinaceae), *Leptospermum* (Myrtaceae), *Melaleuca* (Myrtaceae), *Miconia* (Melastomataceae), *Mimosa* (Fabaceae), *Myrica* (Myricaceae), *Paulownia* (Faba-ceae), *Pittosporum* (Pittosporaceae), *Robinia* (Fabaceae), *Prosopis* (Fabaceae), *Pseudotsuga* (Pinaceae), *Sapium* (Euphorbiaceae), *Salix* (Salicaceae), *Schinus* (Anacardiaceae), and *Tamarix* (Tamaricaeae). Most of the genera in this taxonomically disparate assemblage are useful to humans and have therefore been moved around the world to varying degrees: for example, for erosion control and drift sand stabilization *(Acacia, Tamarix)*, fodder for livestock *(Prosopis)*, fuel wood *(Acacia)*, ornamental use *(Acer, Fraxinus, Elaeagnus, Leptospermum)*, and timber *(Pinus)*.

Partly because of their large size but also because of special features such as the ability to fix atmospheric nitrogen (which many legumes can do), the trees that have become important invasive species have had disproportionately large impacts: for example, *Acacia* and *Pinus* species in South African fynbos, *Melaleuca quinquinervia* in Florida, *Miconia calvesecens* in Tahiti, *Mimosa pigra* in northern Australia, *Myrica faya* in Hawaii, *Prosopis* species in arid parts of South Africa, and *Schinus terebinthifolius* and *Tamarix* species in parts of North America.

Natural experiments are particularly appealing (and necessary) for gaining information on the ecology of tree invasions for several reasons.

- Trees are long-lived organisms, and events at many life-cycle stages affect the likelihood of recruitment, establishment, and persistence (and thus the species' potential impact as an invader). Some of these events are rare occurrences (occurring less frequently than the average life span of a plant) that have profound implications for populations over longer timescales (e.g., Davis 1989). Manipulative experiments required to obtain sufficient data on the range of factors that might affect different stages in the development of the plant tend toward the impossible. As Harper (1977, 600) wrote: "The study of trees is a study of short cuts; the long life and large size of trees makes many of the conventional methods of plant biology impossible or unrealistic."

- Some species of trees are widely used by humans and have been introduced, planted, and managed in many parts of the world. This fact has allowed these species to sample a wide range of habitats and has exposed them to many extrinsic factors that (potentially) mediate their ability to reproduce and spread from sites of cultivation. "The intentional plantings of trees . . . can be dated from time zero and the plantings monitored at intervals to determine the success of the original transplants of seedlings, and their spontaneous progeny, if any. Such instances, while potentially affording the population biologist with ready-made experimental populations, are rarely if at all observed from the demographic standpoint. What exceptional opportunities have been overlooked in the plantations of such tree species such as *Pinus radiata, P. contorta, P. sylvestris, Pseudotsuga menziesii* and many others!" (Kruckeberg 1986, 408).

- Because trees are large and conspicuous and because most alien tree invasions occur in short vegetation, trees are relatively easy to map using various remote-sensing methods (historical photographs, aerial photography, satellite imagery, etc.). Accurate spatial data for large geographical areas is thus relatively easy to acquire. The Geographic Information System (GIS) and geostatistical tools facilitate analysis of such data at many geographical scales.

- Tree distribution and abundance and trajectories of change can often be inferred from historical documents, cultural evidence data from the early stages of invasions, and a range of paleoecological methods (Egan and Howell 2001), all providing important demographic perspectives.

- Many alien trees are planted over large areas. Events and conditions at different parts of the adventive range may effect populations in different ways at different times. Information from many localities greatly improves our knowledge of the factors that drive invasions. Only natural experiments can yield information at such large geographical, temporal, and spatial scales (Figure 12.1).

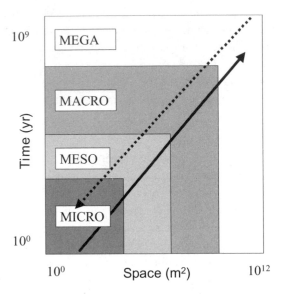

Figure 12.1. The value and potential applicability of
manipulative *(dashed line)* and natural experiments
(solid line) at different spatial and temporal scales.
Value and applicability increase along both lines
in the direction of the arrows. The background
diagram is based on concepts in Delcourt and
Delcourt (1991).

This chapter reviews some approaches toward the study of tree invasions
that involve the innovative use of natural experiments. We discuss the contributions that such studies have made toward generalizations in plant-
invasion ecology, and suggest some potentially profitable avenues for future
work. Many of our examples are from South Africa. This country has probably been invaded by alien trees to a greater extent than any other part of
the world: a recent survey showed that about 10 million ha of the country
had been invaded by the 180 woody species that were mapped (Versfeld, Le
Maitre, and Chapman 1998).

CONCEPTUALIZATION/ABSTRACTION

One important aspect of simplification of concepts is abstraction—the identification of the essence of phenomena or interactions of interest (Pickett,
Kolasa, and Jones 1994). Invasions follow a sequence of events and interac-

tions. Assimilating perspectives by astutely observing invasions in a range of sites, where the introduced taxa have been subjected to a spectrum of mediating factors, is hugely valuable for sketching the game plan for a given invasion. Such an exercise is useful for identifying and characterizing the players, plotting the dimensions of the invasion arena, and identifying the main areas of interaction that drive invasions. At least three levels of conceptualization are important: the naturalization-invasion process, community dynamics, and species-environment interactions that determine invasion success.

The Naturalization–Invasion Process

One way of abstracting a spatial process such as an invasion (*sensu* Richardson, Pyšek, et al. 2000) is to picture an intruder having to negotiate a series of obstacles that hinder its progress toward integration. Richardson, Pyšek, et al. (2000) discuss early uses of this concept in ecology and the various attempts to apply it in invasion ecology. Figure 12.2 represents a framework for defining objective terms for describing the status of an introduced plant species at a given locality. Similar models are easily compiled to explain the current assemblage of alien plants in a given system and the range of levels of success they have achieved. Similarly, collective experience over large spatial and temporal scales facilitates the conceptualization of the interacting role of factors such as human-induced habitat changes, mutualisms, and various feedback loops in mediating invasions (Richardson, Allsopp, et al. 2000, 67).

The widespread planting of many *Pinus* species and the subsequent spread of at least nineteen species in the Southern Hemisphere provide a superb opportunity for deriving important insights on the factors that mediate invasions using "space-for-time substitution" (*sensu* Pickett 1989). "Treatments" involving the introduction and cultivation of pines across a range of sites allow us to evaluate the role of the factors that operate along long temporal scales. For example, the global snapshot of the status of pines as invaders in many parts of the Southern Hemisphere (Richardson and Higgins 1998 and references therein) points strongly to the roles of long residence time, the extent of planting, and moderate levels of disturbance in structuring pine invasions. This "experiment" also sheds light on the relative invasiveness of different taxa. The assumption here is that the many treatments over time and space have subjected the introduced taxa to a wide range of conditions, including rare events that may be extremely influential for the outcome of an invasion (and which would be difficult or impossible to simulate in formal

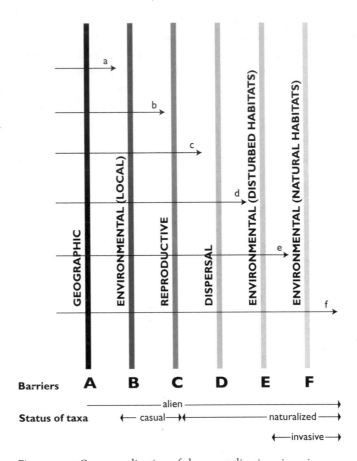

Figure 12.2. Conceptualization of the naturalization–invasion process, illustrating the role of important barriers in mediating invasions (based on Richardson, Pyšek, et al. 2000; *a–f* represent alien taxa). Abstractions such as this are facilitated by natural experiments that enable the observer to determine the role of different factors in mediating invasions.

experiments). From this information we can produce a "score sheet" that identifies features that either facilitate or limit invasions at different sites. Summarizing this data, one can arrive at a qualitative or semiquantitative model of the determinants of pine invasions in the Southern Hemisphere based on species attributes, residence time, extent of planting, ground-cover characteristics, disturbance regime, and the role of resident biota in determining invasibility (Richardson, Williams, and Hobbs 1994).

Community Dynamics

Successful invasion of an ecosystem requires an invader either to be accommodated in the prevailing dynamics of the existing assemblage of organisms occupying the site or to change the prevailing dynamics in such a way as to permit its infiltration and dominance/transformation of a site. In both cases, the careful conceptualization of stages is central to a fundamental understanding of the invasion dynamics. For example, Richardson and Cowling (1992, 175) show how fire-driven cyclical replacement sequences involving native overstory and understory shrubs in fynbos permit the incursion of alien trees such as pines and how these sequences are disrupted to result in a greatly simplified steady state dominated by the invasive trees.

Species–Environment Interactions That Determine Invasion Success

Several species of *Prosopis* are notably weedy within their natural ranges, and several taxa are highly invasive outside their natural range. A review of the many factors known to drive range expansion and increases in population density from studies at many sites can be used to conceptualize population dynamics and to provide at least qualitative abstractions of likely trajectories under a range of scenarios involving, for example, changes in climatic factors and CO_2 levels, altered livestock-stocking rates, and a range of human-related factors (Richardson, Bond, et al. 2000).

DETERMINANTS OF INVASIVENESS

In plant-invasion ecology, the best predictor of the risk of an alien plant becoming invasive is whether that taxon has become invasive elsewhere (references in Rejmánek et al. forthcoming). Despite this generalization's value, several practicalities bedevil its implementation. First, not all potentially invasive species have been moved around the world to the same extent, so not all potential invaders have sampled all potentially invasible sites. Also, not all species have been introduced in the same numbers or over the same period. Consequently, even if there was a readily accessible database with information on all the world's worst invaders, labeling a species as a "low risk" introduction on the basis of empirical evidence would still be risky. This natural experiment is still running. Preliminary results are useful but need to be applied with caution. We clearly need, in addition to such insights, a more mechanistic understanding of invasiveness. Natural experiments are a fundamental prerequisite for such an understanding, partly to provide us with good lists of invasive and non-invasive species—taxa that have been given similar opportunities (over time and space) to invade.

Pinus is a superb (perhaps the best available) subject for exploring the biological characters that distinguish invasive from non-invasive taxa within a group of closely related plants. A large number of pine species have been widely grown for many decades in many habitats well outside their natural ranges. Among these species, we can rank taxa on the basis of their performance as invaders (e.g., with reference to the model in Figure 12.2); there is a reasonably clear distinction between those taxa that are invasive (*sensu* Figure 12.2) and those that have not invaded. Good data are available on the life-history traits of many pines. Rejmánek and Richardson (1996) derived a discriminant function, using only three life-history traits, that separates invasive from non-invasive taxa. Perhaps surprisingly, results thus obtained are useful for explaining the degree of success of all trees as invasive aliens (Rejmánek and Richardson 1996; Rejmánek 2000; Rejmánek et al. forthcoming). Results from this approach suggest priorities for research to elucidate the components of invasiveness, such as genome size (the amount of DNA in one complete set of chromosomes) and seedling growth rate (Grotkopp, Rejmánek, and Rost 2002).

Not all taxa are equally suited for the elucidation of invasiveness through an assessment of life-history traits. For example, we have been unable to find any biological features that separate invasive from non-invasive *Eucalyptus* species (Rejmánek et al. forthcoming; see also below).

INFERRING PROCESS FROM PATTERN

Pinus invasions in South Africa have been studied at scales ranging from small plots to the whole country. For example, studies of spatial patterns of plants, tree rings, and fire scars on *P. halepensis* at the scale of the plant community (10^0 ha) provided information on the role of fire in the recruitment dynamics of alien trees in fynbos (Richardson 1988). Similar studies conducted over much larger areas (landscapes; 10^3 ha) in conjunction with analyses of historical aerial photographs shed light on the role of fire and other disturbances in population growth and also provided useful data on spread rates (Richardson and Brown 1986). Recent studies, utilizing national databases of pine plantations and self-sown pine stands and computer modelling, have examined the roles of a range of environmental factors and propagule pressure in structuring invasions at the national scale (10^8 ha) (Rouget and Richardson 2003b).

These studies have shown that aspects of the invasion process can be inferred from analyzing spatial patterns at various scales. The spatial structure of available information determines which ecological processes can potentially be assessed. For example, we have found it ecologically meaningless to study the role of community dynamics, local disturbance patterns, or even propagule pressure in national-scale assessments since the grain of available data at this scale does not correspond with the scale at which these factors operate.

An Example—Propagule Pressure

Propagule pressure (the spatial mass effect) is an intuitive determinant of invasibility: the greater the number and the higher the rate of introductions, the better the chance of establishment and invasion. A quantitative description of its role has, however, remained illusive (D'Antonio, Levine, and Thomsen 2001). Observed patterns of invasion (or non-invasion) of introduced Myrtaceae in South Africa provide intriguing opportunities for exploring the role of propagule pressure.

About a hundred species of *Eucalyptus* have been introduced to South Africa, where they have been cultivated at different levels of intensity in commercial plantations and arboreta and as amenity plantings for a variety of uses (Poynton 1979). Surprisingly, given their abundant cultivation and long residence time in the country, eucalypts have faired poorly as invaders compared with pines, for example, which have had similar opportunities to invade. Although we have successfully used life-history traits to separate invasive from non-invasive pines, we have been unsuccessful in separating the few invasive species of eucalypts from non-invasive congeners based on any life-history traits or any other feature of biology. However, if we regress the number of records of spontaneous regeneration (from published sources and an intensive review of herbarium specimens) against a crude index of propagule pressure (the number of plantations; Poynton 1979), we see that the extent of spontaneous regeneration (naturalization/invasion) is positively correlated with the number of records of planting (Rejmánek et al. forthcoming). A similar pattern is emerging from a global review of the extent of spontaneous regeneration (from published reports, herbarium specimens, and our observations in many parts of the world for over a decade) for eucalypts introduced to many countries (our primary database contains information on sixty-two eucalypt species planted outside their natural range in Australia, South Africa, California, Florida, New Zealand, the Mediterranean basin, and Hawaii; M. Rejmánek and D. M. Richardson, unpublished data).

Much work remains to be done to explain why eucalypts, so spectacu-

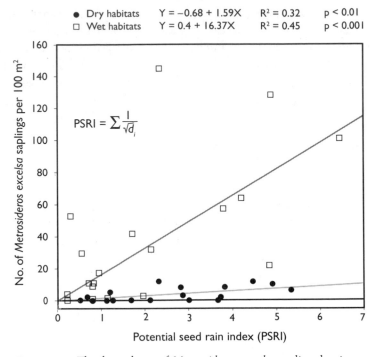

Figure 12.3: The dependence of *Metrosideros excelsa* sapling density on potential seed rain index (PSRI; d_i is the distance (m) to the i-th seed-producing tree within a 300-m radius) in dry and wet fynbos, Betty's Bay, Western Cape, South Africa. Slopes of the two regression lines are significantly different ($p < 0.01$). Fynbos plots were classified as dry (dominated by *Erica ericoides, Elegia filacea,* and *Metalasia muricata*) and wet (dominated by *Cliffortia hirsuta, Osmitopsis asteriscoides, Gleichenia polypodioides,* and *Erica perspicua*) on the basis of results of average linkage clustering using the Jaccard coefficient as a measure of similarity (M. Rejmánek and D. M. Richardson, unpublished).

larly successful in a wide range of habitats in their native range in Australia (Williams and Woinarski 1997) and so widely transplanted across the world (about 14 million ha of plantations at the end of 1993; Turnbull 2000), perform relatively poorly as invaders outside Australia. The fact that propagule pressure is clearly an important driver of invasions (much more so than with pines) suggests that the answer lies in the specialized regeneration requirements of eucalypts. Mycorrhizas and inhibitory microorganisms are probably implicated, but we need formal experiments to elucidate the role of different factors. This concept is an example of how natural experiments can identify crucial questions that can direct manipulative experiments.

Another myrtaceous tree, *Metrosideros excelsa*, which is widely planted as an ornamental tree and hedge plant in the Western Cape of South Africa, provided the opportunity to explore the role of propagule pressure more quantitatively. This species is highly invasive in one small region of the fynbos biome (Richardson and Rejmánek 1999). We mapped all source trees and the entire invasive population in an area of approximately 5 km². We could distinguish, on the basis of native indicator plant species, two distinct habitat types that displayed different invasion patterns—one of these (the wet habitat) was clearly much more suitable for establishment of *M. excelsa* than the other (shown by the much greater number of seedlings). Regression analysis of the number of *Metrosideros* saplings on a potential seed rain index (PSRI) revealed that wet habitats are almost exactly ten times as invasible as dry habitats (Figure 12.3).

GETTING SPATIAL

The above-mentioned approaches are useful, but they give no spatial dimension to invasions. Invasions are spatial phenomena, and plant-invasion ecology must strive for spatially explicit explanations and predictions. The development of spatially explicit modelling and analytical tools offers great opportunities for analyzing invasion dynamics, drawing on information from natural experiments. Spatial dimensions can be effectively added to predictive models developed on the basis of information from natural experiments in various ways. In recent studies we have used "correlative," "semi-mechanistic," and "mechanistic" approaches.

Correlative Approaches

Many studies that address the determinants of plant distribution use correlative methods in which environmental factors (climate, topography, vegetation type, etc.) are assumed to be important drivers of distribution. The invasive spread from plantations of alien trees into natural ecosystems provides wonderful opportunities to investigate the environmental correlates of invasion success. Each invasive tree or stand can be considered as a replicate. If "treatments" occur over a large area relative to the region under consideration, we can assume that the whole range of environmental conditions has been sampled. Using this approach we have studied the determinants of invasive alien trees at different spatial scales in South Africa. We have used regression-tree analysis techniques because they are highly suited to the study of distributions, and they yield better estimates than other classical techniques (De'ath and Fabricius 2000; Rouget et al. 2001). Regression trees revealed that soil pH and vegetation structure were the key determinants of invasive pine distribution at the landscape scale (10^3 ha) in the Kango Valley in the Western Cape (Rouget et al. 2001). A study using a similar approach at the scale of the entire Cape Floristic Region (10^7 ha) showed a different suite of factors (mainly vegetation type and climatic factors) to be the main determinants of invasive trees (Rouget and Richardson 2003b). Correlative methods such as these are useful in conjunction with GIS to produce layers of "potential distribution," which indicate other areas that share the features found to best discriminate invaded from non-invaded sites. For example, for the Cape Floristic Region, this approach suggests that 30% of the area of remaining untransformed land in the region is invasible (Rouget, Richardson, Cowling, Lloyd, and Lombard 2003).

Semimechanistic Approaches

The correlative approach assumes that plants and environment are in equilibrium. In the case of alien plant invasions, such an assumption is probably invalid where introductions and invasions are relatively recent, since the alien species have had insufficient time to sample all potentially suitable sites. We recently developed a set of semimechanistic models (Rouget and Richardson 2003a) for the Agulhas Plain (2160 km^2), a species-rich area of the Cape Floristic Region that is heavily impacted by agriculture and plant invasions. We used extremely detailed maps of the current distribution of thirty-six alien woody plant species to explore determinants of distribution and spread rates of invasive species. Because of the relatively homogenous environment (compared, for example, with many other parts of the Cape Floristic Region), the low environmental heterogeneity, and the recent invasion history of the area, correlative models failed to accurately model the distribution of invasive species.

A null model considered only the distance to invasion foci (the densest/oldest stands) as a predictor of the distribution of alien species. Density of invaders was predicted to decrease with increasing distance from such foci. This approach simulated dispersal, a key biological attribute for understanding and predicting plant invasions at the landscape scale. Environmental factors were then introduced to fine-tune the spread model. Modeling density was considerably improved when we considered environmental factors, which were found to increase or reduce the spread. Spread rate was generally higher in natural areas and lower in cultivated land. Topography was also a key factor for adjusting the spread. In the case of *Pinus pinaster*, the model accurately predicted 85% of the presence/absence and 70% of the species density.

Such models are attractive because they bridge the gap between the complexity of pure mechanistic models and the lack of biological attributes in correlative models. Semimechanistic models are relatively simple to compute and offer reasonable estimates of the invasion process because the key biological attribute—dispersal ability—is accounted for.

Mechanistic Approaches

Over small spatial scales (10^0-10^1 ha), the biological attributes of the invader are of overriding importance and need to be included in models of distribution and spread rates. At this scale, models need to simulate the entire life cycle of the invasive plant, and thus mechanistic models (also called spatially explicit simulation models) become more tractable. This approach integrates space, ecological processes, and stochasticity into a single framework (Czaran and Bartha 1992). Mechanistic models can also trace the spatial locations of modelling entities so that the model predictions can be linked to a geographic information system (GIS) for further analysis. Importantly, these models can be parameterized, calibrated, and validated using information from natural experiments. Individual-based simulation models were built to simulate the spread of alien pine trees from established plantations in natural fynbos ecosystems (Higgins and Richardson 1998). The mean spread rate was close to 10 m yr^{-1}. Results show strong interactions among the ecological processes investigated and emphasize the importance of dispersal ability.

Using information from natural experiments in a variety of ways to model spatial patterns of alien plant invasions has greatly improved our understanding of plant–environment interactions in the invasion process. Most of the studies on the distribution of plant species (including invasive alien plants) have focused on the ecological processes and environmental correlates of importance. However, the different approaches outlined above suggest that the spatial scale strongly influences the types of models that can be used, as well as the outcome of the modelling. Correlative models are well suited for application at the scale of landscapes, regions, or countries but lose their predictive power at smaller spatial scales. Mechanistic models become more tractable and appropriate at smaller spatial scales (Collingham et al. 2000; Rouget and Richardson 2003b). It is important to note that insights derived at one spatial scale cannot be applied at another scale. In our studies, both correlative and mechanistic models yielded inconsistent results when the resolution of the spatial grain was changed. This variation occurs

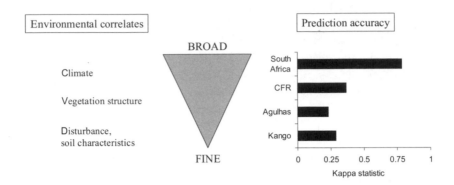

Figure 12.4. Effects of spatial scale on predicting spatial patterns of distribution of invasive alien trees in South Africa. The main environmental factors used to model species distribution are shown on the left, and the accuracy of the modeling prediction is shown on the right. Correlative models were developed for South Africa (1,250,000 km^2), the Cape Floristic Region (CFR; 88,000 km^2), the Agulhas Plain (2160 km^2), and the Kango Valley (25 km^2). The Kappa statistic is an index that compares the agreement against that which might be expected by chance. It ranges from −1 (complete disagreement) through 0 (no agreement above that expected by chance) to 1 (perfect agreement).

partly because of the different processes that operate at different scales but also because of features of the accuracy and precision of available data. Further studies are needed to determine the thresholds of applicability of the different models. Figure 12.4 illustrates the shift in environmental factors used to model and predict invasion processes for alien trees in South Africa from local to national scales. Climatic factors were the main drivers of invasion patterns at a national scale, but these factors were almost irrelevant at a local scale, where soil characteristics were much more important.

These studies show that the recent development of new computational techniques, together with the increased availability of georeferenced data in GIS, have paved the way for new research on ways to better utilize the many opportunities afforded by natural experiments.

APPLICATIONS FOR MANAGEMENT

The approaches outlined above are useful for improving our understanding of the processes of invasion. Three recent applications serve to illustrate how information gleaned from natural experiments can be reformulated for use in management.

Tucker and Richardson (1995) used empirical data on hundreds of introduced woody plants in the fynbos region to compile a system for evaluating the risk that a newly introduced species will become invasive. The system requires fairly detailed information on broad-scale environmental conditions; population characteristics and degree of habitat specialization; dispersal biology; seed production and predation; and life-history traits associated with persistence and proliferation in a fire-prone environment. The system was used to screen a selection of Australian *Banksia* species for their potential to become invasive in fynbos.

Alien species that are desirable and commercially important in parts of the landscape, but damaging invaders in other parts, present a special challenge for managers, planners, and policymakers. Rouget et al. (2002) analyzed the distribution of forestry plantations and invasive (self-sown) stands of *Acacia mearnsii* and *Pinus* spp. in South Africa. The current distribution of the two taxa was subdivided into three groups according to the degree of invasion, the forestry history, and the precision of available data. Regression-tree analyses were used to relate the distribution of invasive stands with environmental variables and to derive habitat-suitability maps for future invasion. Results were used to derive guidelines for policy on alien plant management based on vegetation type, degree of transformation, extent of invasion, and the risk of future alien spread. Areas were identified where these alien species can be grown with the least risk of invasions and where special measures are required to manage spread from plantations.

Higgins, Richardson, and Cowling (2000) describe the development and applications of a landscape model for exploring the relative efficiency of different control options. Options are compared with respect to total costs, time required to clear a given population of invading plants, and also the impacts on three different components of native-plant diversity. Simulations show the importance of prioritizing low-density stands dominated by juvenile plants for control.

CONCLUSIONS

Gotelli and Graves (1996, 30) wrote that the use of natural experiments "is limited only by the imagination of the investigator, not by the constraints of experimental design." Many authors have discussed the benefits and limitations of natural experiments in ecology, and we do not intend to revisit this debate. Much of the current understanding of biological invasions has come from natural experiments. Focusing on alien tree invasions in South Africa, we have shown that information derived from natural experiments has provided the foundation for a reasonable understanding of the ecology of these invasions, and a basis for management.

Many of the approaches available in the toolboxes of Quaternary ecology (e.g., Delcourt and Delcourt 1991; Jackson 1997), historical ecology (Egan and Howell 2001), epidemiology (Lawson 2001), and numerical ecology (ter Braak and Smilauer 1998; Webb 1999) can be applied to contemporary invasions of alien trees for deriving generalizations with immediate relevance in management. Conceptualizations based on natural experiments may extend beyond simple abstractions; they may, for example, be used to parameterize or validate detailed simulation models. They are also useful for fine-tuning hypotheses that could be tested using more rigorous experimental methods. Also, information from natural experiments can be used to parameterize conceptual models. Using computer-modelling techniques, one can readily explore the sensitivity of different factors and, through an iterative process involving model development and validation (e.g., Higgins, Richardson, and Cowling 2001), build models that accurately simulate historical patterns. Although this approach will probably not satisfy the purists, who will question the inference of process from pattern inherent in these methods, the major benefits of adopting this pragmatic approach seem obvious. The utility of such techniques is set to increase rapidly in the future as technological advances make accurate and affordable remote sensing of invasive trees more straightforward and as our understanding of critical factors underpinning environmental modelling (e.g., the importance of scale) improves.

ACKNOWLEDGMENTS

We thank Stephen Jackson and Petr Pyšek for their helpful comments on the manuscript.

REFERENCES

Collingham, Y. C., Wadsworth, R. A., Huntley, B., and Huime, P. E. 2000. Predicting the spatial distribution of non-indigenous riparian weeds: Issues of spatial scale and extent. *J. Appl. Ecol.* 37:13–27.

Cook, T. D., and Campbell, D. T. 1979. *Quasi-experimentation: Design and Analysis Issues for Field Settings.* Chicago: Rand McNally College Publishing Company.

Crawley, M. J. 1987. What makes a community invasible? In *Colonization, Succession, and Stability,* ed. A. J. Gray, M. J. Crawley, and P. J. Edwards, 429–53. Oxford: Blackwell.

Czaran, T., and Bartha, S. 1992. Spatiotemporal dynamic models of plant populations and communities. *Trends Ecol. Evol.* 7:38–42.

D'Antonio, C. M., Levine, J., and Thomsen, M. 2001. Ecosystem resistance to invasion and the role of propagule supply: A California perspective. *J. Medit. Ecol.* 2:233–45.

Davis, M. B. 1989. Retrospective studies. In *Long-Term Studies in Ecology. Approaches and Alternatives,* ed. G. E. Likens, 71–89. New York: Springer-Verlag.

De'ath, G., and Fabricius, K. E. 2000. Classification and regression trees: A powerful yet simple technique for ecological data analysis. *Ecology* 81:3178–92.

Delcourt, H. R., and Delcourt, P. A. 1991. *Quaternary Ecology: A Paleoecological Perspective.* London: Chapman and Hall.

Diamond, J. 1986. Overview: Laboratory experiments, field experiments, and natural experiments. In *Community Ecology,* ed. J. Diamond and T. J. Case, 3–22. New York: Harper and Row.

Egan, D., and Howell, E. A. 2001. *The Historical Ecology Handbook: A Restorationist's Guide to Reference Ecosystems.* Washington, D.C.: Island Press.

Elton, C. S. 1958. *The Ecology of Invasion by Animals and Plants.* London: Methuen.

Gotelli, N. J., and Graves, G. R. 1996. *Null Models in Ecology.* Washington, D.C.: Smithsonian Institution Press.

Grotkopp, E., Rejmánek, M., and Rost, T. L. 2002. Towards a casual explanation of plant invasiveness: Seedling growth and life-history strategies of 29 pine (*Pinus*) species. *Am. Nat.* 159:396–419.

Harper, J. L. 1977. *Population Biology of Plants.* London: Academic Press.

Higgins, S. I., and Richardson, D. M. 1998. Pine invasions in the Southern Hemisphere: Modelling interactions between organism, environment, and disturbance. *Plant Ecol.* 135:79–93.

Higgins, S. I., Richardson, D. M., and Cowling, R. M. 2000. Using a dynamic landscape model for planning the management of alien plant invasions. *Ecol. Appl.* 10:1833–48.

———. 2001. Validation of a spatial simulation model of a spreading alien plant population. *J. Appl. Ecol.* 38:571–84.

Holland, P. 2000. Cultural landscapes as biogeographic experiments: A New Zealand perspective. *J. Biogeog.* 27:39–43.

Jackson, S. T. 1997. Documenting natural and human-caused plant invasions using paleoecological methods. In *Assessment and Management of Plant Invasions,* ed. J. O. Luken and J. W. Thieret, 37–55. New York: Springer.

Kruckeberg, A. R. 1986. The birth and spread of a plant population. *Am. Midl. Nat.* 116:403–10.

Lawson, A. B. 2001. *Statistical Methods in Spatial Epidemiology.* New York: Wiley.

MacArthur, R. H. 1972. *Geographical Ecology: Patterns in the Distribution of Species.* New York: Harper and Row.

Pickett, S.T.A. 1989. Space-for-time substitution as an alternative to long-term studies. In *Long-Term Studies in Ecology: Approaches and Alternatives*, ed. G. E. Likens, 110–35. New York: Springer-Verlag.

Pickett, S.T.A., Kolasa, J., and Jones, C. G. 1994. *Ecological Understanding: The Nature of Theory and the Theory of Nature*. San Diego: Academic Press.

Poynton, R. J. 1979. *Tree Planting in Southern Africa*. Vol. 2, *The Eucalypts*. Pretoria, South Africa: Department of Forestry.

Rejmánek, M. 2000. Invasive plants: Approaches and predictions. *Aust. Ecol.* 25:497–506.

Rejmánek, M., and Richardson, D. M. 1996. What attributes make some plant species more invasive? *Ecology* 77:1655–61.

Rejmánek, M., Richardson, D. M., Higgins, S. I., Pitcairn, M. J., and Grotkopp, E. Forthcoming. Ecology of invasive plants: State of the art. In *Invasive Alien Species: A New Synthesis*, ed. H. A. Mooney, R. N. Mack, J. A. McNeely, L. E. Neville, P. J. Schei, and J. K. Waage. Washington, D.C.: Island Press.

Richardson, D. M. 1988. Age structure and regeneration after fire in a self-sown *Pinus halepensis* forest on the Cape Peninsula, South Africa. *S. Afr. J. Bot.* 54:140–44.

Richardson, D. M., Allsopp, N., D'Antonio, C. M., Milton, S. J., and Rejmánek, M. 2000. Plant invasions: The role of mutualisms. *Biol. Rev.* 75:65–93.

Richardson, D. M., Bond, W. J., Dean, W. R.J., Higgins, S. I., Midgley, G. F., Milton, S. J., Powrie, L. Rutherford, M. C., Samways, M. J., and Schulze, R. E. 2000. Invasive alien organisms and global change: A South African perspective. In *Invasive Species in a Changing World*, ed. H. A. Mooney and R. J. Hobbs, 303–49. Washington, D.C.: Island Press.

Richardson, D. M., and Brown, P. J. 1986. Invasion of mesic mountain fynbos by *Pinus radiata*. *S. Afr. J. Bot.* 52:529–36.

Richardson, D. M., and Cowling, R. M. 1992. Why is mountain fynbos invasible and which species invade? In *Fire in South African Mountain Fynbos*, ed. B. W. Van Wilgen, D. M. Richardson, F. J. Kruger, and H. J. van Hensbergen, 161–81. Berlin: Springer-Verlag.

Richardson, D. M., and Higgins, S. I. 1998. Pines as invaders in the Southern Hemisphere. In *Ecology and Biogeography of* Pinus, ed. D. M. Richardson, 450–73. Cambridge: Cambridge University Press.

Richardson, D. M., Pyšek, P., Rejmánek, M., Barbour, M. G., Panetta, D. F., and West, C. J. 2000. Naturalization and invasion of alien plants—Concepts and definitions. *Diversity Distrib.* 6:93–107.

Richardson, D. M., and Rejmánek, M. 1999. *Metrosideros excelsa* takes off in the fynbos. *Veld & Flora* 85:14–16.

Richardson, D. M., Williams, P. A., and Hobbs, R. J. 1994. Pine invasions in the Southern Hemisphere: Determinants of spread and invadability. *J. Biogeog.* 21:511–27.

Rouget, M., and Richardson, D. M. 2003a. Inferring process from pattern in plant invasions: A semi-mechanistic model incorporating propagule pressure and environmental factors. *Am. Nat.* 162:713–24.

————. 2003b. Understanding actual and potential patterns of plant invasion at different spatial scales: Quantifying the roles of environment and propagule pressure. In *Plant Invasions: Ecological Threats and Management Solutions,* ed. L. E. Child, J. H. Brock, G. Brundu, K. Poach, P. Pysels, P. M. Wade, and M. Williamson, 3–15. Leiden, Netherlands: Backhuys Publishers.

Rouget, M. Richardson, D. M., Cowling, R. M., Lloyd, M. W., and Lombard, A. T. 2003. Current patterns of habitat transformation and future threats to biodiversity in terrestrial ecosystems of the Cape Floristic Region, South Africa. *Biol. Conserv.* 112: 63–85.

Rouget, M., Richardson, D. M., Milton, S. J., and Polakow, D. 2001. Invasion dynamics of four alien *Pinus* species in a highly fragmented semi-arid shrubland in South Africa. *Plant Ecol.* 152:79–92.

Rouget, M., Richardson, D. M., Nel, J. A., and van Wilgen, B. W. 2002. Commercially-important trees as invasive aliens—Towards spatially explicit risk assessment at a national scale. *Biol. Invasions* 4:397–412.

ter Braak, C. J. F., and Smilauer, P. 1998. *CANOCO Reference Manual: User's Guide to CANOCO for Windows, Version 4.* Wageningen, Netherlands: Centre for Biometry.

Tucker, K. C., and Richardson, D. M. 1995. An expert system for screening potentially invasive alien plants in fynbos. *J. Environ. Manage.* 44:309–38.

Turnbull, J. W. 2000. Economic and social importance of eucalypts. In *Diseases and Pathogens of Eucalypts,* ed. P. J. Keane, G. A. Kile, F. D. Podger, and B. N. Brown, 1–9. Collingwood: CSIRO Publishing.

Underwood, A. J. 1997. *Experiments in Ecology: Their Logical Design and Interpretation Using Analysis of Variance.* Cambridge: Cambridge University Press.

Versfeld, D. B., Le Maitre, D. C., and Chapman, R. A. 1998. *Alien Invading Plants and Water Resources in South Africa: A Preliminary Assessment.* Report TT99/98, Water Research Commission, Pretoria, South Africa.

Webb, A. 1999. *Statistical Pattern Recognition.* New York: Oxford University Press.

Whitmore, T. C. 1991. Invasive woody plants in perhumid tropical climates. In *Ecology of Biological Invasions in the Tropics,* ed. P. S. Ramakrishnan, 35–40. New Delhi: International Scientific Publications.

Williams, J. E., and Woinarski, J.C.Z., eds. 1997. *Eucalypt Ecology: Individuals to Ecosystems.* Cambridge: Cambridge University Press.

Williamson, M. 1999. Invasions. *Ecography* 22:5–12.

13   Biological Control in Support of Conservation

Friend or Foe?

Mark S. Hoddle

> So, naturalists observe, a flea
> Hath smaller fleas that on him prey;
> And these have smaller still to bite 'em;
> And so proceed ad infinitum.
>
>   JONATHAN SWIFT,
>   "Poetry, A Rhapsody"

INTRODUCTION

The damage invasive organisms cause to natural and agricultural environments and the potential exotic threats that lurk outside state and country borders are well-documented phenomena that are understood by scientists, politicians, economists, and the lay public. Relatively new environmental concepts such as biological invasions; biodiversity; exotic, alien, and invasive organisms; biotic homogenization; species declines and extinctions; and habitat fragmentation, degradation, and disturbance are mainstream and have precipitated enough public interest to have become the themes of numerous popular nonfiction books, radio, television, and newspaper and magazine articles that describe ecological issues and the science attempting to explain the mechanisms involved.

As E. O. Wilson (1997) noted, both habitat destruction and invasive species—the two great destroyers of the world's biodiversity—are human mediated. The former is brutally destructive, quick, and obvious. The latter stealthily and gradually alter ecosystem functioning after their establishment and subsequent proliferation. It is intriguing and seemingly counterintuitive that the addition of new organisms to a community can have the unintended consequence of biodiversity reduction instead of enhancement (Tenner 1996). Biodiversity reduction by invasive species can occur through the displacement of native species (Kupferberg 1997), modification of trophic structures within communities (Holland 1993), or drastic alteration of ecosystem functioning (Vitousek et al. 1996).

The only habitats that apparently have not suffered obvious degradation by invasive species are the polar ice caps; virtually every other type of terrestrial and aquatic habitat (freshwater and marine) on every continent (excluding Antarctica) and most islands have experienced invasion and modification by some type of metazoan or protozoan (see Simberloff, Schmitz, and Brown 1997 for an extensive review of invasive taxa and the habitats invaded). Recent investigations probing the effects that microbes have on native soil communities and plant biodiversity indicate that not even subterranean habitats are immune to perturbation (Heijden et al. 1998).

Sources and introduction routes of invasive organisms are varied. For example, accidental introductions associated with transportation of people and goods have resulted in the frequent introductions of pestiferous weeds, arthropods, mollusks, and vertebrates into new areas. The brown tree snake, *Boiga irregularis* (Merrem), was transported to Guam on military equipment moved from Indonesia and Papua New Guinea after World War II (Rodda, Fritts, and Chiszar 1997). Zebra mussels, *Dreissena polymorpha* (Pallas), and the water flea, *Cercopagis pengoi* (Ostroumov), were discharged into the Great Lakes with ballast water from Europe (Charlebois, Raffenberg, and Dettmers 2001; Cox 1999).

The aquarium trade has also been responsible for importing noxious plants, algae, mollusks, crustaceans, vertebrates, and aquatic pathogens into new areas (Bright 1998). One highly publicized aquarium escapee is the green marine alga *Caulerpa taxifolia* (Vahl) C. Agardh, a Caribbean native that occupies more than 6000 acres of seafloor in the Mediterranean Sea (Meinesz 1999). This aggressive Mediterranean-adapted strain of *C. taxifolia* recently invaded the coast of California with wastewater discharged from commercial aquariums (Jousson et al. 2000).

Plants with "weedy" characteristics are popular nursery plants because they are hardy and easy to grow, but these attributes also promote escape from cultivation. *Clematis vitalba* L., euphemistically referred to as old man's beard, is a vine of European origin that escaped from gardens in the 1930s to become one of New Zealand's worst forest weeds (Ogle et al. 2000).

The importation and redistribution of novel horticultural plants by nurseries and botanical gardens may also assist the spread of adventive insects, mites, slugs, and pathogens that attack or infest leaves and stems (Bright 1998; Guy, Webster, and Davis 1998). Exotic organisms can also spread in the soil of transported plants. The New Zealand flatworm, *Artioposthia triangulata* (Dendy), has been associated with earthworm declines in Scotland

and Ireland following its introduction in the 1950s in potting soil (Christensen and Mather 1995).

The pet trade is another major biotic conduit for introducing exotic vertebrates (e.g., amphibians, birds, mammals, and reptiles) and invertebrates (e.g., spiders, cockroaches, millipedes, and scorpions) into areas outside of their natural range. The monk parakeet, *Myiopsitta monachus* (Boddeart), a South American native that became naturalized in New York City in 1967 (Todd 2001), has now spread to fifteen states and threatens to become an agricultural pest (Cox 1999). The wild-bird trade is responsible for the introduction of at least nine species of exotic parrots that are now established in the United States (Cox 1999).

Exotic game and sport fish have been intentionally imported, established, and redistributed by acclimatization societies and private individuals for recreational pursuits (i.e., hunting and fishing). Reproducing populations—some of which may be periodically augmented with mass-reared individuals (e.g., hatchery-raised fish)—of deer, goats, pigs, chamois, brush-tail possums, rabbits, tahr, wallabies, pheasants, quail, ducks, geese, trout, and bass can have profound detrimental impacts on native vegetation and can compete with native animals for food and habitat (Cox 1999; Hoddle 1999; King 1990a). These populations may also inadvertently spread pathogens lethal to native wildlife (Kiesecker, Blaustein, and Miller 2001) and domesticated animals.

The liberation of upper-trophic-level organisms—known as biological control agents, or natural enemies—is a deliberate importation-and-release practice that attempts to establish permanent exotic populations that alter community structure and reduce densities of target pest species. (Professor Harry S. Smith, an entomologist at University of California, Riverside, coined the phrase *biological control* in 1919 [Smith 1919].) Introduction of a biological control agent either enriches an existing guild or adds a guild that was previously lacking in the target community. The introduction of an efficient natural enemy can substantially reduce pest densities and can free resources for use by competing organisms (e.g., endemic wildlife) in the same trophic level or in lower trophic levels (Bellows 2001). The invasive Brazilian floating aquatic weed *Salvinia molesta* D. S. Mitchell smothers slow-moving waterways by forming impenetrable mats that cut off light to submerged plants and animals, and the weed promotes deoxygenation of water that kills submerged fauna. Weed mats also affect flood regimens, hydroelectric production, and surface transportation (Room and Thomas 1985). The absence of specialized herbivores (i.e., natural enemies) allowed *S. molesta* to attain unacceptable population densities in tropical and subtropical areas outside its natural range (Room et al. 1981). Release of the weevil *Cyrtobagous salviniae* Calder and Sands, a specialized *S. molesta* herbivore native to Brazil, destroyed 18,000 tonnes of this plant in the heavily infested Lake Moondarra in Australia over a 15-month period (Forno and Julien 2000). The introduction of this weevil and other host-specific insects filled a largely unoccupied niche and reduced the density of *S. molesta* in its adventive range to levels that are now similar to those found in Brazil. The establishment of this new guild of herbivores on *S. molesta* was solely responsible for the recovery of affected waterways and allowed local economies and wildlife to recover (Room 1990). The threat that invasive organisms pose to natural environments is immense, and given the propensity for humans to move plants and wildlife easily and quickly around the planet, this chronic problem will require constant vigilance if exotic pest threats are to be identified and dealt with expeditiously.

Once problematic invasive organisms are identified, management plans can follow two distinct routes: proactive strategies to mitigate potential problems before they arrive or reactive management strategies to deal with incipient or existing problems. Policy implementation—which provides a legal and guiding infrastructure to prevent pest establishment (e.g., exclusion, border interception, or quarantine policies) or mitigate spread and impact after detection (e.g., localized eradication efforts)—can be viewed as proactive. Direct hands-on approaches for pest management that may use any or all the following strategies—habitat restoration, containment, regular and persistent cultural and chemical management, and biological control—can be classified as reactive management strategies. The practice of biological control, that is, the deliberate scientific use of natural enemies to reduce pest densities to levels that are no longer damaging, has been a widely practiced pest management option for exotic agricultural pests, especially insects and weeds. Historically, the application of biological control in support of conservation programs has not been widely practiced, but its use has increased since the mid-1980s. Because of the large number of well-financed and highly publicized "traditional" programs focusing on arthropods and weeds of agricultural importance, program managers dealing with invasive organisms may be unfamiliar with biological control or misled into assuming that it is not an appropriate conservation technology. The history, theory, application, benefits, and risks of biological control; the identification of nontraditional pest targets of conservation importance that may be amenable to biological control; and experimental investigations of factors influencing the invasion success of natural enemies will be the focus of this chapter.

HISTORY AND THEORY OF BIOLOGICAL CONTROL

Biological control is the intentional use by humans of parasitoid, predator, pathogen, antagonist, or competitor populations to suppress a pest population and thereby make the pest less abundant and damaging (DeBach 1964, 1974; Van Driesche and Bellows 1996). Most applied biological control programs involve the addition of new species that fill an unoccupied niche (i.e., that use the pest as a resource). Often, adventive pests have very few native natural enemies, and those that exist are native generalist predators or herbivores that utilize other resources in the environment. Introductions of specialized natural enemies from the home range of the pest fill niches not occupied by local generalists. In most cases, native organisms in the upper trophic level are not adversely affected by the introduction of specialized natural enemies as they can utilize other resources for survival. With successful biological control programs, both the pest and the natural enemy decline in abundance through time (Bellows 2001). The process of natural regulation in the pest's home range and the subsequent importation of specialized natural enemies from the home range of the pest for use in a biological control program are illustrated in Figure 13.1. Biological control has been used as a management tool for the control of crop and forest pests and for the restoration of natural systems affected by adventive pests (Van Driesche and Bellows 1996).

The development of the concept of biological control occurred with the gradual accumulation of biological and ecological knowledge of the natural world by humans. The first orchestrated biological control programs were run by Chinese and Yemenite farmers who encouraged native predatory ants to protect citrus and date trees from insect pests (DeBach 1974). By the 1700s, observations on the peculiar and mysterious production of flies and wasps from the larvae and pupae of various Lepidoptera fostered an understanding of insect parasitism by hymenopterous and dipteran parasitoids. Investigations of diseases affecting economically important insects like silkworms were burgeoning at this time also (DeBach 1974). By the 1800s, the accrued body of knowledge on insect pathogen transmission and disease control, and observations of natural epizootics, had scientists in Europe and the United States promoting the idea of mass rearing fungi for release against agricultural pests. The first field releases of mass-reared fungi for control of weevils attacking sugar beet occurred in Russia in 1884. Around 1855, U.S. entomologists suggested that insects be imported from Europe to control European weeds that were taking over valuable pastureland and that lacked herbivorous arthropods known to feed on these plants in Europe (DeBach 1974).

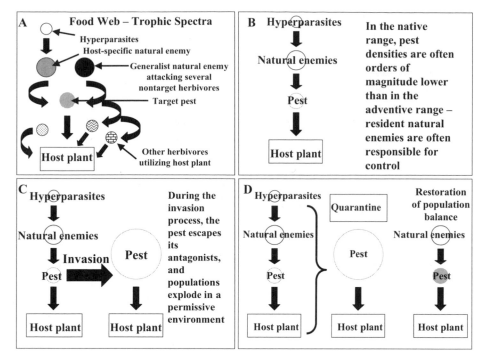

Figure 13.1. *(A)* In the target pest's home range, several natural enemies use the target as a resource. Some of these natural enemies are host specific and use only the pest, whereas others are polyphagous and may utilize several other herbivores on the host plant. These natural enemies may in turn be exploited by hyperparasites, which constitute a fourth trophic level after the plant, target herbivore, and the herbivore's natural enemies. *(B)* In the native range of the pest, it is this trophic organization that, in part (other factors such as climate and interspecific competition may also limit pest population growth), often holds pest densities at much lower levels than those observed in the adventive range. *(C)* When the target pest successfully invades and establishes in the absence of its antagonists in a permissive environment (i.e., good climate, lots of food, and no natural enemies), populations explode and cause severe economic and ecological damage. *(D)* Biological control is the deliberate importation of host-specific natural enemies of the target pest for its home range. These natural enemies are passed through a quarantine facility to eliminate disease and hyperparasites and to run host-specificity tests prior to release. After the natural enemies are released from quarantine (federal, state, and Fish and Wildlife Service permission is required) and established, the pest population densities may be restored to levels similar to those observed in the home range. Figure adapted from Bellows 2001.

By the 1880s, biological and ecological knowledge of natural enemies (i.e., predators, parasitoids, and pathogens) attacking insects and weeds helped form the concept of bug versus bug, or the parasite method, for pest management. Organized biological control programs sponsored by fledgling institutions such as the U.S. Department of Agriculture (USDA, established in 1862) were launched. The first of these programs, the biological control campaign against the cottony cushion scale, *Icerya purchasi* Maskell, a devastating pest of California-grown citrus, was launched in 1886 and was a spectacular success. This program is a premier example of pest regulation by upper-trophic-level organisms (Sawyer 1996).

Icerya purchasi, an insect native to Australia and New Zealand, had been accidentally imported into California on acacia trees in 1868. It rapidly spread to citrus trees, causing massive crop losses and tree death. The nonpest status of *I. purchasi* in Australia supported the idea that in its home country this pest was controlled by natural enemies and that these antagonists could provide similar pest suppression if they could be established in California. Foreign exploration (i.e., prospecting for natural enemies that coevolved with the pest in its area of origin) in Australia and New Zealand for natural enemies of *I. purchasi* resulted in the importation of a predatory beetle *(Rodolia cardinalis* [Mulsant]) and a parasitic fly (*Cryptochaetum iceryae* [Williston]). By 1888, *I. purchasi* was permanently reduced to densities that no longer caused economic damage, and these natural enemies were subsequently distributed to other countries with cottony cushion scale problems (Caltagirone and Doutt 1989). Biological control of *I. purchasi* put California on an economic and agricultural trajectory that led to unprecedented prosperity, and pest control with natural enemies was enthusiastically endorsed by California citrus growers, who backed research programs with commodity funds (Sawyer 1996). If the citrus industry had failed in California because of *I. purchasi*, the history and economic development of the state would have been radically different.

The cottony cushion scale project in California is considered the first scientific and institutionally backed biological control program, and it illustrates the concepts of (1) adventive pest identification; (2) foreign exploration for specialized natural enemies in the pest's home range; (3) importation and mass rearing of natural enemies; and (4) establishment, redistribution, and impact monitoring of imported biological control agents. These basic practices, although substantially refined, are the foundation of current biological control programs (Bellows and Fisher 1999; DeBach 1964; Van Driesche and Hoddle 2000; Van Driesche and Bellows 1993).

The addition of exotic natural enemies to the environment has three impacts on the target system. First, the number and function of food web links that connect the pest to other members of the community are permanently changed. Second, natural enemies can radically alter pest densities and population dynamics. Third, reduction in pest density by natural enemies changes the community structure of the affected system. In successful biological control programs, reduction of pest densities and recovery of adversely affected flora or fauna lead toward a system with better ecological balance and community structure (Bellows 2001). Once established, successful natural enemies can provide enduring pest control, they can replicate and disperse without continued human management, and they can persist when pest populations are stabilized at very low densities. In addition, the technology is often relatively cheap to implement. Successful biological control programs against noxious insects, mites, weeds (aquatic and terrestrial), plant pathogens, and vertebrates are extensive; and well-documented examples are covered by Bellows and Fisher (1999), and Gurr and Wratten (2000).

POTENTIAL ECONOMIC AND ENVIRONMENTAL BENEFITS OF BIOLOGICAL CONTROL

Reduced expenditure for pesticides, labor, and specialized equipment and, potentially, a permanent return to ecological conditions similar to those seen before the arrival of the pest are compelling motivations for the adoption of biological control. Economic analyses indicate that benefit:cost ratios for successful biological control of arthropod pests are high and can exceed 145:1 (Norgaard 1988; Pickett et al. 1996; Jetter, Klonsky, and Pickett 1997), and potential benefit:cost ratios overwhelmingly favor support for biological control programs as an option for pest control (Gutierrez, Caltagirone, and Meikle 1999). Estimates of economic benefits from successful biological control programs tend to be conservative, and profits continue to accrue annually with little or no additional management of the system (Norgaard 1988). Comparisons of costs for biological control programs indicate that benefits amassed from successful projects outweigh the combined costs of unsuccessful projects, even though failures are more numerous. For example, just 10% of arthropod biological control programs have provided full control of the target pest (Gurr et al. 2000). For weed programs, less than 30% of projects have resulted in either total or partial control of the target (Syrett, Briese, and Hoffmann 2000), although in some instances evaluations may have been conducted too early to determine final outcomes (McFadyen 1998). Projects sponsored by the Australian Center for International Agricultural Research had a benefit:cost ratio of 13.4:1 for ten projects that ran from 1983 to 1996, even though just four of these projects were documented successes (Lubulwa and McMeniman 1998).

Biological control of agricultural pests can indirectly benefit native wildlife through the reduction of pesticides released into the environment, because natural enemies suppress economically important targets. The acute impact of insecticides on wildlife because of aerosol drift from agricultural areas, runoff into waterways, food chain accumulation, or indiscriminant application was first brought to public attention by Rachel Carson (1962). An insidious, chronic side effect from pesticide use that has been recently postulated is the potential ability of synthetic chemical pollutants in the environment to accumulate in the bodies of vertebrates. These sequestered compounds may mimic or block the actions of endogenous hormones (Colborn, Dumanoski, and Peterson Myers 1997). The environmental endocrine hypothesis has been used as a unifying theory linking wildlife declines, reproductive ailments, behavioral abnormalities (e.g., reproductive and antipredator), and gross physical deformities with agricultural pesticides, pharmaceuticals, and other industrial chemicals that mimic or obstruct hormonal activity in animals (Ankley et al. 1998; Krimsky 2000; Nagler et al. 2001; Park, Hempleman, and Propper 2001; Pelley 1997; Relyea and Mills 2001; Souder 2000) and humans (Schettler, Soloman, and Huddle 1999).

BIOLOGICAL CONTROL DISASTERS

Biological control, when practiced correctly, is a carefully orchestrated scientific endeavor that alters community structure through the deliberate manipulation of upper-trophic-level organisms that use the targeted pest as a resource. Therefore, the practice of classical biological control (i.e., the importation of specialized natural enemies from the pest's home range) intimately links this pest management strategy to the science of population ecology and supports the paradigm of top-down regulation by host-specific biological control agents. Consequently, high levels of host specificity ensure strong links and maximal impact by natural enemies on the target while ensuring weak links and minimal impacts to nontargets. Theoretical community-assembly studies indicate risks to nontargets are increased if only moderately effective natural enemies are established, because they maintain high numbers but do not substantially reduce pest densities, which ultimately causes their own population decline. Consequently, susceptible native species are subject to constant attack by these mediocre natural enemies that maintain moderate densities of the target pest. Additionally, if exotic natural enemies provide an abundant resource for generalist resident predators or parasites, thereby promoting an increase in their density, attacks on preferred native prey may occur more frequently as a consequence (Holt and Hochberg 2001).

When biological control projects stray from the fundamental ecological principle of using natural enemies with high host specificity or when the technology is applied without ecological justification to poorly chosen pest targets (e.g., neoclassical biological control of native pests [see Hokkanen and Pimental 1989]), then undesired outcomes such as nontarget impacts and lack of control are more likely to occur. This risk of unintended consequences is further amplified by releases of generalist natural enemies, that is, biological control agents that are polyphagous and can attack many hosts. Generalist natural enemies, which by definition lack high levels of host and habitat specificity, are frequently cited as examples of the inherent and unpredictable risks associated with releasing biological control agents because of their adverse effects on native organisms and lack of impact on the pestiferous target (Elliot, Kieckhefer, and Kauffman 1996; Henneman and Memmott 2001; Howarth 1983, 1991; Simberloff and Stiling 1996; Stiling and Simberloff 2000). Biological control introductions have also been criticized for diluting endemic biodiversity and contributing to the homogenization of global biota. In New Zealand, 13% of the country's insects are exotic. Of these, 2.5% have been intentionally introduced for biological control, and these natural enemies comprise 0.33% of New Zealand's total insect fauna. The number of intentional introductions of natural enemies is negligible compared to the number of adventive insect pests and weeds that have become established in New Zealand, and biological control agents are not considered a major source of biological pollution diluting native biodiversity (Emberson 2000).

Examination of the commonly cited "rogue" biological control agents presented in Table 13.1 clearly demonstrates that these biological control projects were ill conceived, not necessarily because the pests were unsuitable targets but primarily because the natural enemies selected had broad host ranges and substantial nontarget impacts should have been predicted. Some of the projects listed in the table were carried out by agricultural interest groups (e.g., sugarcane growers and farmers) with little or no scientific grounding, and government oversight was lax either because of noninvolvement or because there was no regulatory infrastructure (i.e., governing legislation) for identifying suitable targets and assessing the safety of imported natural enemies before their release or subsequent redistribution following establishment.

The parasitic fly *Compsilura concinata* Meigen illustrates the environ-

TABLE 13.1 *Rogue biological control agents*

Biological Control Agent (native range)	Target Pest	Country, Year of First Introduction
European red fox, *Vulpes vulpes* L. (palaearctic regions)	Rabbits, *Oryctolagus cuniculus* L	Australia, 1860s Originally released for hunting.
Stoat, *Mustela erminea* L. (Eurasia and North America)	Rabbits	New Zealand, 1884
Ferret, *Mustela furo* L. (central Europe and the Mediterranean)	Rabbits	New Zealand, 1879
Weasel, *Mustela nivalis vulgaris* Erxleben (Eurasia and North America)	Rabbits	New Zealand, 1884
Small Indian mongoose, *Herpestes javanicus* (Saint-Hilaire), (= *auropunctatus* [Hodgson]) (Iraq to the Malay Peninsula)	Rats, *Rattus* spp.	Trinidad, 1870; Jamaica, 1872; Cuba, 1886; Puerto Rico, 1877; Barbados, 1877; Hispaniola, 1895; St. Croix, 1884; Surinam, 1900; Hawaii, 1883
Cane toad, *Bufo marinus* L. (northwestern Mexico through southern Brazil)	White grubs, *Phyllophaga* spp.; sweet potato hawk moth, *Agrius convolvuli* L.; greyback cane beetle, *Dermolepida albohirtum* (Waterhouse)	Jamaica, 1844; Bermuda, 1855; Puerto Rico, 1920; Hawaii, 1932; Australia, 1935; Fiji, 1936; Guam, 1937; New Guinea,1937; Philippines, 1934
Mosquitofish, *Gambusia affinis* (Baird & Girard) (eastern U.S. and Mexico)	Worldwide dissemination for control of mosquito larvae promoted by the World Health Organization until 1982.	Intensive releases began worldwide; around 1900, ~70 countries now have permanent G. affinis populations, including Afghanistan, Australia, Canada, China, Ethiopia, Grand Cayman, Greece, Hawaii, Iran, Korea, New Zealand, Somalia, Turkey, Ukraine

TABLE 13.1 *(continued)*

Nontarget Impacts	Reference
Eats native marsupials, birds, and lambs	Saunders et al. 1995
Eats native birds, insects, and lizards	King 1990b
Eats native birds	Lavers and Clapperton 1990
Eats native birds, insects, and lizards	King 1990c
Eats native birds and reptiles	Hinton and Dunn 1967; Loope, Hamann, and Stone 1988
Eats native insects, amphibians, and reptiles; toxic to native wildlife that consume it; *B. marinus* outcompetes native amphibians for shelter and breeding sites	Easteal 1981; Freeland 1985
Substantial nontarget attacks on native aquatic invertebrates and vertebrates outside its native range	Diamond 1996; Gamradt and Kats 1996; Legner 1996; Meisch 1985; Rupp 1996; Walton and Mulla 1991

(continued)

TABLE 13.1 *(continued)*

Biological Control Agent (native range)	Target Pest	Country, Year of First Introduction
Predatory snail, *Euglandina rosea* (Ferrusac) (southeastern U.S.)	Giant African snail, *Achatina fulica* Bowdich	Hawaii, 1955; Tahiti, 1974; Moorea, 1977; New Caledonia, 1978; Guam, 1957; Vanuatu, 1973; Papua New Guinea, 1952; Japan 1958; Taiwan, 1960; Madagascar, 1962; Seychelles, 1960; Mauritius, 1961; Réunion, 1966; India, 1968; Bermuda, 1958
Tachinid fly, *Compsilura concinata* (Meigen) (Europe)	Gypsy moth, *Lymantria dispar* (L.) and other lepidopteran pests	U.S., 1906
Tachinid fly, *Bessa* (=*Ptychomyia*) *remota* (Aldrich) (Native range Indo-Malay archipelago)	Coconut moth, *Levuana iridescens* Bethune–Baker	Fiji, 1925
Seven-spotted lady beetle, *Coccinella septempunctata* L. (palearctic regions)	Pestiferous aphid spp.	U.S., 1957
Cactus moth, *Cactoblastis cactorum* (Bergroth) (northern Argentina, Uruguay, Paraguay, and southern Brazil)	*Opuntia* spp. of cacti	Australia, 1925; Caribbean, 1957; Hawaii, 1950; Mauritius, 1950; South Africa, 1932; U.S. (accidental), 1989
Flowerhead weevil, *Rhinocyllus conicus* Fröelick (Eurasia)	*Carduus* spp. thistles	U.S., 1968

TABLE 13.1 *(continued)*

Nontarget Impacts	Reference
Predation of native snails (e.g., native (*Achatinella* spp. and *Partula* spp.), probably leading to extinction of some native species	Clarke, Murray, and Johnson 1984; Davis and Butler 1964; Griffiths, Cook, and Wells 1993; Kinzie 1992; Murray et al. 1988; Simmonds and Hughes 1963
Regional declines of native saturniid moths because of heavy parasitism of larvae	Boettner, Elkinton, and Boettner 2000
Possible extirpation of the native Fijian zygaenid *Heteropan dolens* and reduction in abundance of other native zygaenids	Howarth 1991; Robinson 1975; Sands 1997; Tothill, Taylor, and Paine 1930
Competitive displacement of native aphidophagous coccinellids in agricultural crops, and nontarget predation of native lepidopterans	Elliot, Kieckhefer, and Kauffman 1996; Obrycki, Elliot, and Giles 2000
Invaded mainland U.S. from the Caribbean in 1989; attacks native *Opuntia* spp. thereby threatening the survival of endangered native spp.	Bennett and Habeck 1992; Holloway 1964; Pemberton 1995
Attacks on seed heads of native *Cirsium* spp. thistles, potentially limiting regeneration of plants and displacing native thistle-head feeders	Gassmann, and Louda 2001; Louda et al. 1997; Strong 1997

mental risk posed by natural enemies that are extreme generalists. This fly was first released in the United States by government authorities in the early 1900s for suppression of the gypsy moth (a severe defoliating forest pest in the northeastern United States) and for control of thirteen other pest species. Following its establishment, *C. concinata* attacked more than 180 species, including native moths, beetles, and sawflies (Boettner, Elkinton, and Boettner 2000). When this project was initiated, entomologists saw the broad host range of *C. concinata* as a desirable attribute because it enabled the fly to maintain high population densities by using nontarget species at times or in places where gypsy moth larvae were few or non-existent, thereby providing more persistent pest suppression when gypsy moth densities rebounded (Van Driesche and Hoddle 1997).

Similarly, the flowerhead weevil, *Rhinocyllus conicus* Fröelick, a generalist thistle-feeding species, was released in the United States in 1968 by the USDA for control of exotic musk thistles. This insect moved onto native *Cirsium* spp. thistles, as predicted by host-range tests performed by scientists prior to weevil releases from quarantine facilities and confirmed from early postrelease observations (Gassmann and Louda 2001). These results clearly demonstrate the usefulness and predictivness of these prerelease studies. Use of native thistles by *R. conicus* is currently viewed as an unacceptable outcome of this biological control project because the regeneration potential of these rare plants is potentially compromised, and native insects utilizing these resources are also threatened (Louda et al. 1997; Strong 1997; Louda 2000). In countries that lack native thistles but suffer heavy infestations of exotic thistle species, such as New Zealand, use of this weevil for thistle biological control has not resulted in attacks on native plants, and weevil impact on thistles is significant (Kelly et al. 1990). Furthermore, over recent years, the value of indigenous flora and fauna has increased as the public's conservation awareness has become more acute; and as fluid social perceptions of "acceptable risk" change, the nontarget impact of biological control agents such as *R. conicus* on native flora is viewed differently. In the 1960s, attacks on native thistles by exotic insects were considered acceptable by society, but in the late 1990s, they were considered unacceptable because of the high value placed on protecting native flora and fauna. In the current prevailing scientific and social climate, oligophagous natural enemies such as *C. concinata* and *R. conicus* are unlikely to be approved for release by regulatory authorities in the United States.

The potential for introduced natural enemies to adversely alter food-web linkages has been illustrated with *R. conicus,* and there are concerns that community-wide effects can result from competitive interactions between natural enemies and indigenous predators, parasitoids, and herbivores (Strong and Pemberton 2001). Although nontarget attacks by natural enemies, in particular parasitoids, have been documented (Boettner, Boettner, and Elkinton 2000; Henneman and Memmott 2001), the exact ecological impact on native invertebrate populations is uncertain, and some are arguing for detailed studies on trophic spectra to determine natural-enemy impacts on the communities into which they are introduced (Strong and Pemberton 2001).

Selection of natural enemies with narrow host ranges protects nontarget species because physiological, behavioral, ecological, or geographical attributes make native organisms unsuitable for exploitation by natural enemies (Strand and Obrycki 1996; Frank 1998). Furthermore, high levels of host specificity by natural enemies ensure that as social climates change, the perceived benefit of the natural enemy for pest control will not wane unless the perceived value of the pest changes. Host-range expansion by specialized natural enemies to exploit novel hosts through evolutionary adaptation is considered rare (Onstad and McManus 1996; Nechols, Kauffman, and Schaefer 1992). Although host shifts and host-range expansions by some types of natural enemies are rare, such possibilities warrant consideration, and increased research effort is needed to aid in predicting the likelihood of such events (Howarth 2000; Roderick 1992; Secord and Kareiva 1996; Simberloff and Stiling 1998).

SAFEGUARDS AGAINST
BIOLOGICAL CONTROL ACCIDENTS

The increasing demand for greater use of biological control to suppress invasive species, coupled with recent criticisms by reputable biologists that biological control may not always be a safe alternative to pesticides, has resulted in the development of evaluation protocols and legislation to regulate the importation and release of exotic natural enemies. Laws governing biological control vary by country, or they may not exist at all. In the United States, the USDA and the Department of the Interior are required by Executive Order 11987 (Exotic Organisms; 1977) to restrict the introduction of exotic species into natural ecosystems unless it had been shown that there will be no adverse effects (Follett et al. 2000). Biological control in the United States has been facilitated by the recent Executive Order 13112 (Invasive Species; 1999), which established the cabinet-level Invasive Species Council to provide guidance on rational and cost-effective control measures of exotic pests. The Animal and Plant Heath Inspection Service (an arm of the USDA) is charged with examining the potential environmental impacts of introduced biological control agents before authorizing their release to comply with statutes such as the National Environmental Policy Act (1969) and the Endangered Species Act (1973). Despite these regulations, the United States does not have an encompassing biological control law, and there is no legal mandate or agency to explicitly oversee the importation and release of exotic organisms (Howarth 2000). This is in stark contrast with current legislation enacted in New Zealand and Australia (see below). Interestingly, the importation of candidate biological control agents by scientists into the United States is regulated: highly secure quarantine facilities are used to screen and test natural enemies prior to release, and federal- and state-level clearances are needed to move organisms from quarantine facilities. However, such procedures are generally not required for the importation of exotic and potentially invasive aquatic, terrestrial, and arboreal species that constitute the pet, nursery, and aquarium trades in the United States (Van Driesche and Van Driesche 2000). Strong and Pemberton (2001) suggest that native invertebrates are inadequately protected from biological control agents under current U.S. legislation, and they recommend that a review process similar to the one currently in place for biological control of weeds be applied to invertebrate targets to reduce the risks of collateral damage to nontarget species. Host-specificity testing protocols are being developed and evaluated for arthropod biological control agents, and current protocols are following systems developed for determining the host specificity of weed biological control agents (see Withers, Browne, and Stanley 1999; Van Driesche et al. 1999).

In Australia, the importation of exotic organisms is controlled by two legislative acts: the Quarantine Act (1985), designed to prevent the introduction of agricultural pests as well as diseases of humans, and the Wildlife Protection Act (1982), intended to control trade in endangered wildlife (McFadyen 1997). A third piece of related legislation is the Biological Control Act (1984), the purpose of which is to resolve conflicts of interest that arise when a biological control target is classified both as a pest and a beneficial organism. For example, vast monotypic stands of invasive weeds that provide nectar and pollen for commercially managed bees are seen as beneficial by beekeepers but not by conservationists or rangeland managers (Cullen and Delfosse 1985). Permits for the importation and release of biological control agents are granted by the Australian Quarantine Inspection Service and involve wide consultation with interested parties within Australia before a consensus on the outcome of the application for release is achieved.

New Zealand has one of the most stringent legislative requirements for importation of potential biological control agents. The Hazardous Substances and New Organisms Act (1996) has greatly increased the obligations incumbent on proponents of new biological control agents, requiring them to provide adequate data on which approvals for importation and release can be based (Fowler, Syrett, and Jarvis 2000). This legislation provides a solid framework within which risks and benefits of proposed natural-enemy introductions can be weighed and decisions can be made in accordance with presented data. The Environmental Risk Management Authority administers the review process for the importation and release of biological control agents in New Zealand.

International agreements designed to prevent the introduction of biological control agents from causing economic and environmental damage may lead to increased restrictions on the release of biological control agents in other parts of the world. The Food and Agriculture Organization (FAO) Code for the Import and Release of Exotic Biological Control Agents was approved by all member states in 1995, and these guidelines should be adopted worldwide. Under these guidelines, not only must the government of the importing country approve the introduction but other countries in the region must also be consulted, as natural enemies may cross international boundaries. The host range of the proposed biological control agent must be adequately measured before release, and the impact of the organism must be evaluated following its establishment (Code of conduct 1997; Food and Agriculture Organization 1996; McFadyen 1997).

Adoption and enforcement of regulatory acts could significantly reduce the likelihood of egregious harm to native biota and could prevent foolish projects such as the introduction of *Englandina rosea* to the Pacific island of Moorea, which resulted in the extinction of *Partula* spp. snails (Clarke, Murray, and Johnson 1984), or the casual importation of the crazy ant, *Paratrechina fulva* (Mayr), from Brazil into Colombia for the control of poisonous snakes (de Polania and Wilches 1992). This snake control project resulted in the disappearance of thirty-six of thirty-eight species of ants indigenous to the invaded area, seven other soil-dwelling insect species, one species of snake, and three lizard species.

Despite the best-intentioned laws, flagrant disregard of legislation by stakeholders who feel disenfranchised by the regulatory bureaucracy can result in the illegal importation of biological control agents. Rabbit calicivirus disease (RCD) was probably smuggled into the South Island of New Zealand from Australia by high-country farmers in August 1997 for the control of rabbits. The virus was illegally disseminated by feeding rabbits carrots and oats saturated with contaminated liquefied livers extracted from rabbits that died from RCD. A network of cooperators spread the virus over large areas of the South Island, and its subsequent spread (human assisted through the movement of carcasses, baiting, and insect vectors) made containment and eradication of the disease impossible. Such actions by farmers clearly violated New Zealand's Biosecurity Act (1994), which was enacted in part to protect agriculture and the country's unique flora and fauna from unwanted introductions of pests. A small group of New Zealand farmers justified their actions because they felt that the New Zealand government was not moving rapidly enough on the importation of biological control agents for rabbits (the myxoma virus, which causes the lethal rabbit disease myxomatosis, is not present in New Zealand). The New Zealand government has now sanctioned controlled virus releases into new areas, and the short-term impact of RCD on New Zealand rabbit populations has resulted in 47–66% mortality in central Otago. Large-scale, long-term field studies of RCD on rabbits are planned. Fortunately, native New Zealand birds and mammals, groups identified to be at high risk from RCD, do not appear to be affected by this virus (Buddle et al. 1997).

BIOLOGICAL CONTROL PROGRAMS IN SUPPORT OF CONSERVATION

Biological control is perhaps the best and, in some instances, the only technology available for the management and restoration of ecosystems degraded by exotic invasives (Headrick and Goeden 2001). Suppression of weeds in natural areas is the most prominent application of biological control in support of conservation. The practice has grown out of the earlier use of weed biological control for range and agricultural management (McFadyen 1998). In the Everglades World Heritage Park in Florida, biological control programs utilizing insects are currently targeting the thick stands of Australian melaleuca tree (*Melaleuca quinquenervia* [Cavanilles]) that are adversely affecting water table levels and displacing native plants and wildlife (Center, Frank, and Dray 1997; Goolsby, Makinson, and Purcell 2000). Similar programs are being conducted in New Zealand's Tongariro National Park, a World Heritage Area, where an exotic European heather, *Calluna vulgaris* (L.), is being targeted with heather beetles, *Lochmaea suturalis* (Thomson), that feed exclusively on this weed (Syrett et al. 2000). Other weeds of conservation importance in the United States that are current targets of biological control programs are purple loosestrife, *Lythrum salicaria* L. (Hight and Drea 1991; Malecki et al. 1993); Brazilian peppertree,

Schinus terebinthfolius Raddi (Medal et al. 1999); and salt cedar, *Tamarix* spp. (DeLoach et al. 1996).

Biological control has been used against arthropods that are threatening native flora and fauna in natural areas. Introduced scale insects, *Carulaspis minima* (Targioni–Tozzetti) and *Insulaspis pallida* (Maskell), caused extreme declines of Bermuda cedar, *Juniperus bermudiana* L.; and natural enemies were used in a control program for this pest (Cook 1985). Ensign scale, *Orthezia insignis* Browne, threatened the survival of the endemic gumwood tree, *Commidendrum robustum*, on the isolated island of St. Helena until it was brought under successful biological control with the coccinellid *Hyperaspis pantherina* Fürsch (Booth et al. 1995). The hemlock woolly adelgid, *Adelges tsugae* Annand, an adventive pest from Asia, is killing native eastern hemlock trees, *Tsuga canadensis* (L.), in the northeastern United States at an unprecedented rate. A biological control program using coccinellids (*Scymnus sinuanodulus* Yu et Yao and *Pseudoscymnus tsugae* Sasaji and McClure) and oribatid mites (*Diapterobates humeralis* [Hermann]) that attack adelgids is under way (Lu and Montgomery 2001). There are no known hymenopterous parasitoids that attack adelgids.

An exotic Mexican weevil, *Metamasius callizona* (Chevrolat), is attacking and killing threatened species of epiphytic bromeliads in Florida. This pest was introduced through the commercial bromeliad trade, and biological control with a tachinid fly, *Admontia* sp., may be the only feasible solution for controlling this pest in environmentally sensitive areas (Frank and Thomas 1994; Frank 1999; Salas and Frank 2001). New Zealand's native nectar-feeding birds are being outcompeted for beech scale honeydew in South Island forests by highly aggressive yellow jackets (*Vespula vulgaris* L.). A specialized ichneumonid parasitoid (*Sphecophaga vesparum vesparum* [Curtis]) that attacks immature yellow jackets in nests has been established in New Zealand in an attempt to reduce the inordinate numbers of these wasps in forests (Barlow, Moller, and Beggs 1996).

Biological control projects involving arthropod pests of natural areas have not been evaluated in terms of economic costs and sociological benefits to conservation. This situation will undoubtedly change as the success of recently initiated projects is evaluated and the number of new projects increases. Natural enemies are not used against nonmammalian vertebrates (e.g., snakes), freshwater and marine crustaceans and mollusks, and platyhelminths, because there are no precedent-setting examples for these groups and perhaps because specialists studying these invasive organisms may be unfamiliar with the concept of biological control and the benefits it offers. In many instances, biological control may have an irreplaceable role in protecting precious and unique ecosystems from degradation by invasive pests that threaten flora and fauna of conservation importance (Van Driesche 1994).

NONTRADITIONAL PEST TARGETS OF CONSERVATION IMPORTANCE AND NOVEL BIOLOGICAL CONTROL TECHNIQUES

The brown tree snake is a pest of conservation importance that is a suitable but nontraditional target for biological control. *Boiga irregularis* is the proximate cause of twelve native-bird extinctions on Guam following its accidental introduction after World War II on military equipment (Pimm 1987; Savidge 1987; Jaffe 1994; Rodda, Fritts, and Chiszar 1997). The snake has also caused declines of native-reptile and small mammal populations, and it will enter houses and attack sleeping human infants (Rodda, Fritts, and Chiszar 1997). Additionally, the brown tree snake has caused economic loses by adversely affecting domestic animals (e.g., chickens and pets), and high densities of snakes on power lines regularly cause short circuits that interrupt electrical supplies and necessitate repairs. Control of the brown tree snake has been attempted through trapping, but the snake's extreme preference for live bait over artificial lures has made this approach impractical (Rodda, Fritts, and Chiszar 1997).

The brown tree snake, a member of the family Colubridae, is native to eastern Indonesia, the Solomon Islands, New Guinea, and northeastern Australia, and it is the only member of this family on Guam. There is one native species of snake on Guam, the blind snake, *Rhamphotyphlopys braminus* (Daudin), which belongs to the family Typhlopidae and is the only snake occurring on many islands in the central Pacific region (T. Fritts, pers. comm. 1998). The brown tree snake has extended its range: it is now established on the previously snake-free island of Saipan and has been intercepted in Hawaii, Texas, and Spain (Rodda, Fritts, and Chiszar 1997). Given the brown tree snake's propensity to be dispersed to new habitats by cargo planes and ships, the major social, economic, and ecological problems that occur on islands after colonization and the snake's lack of close taxonomic relationship to snakes common to Pacific islands make it an excellent target for biological control.

The taxonomic distance between colubrids and typhlopids may simplify the task and reduce the cost of finding natural enemies unique to the brown tree snake. Parasites or pathogens that are host specific at the family (i.e., Colubridae) or genus (i.e., *Boiga*) level may be safe to nontarget snakes (e.g., typhlopids) if these organisms have not evolved the ability to cause disease in distantly related hosts.

In general, vertebrate pests, especially mammals, have proven to be notoriously difficult to control because of their ability to learn and their generally secretive habits, and the use of vertebrate predators to control vertebrate pests has exacerbated many problems as the biological control agents have in turn become problematic (see Table 13.1). Many vertebrate species that become pests are distinguished from nonpestiferous species by their higher intrinsic rates of increase. Agents that reduce vertebrate reproductive rates without causing mortality are receiving increased attention because this approach may present fewer risks to nontarget species. For example, sexual transmission of diseases may guarantee host specificity in biological control programs against vertebrate pests, and the potential of genetically engineering sexually transmitted viruses to sterilize infected hosts without killing them is being investigated as a novel biological control technology (Barlow 1994).

Viruses that have antigens from the host sperm or from the zona pellucida around host eggs engineered into the genome can provoke an immune system response that renders the recipient sterile. Immunocontraception (also referred to as immunosterilization) as a means to control noxious vertebrates (e.g., foxes, rabbits, and mice) is being actively pursued in Australia and New Zealand (McCallum 1996). An alternative approach to immunocontraception is the use of genetically modified pathogens to prevent lactation in females so that juveniles are not successfully weaned or to interfere with hormonal control of reproduction (Cowan 1996; Jolly 1993; Rodger 1997).

Genetically engineered pathogens that sterilize pest animals offer the possibility of humane control without killing or causing animals to suffer the effects of debilitating disease. As a form of biological control, immunocontraception may reduce the need for broadcast distribution of toxins for pest suppression, thereby reducing environmental contamination and nontarget mortality. This is of special concern when pests inhabit suburbs, urban parks, government and state campuses, nature reserves, military bases, or other areas where lethal controls may be illegal or unsafe (Kirkpatrick et al. 1997; Williams 1997). The concept of virally mediated immunocontraception has generated considerable debate on legal and ethical issues regarding releases of engineered microorganisms into the environment. Once contagious recombinant agents that cause permanent sterilization in animals are released into the environment, they cannot be recalled (Tyndale-Biscoe 1995). A sterilizing agent that does not cause painful disease symptoms is an ethically acceptable form of pest control that is justifiable from animal rights perspectives, because it does not cause the suffering typical of current lethal methods (e.g., trapping shooting, poisoning, and introducing disease) (Oogjes 1997; Singer 1997). Limited field trials with sterilizing microorganisms are unlikely before 2005 (Anderson 1997).

EXPERIMENTAL APPROACHES IN CONSERVATION BIOLOGY: CONTRIBUTIONS TO THEORY FROM BIOLOGICAL CONTROL

Many questions that are of importance to conservation biologists are also of concern to biological control practitioners. For example, what are the factors influencing the establishment and maintenance of viable populations? This question is relevant for establishing new natural-enemy species outside of their natural ranges, for introducing native organisms threatened by extinction into reclaimed habitat (e.g., releases of native New Zealand birds on offshore islands that have been cleared of rats and mice), and for assessing the likelihood of invasive-species establishment.

Retrospective studies of biological control programs have sought to determine the factors affecting the establishment of released organisms (Stiling 1990). Aspects thought to influence establishment rates include (1) the number organisms released (propagule size) and the frequency of releases (Beirne 1975, 1985; Memmott, Fowler, and Hill 1998; Williamson and Fitter 1996); (2) biogeographical influences such as the ready accommodation by islands for new species as opposed to the resistance exhibited by continents (Hall and Ehler 1979; Sailer 1978); (3) the geographic origin of the agent and the climatic suitability of the new range (Goolsby, Makinson, and Purcell 2000; Sutherst and Maywald 1985; Sutherst, Floyd, and Maywald 1995); (4) the taxonomic order of the agent and its propensity to establish and proliferate (i.e., wasps versus flies) (Hall and Ehler 1979; Sailer 1978); (5) environmental (e.g., climatic events and catastrophes [storms and volcanoes]) and demographic stochasticity (Allee effects leading to extinction because of low population densities) (Beirne 1975; Syrett, Smith, Bourner, Fowler, and Wilcox 2000); (6) biotypes and genetic constitution of founding populations (Roush 1990); and (7) the impact of resident organisms on population growth of exotics through either direct attack or competitive exclusion (Goeden and Louda 1976; Cornell and Hawkins 1993).

Once exotic organisms (e.g., natural enemies) are established, there is a general need to understand their population phenology, the factors promoting and retarding population growth, and the subsequent range expansions and contractions. Biological control research has made major contributions to current paradigms on population regulation that were first posited by Nicholson and Bailey (1935) to explain the observed "balance of nature" and the population cycles between prey and predators that can result from density-dependent mortality as defined by Smith (1935). The mathematical modeling (Nicholson–Bailey and Lotka–Volterra models) of experimental and field systems has become increasingly sophisticated to incorporate natural-enemy functional responses, metapopulation dynamics, and spatial effects to explain the long-term persistence of natural enemies and their prey (Barlow and Wratten 1996). These models have been useful in estimating the impact of generalist predators (e.g., foxes and ferrets) on rabbit populations and in determining critical rabbit densities below which these predators will preferentially attack vulnerable native animals (e.g., birds and marsupials). Detection of declining primary-prey populations can initiate proactive culling of predators before attacks on native wildlife begin (Newsome 1990). Experimental investigations of population cycles and factors affecting predator–prey oscillations have identified the importance of food supplies, natality, and mortality, habitat stability, and climate as components of complex links that form food webs in natural systems (Sinclair, Olsen, and Redhead et al. 1990). Perturbations to nonresilient species links in these webs either because of habitat destruction, global warming, or invasive species can threaten the existence of some of these best-studied population phenomena (Krebs et al. 2001).

CONCLUSIONS

The rate at which invasive species degrade valued habitats continues to accelerate. In many instances, biological control may be the only sustainable, cost-effective way to manage these problematic organisms. Caution should guide biological control programs to enhance project safety and ensure appropriate selection of targets and natural enemies. Potential targets must be carefully selected because not every established alien species is a threat. Biological control may not be the most appropriate response to every pest, especially for native species that lack effective natural enemies or could be controlled with generalist natural enemies from outside of the endemic pest's native range. Judicious importation and release of host-specific natural enemies is needed. The shotgun approach—that is, the importation of multiple natural-enemy species for mass rearing and release with the expectation that environmental winnowing will select an effective biological control agent—should not be practiced, as it increases both the probability of establishment by generalist natural enemies and of attacks on nontargets. These two issues, identification of appropriate targets and natural-enemy selection, can be addressed within sensible regulatory frameworks, and legislation in New Zealand and Australia and the FAO Code of Conduct provide excellent working models for the design and implementation of protocols governing importation, evaluation, and releases of exotic natural enemies.

Professional scientists with explicit training and a proven track record in biological control research should conduct biological control projects. Location, selection, and evaluation of natural enemies are difficult and require the expertise of many collaborators and elaborate research facilities, and projects can span many years before suitable natural enemies are located and cleared for release. Consequently, the desire for quick fixes to pest problems has prompted agriculturists and land mangers to sidestep these rigorous procedures and import generalist natural enemies (e.g., foxes and predatory snails) that have high probabilities of establishment, voracious appetites, and weak associations with the target pest. Such reckless practices cannot be considered science-based biological control programs. As seen with the rabbit biological control program in Australia, highly specific pathogens may exist for nontraditional pest targets, and disease-causing organisms can be effective natural enemies. Researchers are now investigating immunocontraception to further strengthen the rabbit biological control program. Obviously, the expertise and the time required to work with such minute and highly specialized natural enemies and to assess the risks and benefits of biological control projects is not the domain of nonspecialists seeking a rapid solution to an environmental problem.

Biological control can greatly benefit conservation efforts through the practical suppression of pests. Biological control programs based on carefully planned natural-enemy introductions and manipulative experiments can enhance our theoretical understanding of critical issues concerning the dynamics of invasion, genetic heterogeneity and lag periods, propagule size and establishment rates, and factors influencing population dynamics and range extension. Additional incentives to use natural enemies as cornerstones for integrated pest management programs arise from government legislation mandating reduced pesticide use. Despite the promise that classical biological control can reduce pesticide use and other detrimental environmental impacts from conventional pest management, the technology is under intense scrutiny because of concerns over nontarget impacts and intentional homogenization of the world's biota. Ultimately, when faced with the devastating impacts of an ever increasing multitude of invasive organisms, the risks of the do-nothing option are unacceptably high. Well-targeted biological control projects can make a real difference in the battle against the stealth destroyers of the world's wilderness areas. Biological control is unreservedly an ally of conservation.

REFERENCES

Anderson, I. 1997. Alarm greets contraceptive virus. *New Scientist* 154:4.

Ankley, G. T., Tietge, J. E., DeFoe, D. L., Jensen, K. M., Holcombe, G. W., Durhan, E. J., and Diamond, S. A. 1998. Effects of ultraviolet light and methoprene on survival and development of *Rana pipiens*. *Environmental Toxicology and Chemistry* 17:2530–42.

Barlow, N. D. 1994. Predicting the effect of a novel vertebrate biocontrol agent: A model for viral-vectored immunocontraception of New Zealand possums. *Journal of Applied Ecology* 31:454–62.

Barlow, N. D., Moller, H., and Beggs, J. R. 1996. A model for the effect of *Sphecophaga vesparum vesparum* as a biological control agent of the common wasp in New Zealand. *Journal of Applied Ecology* 33:31–44.

Barlow, N. D., and Wratten, S. D. 1996. Ecology of predator-prey and parasitoid-host systems: Progress since Nicholson. In *Frontiers of Population Ecology,* ed. R. B. Floyd, A. W. Sheppard, and P. J. De Barro, 217–43. Melbourne, Aust.: CSIRO Publishing.

Beirne, B. P. 1975. Biological control attempts by introductions against insect pests in the field in Canada. *Canadian Entomologist* 107:225–36.

———. 1985. Avoidable obstacles to colonization in classical biological control of insects. *Canadian Journal of Zoology* 63:743–47.

Bellows, T. S., Jr. 2001. Restoring population balance through natural enemy introductions. *Biological Control* 21:199–205.

Bellows, T. S., Jr., and Fisher, T. W., eds. 1999. *Handbook of Biological Control.* San Diego, Calif.: Academic Press.

Bennett, F. D., and Habeck, D. H. 1992. *Cactoblastis cactorum:* A successful weed control agent in the Caribbean, now a pest in Florida? In *Proceedings of the Eighth International Symposium on Biological Control of Weeds,* ed. E. S. Delfosse and R. R., 21–26. Melbourne, Australia: CSIRO Publishing.

Boettner, G. H., Elkinton, J. S., and Boettner, C. J. 2000. Effects of biological control introduction on three nontarget native species of saturniid moths. *Conservation Biology* 14:1798–806.

Booth, R. G., Cross, A. E., Fowler, S. V., and Shaw, R. H. 1995. The biology and taxonomy of *Hyperaspis pantherina* (Coleoptera: Coccinellidae) and the classical biological control of its prey, *Orthezia insignis* (Homoptera: Ortheziidae). *Bulletin of Entomological Research* 85:307–14.

Bright, C. 1998. *Life Out of Bounds: Bioinvasion in a Borderless World.* New York: W. W. Norton.

Buddle, B. M., de Lisle, G. W., McColl, K., Collins, B. J., Morrissy, C., and Westbury, H. A. 1997. Response of the North Island brown kiwi, *Apteryx australis mantelli*, and the lesser short-tailed bat, *Mystacina tuberculata*, to a measured dose of rabbit haemorrhagic disease virus. *New Zealand Veterinary Journal* 45:109–13.

Caltagirone, L. E., and Doutt, R. L. 1989. The history of the vedalia beetle importation to California and its impact on the development of biological control. *Annual Review of Entomology* 34:1–16.

Carson, R. 1962. *Silent Spring*. Boston, Mass.: Houghton Mifflin.

Center, T. D., Frank, J. H., and Dray, F. A. 1997. Biological control. In *Strangers in Paradise: Impact and Management of Nonindigenous Species in Florida*, ed. D. Simberloff, D. C. Schmitz, and T. C, Brown, 245–63. Washington, D.C.: Island Press.

Charlebois, P. M., Raffenberg, M. J., and Dettmers, J. M. 2001. First occurrence of *Cercopagis pengoi* in Lake Michigan. *Journal of Great Lakes Research* 27:258–61.

Christensen, O. M., and Mather, J. G. 1995. Colonisation by the land planarian *Artioposthia triangulata* and impact on lumbricid earthworms at a horticultural site. *Pedobiologia* 39:144–54.

Clarke, B., Murray, J., and Johnson, M. S. 1984. The extinction of endemic species by a program of biological control. *Pacific Science* 38:97–104.

Code of conduct for the import and release of exotic biological control agents. 1997. *Biocontrol News and Information* 18:119N–124N.

Colborn, T., Dumanoski, D., and Peterson Myers, J. 1997. *Our Stolen Future: Are We Threatening Our Fertility, Intelligence, and Survival? A Scientific Detective Story*. New York: Penguin Books.

Cook, M. J. W. 1985. *A Review of Biological Control of Pests in the Commonwealth Caribbean and Bermuda up to 1982*. Commonwealth Agricultural Bureau Technical Communication no. 9. Surrey, U.K.: Unwin Brothers.

Cornell, H. V., and Hawkins, B. A. 1993. Accumulation of native parasitoid species on introduced herbivores: A comparison of hosts as natives and hosts as invaders. *American Naturalist* 141:847–65.

Cowan, P. E. 1996. Possum biocontrol: Prospects for fertility control. *Reproduction, Fertility, and Development* 8:655–60.

Cox, G. W. 1999. *Alien Species in North America and Hawaii*. Washington, D.C.: Island Press.

Cullen, J. M., and Delfosse, E. S. 1985. *Echium plantagineum*: Catalyst for conflict and change in Australia. In *Proceedings of the 6th International Symposium on Biological Control of Weeds*, ed. E. S. Delfosse, 249–92. Vancouver, Canada: Agriculture Canada.

Davis, C. J., and Butler, G. D. 1964. Introduced enemies of the giant African snail, *Achatina fulica* Bowdich, in Hawaii (Pulmonta: Achatinidae). *Proceedings of the Hawaiian Entomological Society* 18:377–89.

DeBach, P., ed. 1964. *Biological Control of Insect Pests and Weeds*. New York: Reinhold Publishing.

————. 1974. *Biological Control by Natural Enemies.* Cambridge: Cambridge University Press.

DeLoach, C. J., Gerling, D., Fornasari, L., Sobhian, R., Myartseva, S., Mityaev, I. D., and Lu, Q. G. 1996. Biological control program against salt cedar (*Tamarix* spp.) in the United States of America: Progress and problems. In *Proceedings of the IX International Symposium on Biological Control of Weeds,* ed. V. C. Moran and J. H. Hoffmann, 253–60. Stellenbosch, South Africa: University of Cape Town.

de Polania, I. Z., and Wilches, O. M. 1992. Impacto ecologico de la hormiga loco, *Paratrechina fulva* (Mayr), en el municipio de Cimitarra (Santander). *Revista Colombiana de Entomología* 18:14–22.

Diamond, J. M. 1996. A-bombs against amphibians. *Nature* 383:386–87.

Easteal, S. 1981. The history of introductions of *Bufo marinus* (Amphibia: Anura); a natural experiment in evolution. *Biological Journal of the Linnean Society* 16:93–113.

Elliot, N., Kieckhefer, R., and Kauffman, W. 1996. Effects of an invading coccinellid on native coccinellids in an agricultural landscape. *Oecologia* 105:537–44.

Emberson, R. 2000. Endemic biodiversity, natural enemies, and the future of biological control. In *Proceedings of the Tenth International Symposium on Biological Control of Weeds,* ed. R. Spencer, 875–80. Bozeman: Montana State University.

Follett, P. A., Duan, J. J., Messing, R. H., and Jones, V. P. 2000. Parasitoid drift after biological control introductions: Re-examining Pandora's box. *American Entomologist* 46:82–94.

Food and Agriculture Organization of the United Nations/Secretariat of the International Plant Protection Convention. 1996. *Code of Conduct for the Import and Release of Exotic Biological Control Agents.* International Standards for Phytosanitary Measures, no. 3.

Forno, I. W., and Julien, M. H. 2000. Success in biological control of aquatic weeds by arthropods. In *Biological Control: Measures of Success,* ed. G. Gurr and S. Wratten, 159–87. Dordrecht, Netherlands: Kluwer Academic Publishers.

Fowler, S. V., Syrett, P., and Jarvis, P. J. 2000. Will expected and unexpected non-target effects, and the new Hazardous Substances and New Organisms Act, cause biological control of broom to fail in New Zealand? In *Proceedings of the Tenth International Symposium on Biological Control of Weeds,* ed. R. Spencer, 173–86. Bozeman: Montana State University.

Frank, J. H. 1998. How risky is biological control? Comment. *Ecology* 79:1829–34.

————. 1999. Bromeliad feeding weevils. *Selbyana* 20:40–48.

Frank, J. H., and Thomas, M. C. 1994. *Metamasius callizona* (Chevrolat) (Coleoptera: Curculionidae), an immigrant pest, destroys bromeliads in Florida. *Canadian Entomologist* 126:673–82.

Freeland, W. J. 1985. The need to control cane toads. *Search* 16:211–15.

Gamradt, S. C., and Kats, L. B. 1996. Effect of introduced crayfish and mosquitofish on California newts. *Conservation Biology* 10:1155–62.

Gassmann, A., and Louda, S. V. 2001. *Rhinocyllus conicus:* Initial evaluation and subsequent ecological impacts in North America. In *Evaluating Indirect Ecological Effects of Biological Control,* ed. E. Wajnberg, J. K. Scott, and P. C. Quimby, 147–83. Wallingford, U.K.: CABI Publishing.

Goeden, R. D., and Louda, S. 1976. Biotic interference with insects imported for weed biological control. *Annual Review of Entomology* 21:325–42.

Goolsby, J. A., Makinson, J., and Purcell, M. 2000. Seasonal phenology of the gall making fly *Fergusonnia* sp. (Diptera: Fergusonnidae) and its implications for biological control of *Melaleuca quinquenervia. Australian Journal of Entomology* 39:336–43.

Griffiths, O., Cook, A., and Wells, S. M. 1993. The diet of the introduced carnivorous snail *Euglandina rosea* in Mauritius and its implications for threatened island gastropod faunas. *Journal of Zoology* 229:79–89.

Gurr, G. M., Barlow, N. D., Memmott, J., Wratten, S. D., and Greathead, D. J. 2000. A history of methodical, theoretical, and empirical approaches to biological control. In *Biological Control: Measures of Success,* ed. G. Gurr and S. Wratten, 3–37. Dordrecht, Netherlands: Kluwer Academic Publishers.

Gurr, G., and Wratten, S., eds. 2000. *Biological Control: Measures of Success.* Dordrecht, Netherlands: Kluwer Academic Publishers.

Gutierrez, A. P., Caltagirone, L. E., and Meikle, W. 1999. Evaluation of results: Economics of biological control. In *Handbook of Biological Control,* ed. T. S. Bellows Jr. and T. W. Fisher, 243–52. San Diego, Calif.: Academic Press.

Guy, P. L., Webster, D. E., and Davis, L. 1998. Pests of non-indigenous organisms: Hidden costs of introduction. *Trends in Evolution and Ecology* 13:111.

Hall, R. W., and Ehler, L. E. 1979. Rate of establishment of natural enemies in classical biological control. *Entomological Society of America Bulletin* 25:280–82.

Headrick D. H., and Goeden, R. D. 2001. Biological control as a tool for ecosystem management. *Biological Control* 21:249–57.

Heijden, M.G.A., Klironomos, J. N., Ursic, M., Moutoglis, P., Streitwolf-Engel, R., Boller, T., Wiemken, A., and Sanders, I. R. 1998. Mycorrhizal fungal biodiversity, ecosystem variability, and productivity. *Nature* 396:69–72.

Henneman, M. L., and Memmott, J. 2001. Infiltration of a Hawaiian community by introduced biological control agents. *Science* 293:1314–16.

Hight, S., and Drea, J. J., Jr. 1991. Prospects for a classical biological control project against purple loosestrife (*Lythrum salicaria* L.). *Natural Areas Journal* 11:151–57.

Hinton, H. E., and Dunn, A.M.S. 1967. *Mongooses: Their Natural History and Behaviour.* Berkeley: University of California Press.

Hoddle, M. S. 1999. Biological control of vertebrate pests. In *Handbook of Biological Control,* ed. T. S. Bellows Jr. and T. W. Fisher, 955–74. San Diego, Calif.: Academic Press.

Hokkanen, H., and Pimental, D. 1989. New associations in biological control: Theory and practice. *Canadian Entomologist* 121:829–40.

Holland, R. E. 1993. Changes in planktonic diatoms and water transparency in Hatchery Bay, Bass Island area, western Lake Erie, since the establishment of the zebra mussel. *Journal of Great Lakes Research* 19:617–24.

Holloway, J. K. 1964. Projects in biological control of weeds. In *Biological Control of Insect Pests and Weeds*, ed. P. DeBach, 650–70. New York: Reinhold Publishing Corporation.

Holt, R. D., and Hochberg, M. E. 2001. Indirect interactions, community modules, and biological control: A theoretical perspective. In *Evaluating Indirect Ecological Effects of Biological Control*, ed. E. Wajnberg, J. K. Scott, and P. C. Quimby, 13–37. Wallingford, U.K.: CABI Publishing.

Howarth, F. G. 1983. Classical biological control: Panacea or Pandora's box? *Proceedings of the Hawaiian Entomological Society* 24:239–44.

———. 1991. Environmental impacts of biological control. *Annual Review of Entomology* 36:485–509.

———. 2000. Non-target effects of biological control agents. In *Biological Control: Measures of Success*, ed. G. Gurr and S. Wratten, 369–403. Dordrecht, Netherlands: Kluwer Academic Publishers.

Jaffe, M. 1994. *And No Birds Sing: The Story of an Ecological Disaster in a Tropical Paradise*. New York: Simon and Schuster.

Jetter, K., Klonksy, K., and Pickett, C. H. 1997. A cost/benefit analysis of the ash whitefly biological control program in California. *Journal of Arboriculture* 23:65–72.

Jolly, S. E. 1993. Biological control of possums. *New Zealand Journal of Zoology* 20:335–39.

Jousson, O., Pawlowski, J., Zaninetti, L., Zechman, F. W., Dini, F., Di Guiseppe, G., Woodfield, R., Millar, A., and Meinesz, A. 2000. Invasive alga reaches California. *Nature* 408:157.

Kelly, D., McCallum, K., Schmidt, C. J., and Scanlan, P. M. 1990. Seed predation in nodding and slender winged thistles by nodding thistle receptacle weevil. In *Proceedings of the 43rd New Zealand Weed and Pest Control Conference*, ed. A. J. Popay, 212–15. Dunedin, New Zealand, n.p..

Kiesecker, J. M., Blaustein, A. R., and Miller, C. I. 2001. Transfer of a pathogen from fish to amphibians. *Conservation Biology* 15:1064–70.

King, C. M., ed. 1990a. *The Handbook of New Zealand Mammals*. Auckland, New Zealand: Oxford University Press.

———. 1990b. Stoat. In *The Handbook of New Zealand Mammals*, ed. C. M. King, 228–312. Auckland, New Zealand: Oxford University Press.

———. 1990c. Weasel. In *The Handbook of New Zealand Mammals*, ed. C. M. King, 313–20. Auckland, New Zealand: Oxford University Press.

Kinzie, R. A., III. 1992. Predation by the introduced carnivorous snail *Euglandina rosea* (Ferussac) on endemic aquatic lymnaeid snails in Hawaii. *Biological Conservation* 60:149–55.

Kirkpatrick, J. F., Turner, J. W., Jr., Liu, I.K.M., Fyrer-Hosken, R., and Rutberg,

A. T. 1997. Case studies in wildlife immunocontraception: Wild and feral equids and white-tailed deer. *Reproduction, Fertility, and Development* 9:105–10.

Krebs, C. J., Boonstra, R., Boutin, S., and Sinclair, A.R.E. 2001. What drives the 10-year cycle of snowshoe hares? *BioScience* 51:25–35.

Krimsky, S. 2000. *Hormonal Chaos: The Scientific and Social Origins of the Environmental Endocrine Hypothesis.* Baltimore, Md.: The Johns Hopkins University Press.

Kupferberg, S. J. 1997. Bullfrog (*Rana catesbeiana*) invasion of a California river: The role of larval competition. *Ecology* 78:1736–51.

Lavers, R. B., and Clapperton, B. K. 1990. Ferret. In *The Handbook of New Zealand Mammals,* ed. C. M. King, 320–30. Auckland, New Zealand: Oxford University Press.

Legner, E. F. 1996. Comments on adverse assessments of *Gambusia affinis. Journal of the American Mosquito Control Association* 12:161.

Loope, L., Hamann, O., and Stone, C. P. 1988. Comparative conservation biology of oceanic archipelagos. *BioScience* 38:272–82.

Louda, S. M. 2000. Negative ecological effects of the musk thistle biological control agent, *Rhinocyllus conicus.* In *Nontarget Effects of Biological Control,* ed. P. A. Follet and J. J. Duan, 215–43. Boston, Mass.: Kluwer Academic Publishers.

Louda, S. M., Kendall, D., Connor, J., and Simberloff, D. 1997. Ecological effects of an insect introduced for the biological control of weeds. *Science* 277:1088–90.

Lu, W., and Montgomery, M. E. 2001. Oviposition, development, and feeding of *Scymnus (Neopullus) sinuanodulus* (Coleoptera: Coccinellidae): A predator of *Adelges tsugae* (Homoptera: Adelgidae). *Annals of the Entomological Society of America* 94:64–70.

Lubulwa, G., and McMeniman, S. 1998. ACIAR-supported biological control projects in the South Pacific (1983–1996): An economic assessment. *Biocontrol News and Information* 19 (3): 91N–98N.

Malecki, R. A., Blossey, B., Hight, S. D., Schroeder, D., Kok, L. T., and Coulson, J. R. 1993. Biological control of purple loosestrife. *BioScience* 43:680–86.

McCallum, H. I. 1996. Immunocontraception for wildlife population control. *Trends in Ecology and Evolution* 11:491–93.

McFadyen, R. 1997. Protocols and quarantine procedures for importation and release of biological control agents. In *Biological Control of Weeds: Theory and Practical Application,* ed. M. Julien and G. White, 63–69. Canberra, Australia: ACIAR Monograph Series.

———. 1998. Biological control of weeds. *Annual Review of Entomology* 43:369–93.

Medal, J. C., Vitorino, M. D., Habeck, D. H., Gillmore, J. L., Pedrosa, J. H., and De Sousa, L. P. 1999. Host specificity of *Heteroperreyia hubrichi* Malaise (Hymenoptera: Pergidae), a potential biological control agent of Brazilian peppertree (*Schinus terebinthifolius* Raddi). *Biological Control* 14:60–65.

Meinesz, A. 1999. *Killer Algae: The True Tale of a Biological Invasion.* Chicago: University of Chicago Press.

Meisch, M. V. 1985. *Gambusia affinis affinis.* In *Biological Control of Mosquitoes*, ed. H. C. Chapman, 3–17. American Mosquito Control Association, bull. no. 6, American Mosquito Control Association, Fresno, Calif.

Memmott, J., Fowler, S. V., and Hill, R. L. 1998. The effect of release size on the probability of establishment of biological control agents: Gorse thrips (*Sericothrips staphylinus*) released against gorse *(Ulex europaeus)* in New Zealand. *Biocontrol Science and Technology* 8:103–15.

Murray, J., Murray, E., Johnson, M. S., and Clarke, B. 1988. The extinction of *Partula* on Moorea. *Pacific Science* 42:150–53.

Nagler, J. J., Bouma, J., Thorgaard, G. H., and Dauble, D. D. 2001. High incidence of a male-specific genetic marker in phenotypic female Chinook salmon from the Columbia River. *Environmental Health Perspectives* 109:67–69.

Nechols, J. R., Kauffman, W. C., and Schaefer, P. W. 1992. Significance of host specificity in classical biological control. In *Selection Criteria and Ecological Consequences of Importing Natural Enemies*, ed. W. C. Kauffman and J. R. Nechols, 41–52. Lanham, Md.: Entomological Society of America.

Newsome, A. 1990. The control of vertebrate pests by vertebrate predators. *Trends in Ecology and Evolution* 5:187–91.

Nicholson, A. J., and Bailey, V. A. 1935. The balance of animal populations. Part I. *Proceedings of the Zoological Society of London* 3:551–98.

Norgaard, R. B. 1988. Economics of the cassava mealybug *(Phaenacoccus manihoti;* Hom.: Pseudococcidae) biological control program in Africa. *Entomophaga* 33:3–6.

Obrycki, J. J., Elliot, N. C., and Giles, K. L. 2000. Coccinellid introductions: Potential for and evaluation of nontarget effects. In *Nontarget Effects of Biological Control*, ed. P. A. Follet and J. J. Duan, 127–45. Boston, Mass.: Kluwer Academic Publishers.

Ogle, C. C., Cock, G.D.L., Arnold, G., and Mickleson, N. 2000. Impact of an exotic vine *Clematis vitalba (F. Ranunculaceae)* and of control measures on plant biodiversity in indigenous forest, Taihape, New Zealand. *Australian Ecology* 25:539–51.

Onstad, D. W., and McManus, M. L. 1996. Risks of host range expansion by parasites of insects. *BioScience* 46:430–35.

Oogjes, G. 1997. Ethical aspects and dilemmas of fertility control of unwanted wildlife: An animal welfarist's perspective. *Reproduction, Fertility, and Development* 9:163–67.

Park, D., Hempleman, S. C., and Propper, C. R. 2001. Endosulfan exposure disrupts pheromonal systems in the red-spotted newt: A mechanism for subtle effects of environmental chemicals. *Environmental Health Perspectives* 109:669–73.

Pelley, J. 1997. Deformities in Minnesota frogs linked to water in new study. *Environmental Science and Technology* 31 (12): 552A.

Pemberton, R. W. 1995. *Cactoblastis cactorum* (Lepidoptera: Pyralidae) in the

United States: An immigrant biological control agent or an introduction of the nursery trade? *American Entomologist* 41:230–32.

Pickett, C. H., Ball, J. C., Casanave, K. C., Klonksy, K., Jetter, K. M., Bezark, L. G., and Schoenig, S. E. 1996. Establishment of the ash whitefly parasitoid *Encarsia inaron* (Walker) and its economic benefits to ornamental street trees in California. *Biological Control* 6:260–72.

Pimm, S. L. 1987. The snake that ate Guam. *Trends in Ecology and Evolution* 2:293–95.

Relyea, R. A., and Mills, N. 2001. Predator-induced stress makes the pesticide carbaryl more deadly to gray treefrog tadpoles (*Hyla versicolor*). *Proceedings of the National Academy of Sciences* 98:2491–96.

Robinson G. S. 1975. *Macrolepidoptera of Fiji and Rotuma*. Oxon, U.K.: Farringdon Classey.

Rodda, G. H., Fritts, T. H., and Chiszar, D. 1997. The disappearance of Guam's wildlife. *BioScience* 47:565–74.

Roderick, G. K. 1992. Postcolonization evolution of natural enemies. In *Selection Criteria and Ecological Consequences of Importing Natural Enemies*, ed. W. C. Kauffman and J. E. Nechols, 71–86. Lanham, Md.: Entomological Society of America.

Rodger, J. C. 1997. Likely targets for immunocontraception in marsupials. *Reproduction, Fertility, and Development* 9:131–36.

Room, P. M. 1990. Ecology of a simple plant-herbivore system: Biological control of *Salvinia*. *Trends in Ecology and Evolution* 5:74–79.

Room, P. M., Harley, K.L.S., Forno, I. W., and Sands D.P.A. 1981. Successful biological control of the floating weed *Salvinia*. *Nature* 294:78–80.

Room, P. M., and Thomas, P. A. 1985. Nitrogen and establishment of a beetle for biological control of the floating weed *Salvinia* in Papua New Guinea. *Journal of Applied Ecology* 22:139–56.

Roush, R. T. 1990. Genetic variation in natural enemies: Critical issues for colonization in biological control. In *Critical Issues in Biological Control*, ed. M. MacKauer, L. E. Ehler, and J. Roland, 263–88. Andover, U.K.: J. Intercept.

Rupp, H. R. 1996. Adverse assessments of *Gambusia affinis*: An alternate view for mosquito control practitioners. *Journal of the American Mosquito Control Association* 12:155–66.

Sailer, R. I. 1978. Our immigrant insect fauna. *Entomological Society of America Bulletin* 24:3–11.

Salas, J., and Frank, J. H. 2001. Development of *Metamasius callizona* (Coleoptera: Curculionidae) on pineapple stems. *Florida Entomologist* 84:123–26.

Sands, D.P.A. 1997. The safety of biological control agents: Assessing their impact on beneficial and other non-target hosts. *Memoirs of the Museum of Victoria* 56:611–16.

Saunders, G., Coman, B., Kinnear, J., and Braysher, M. 1995. *Managing Vertebrate Pests: Foxes*. Canberra: Australian Government Printing Service.

Savidge, J. A. 1987. Extinction of an island forest avifauna by an introduced snake. *Ecology* 68:660–68.

Sawyer, R. C. 1996. *To Make a Spotless Orange: Biological Control in California*. Ames: Iowa State University Press.

Schettler, T., Soloman, G., Valenti, M., and Huddle, A. 1999. *Generations at Risk: Reproductive Health and the Environment*. Cambridge: MIT Press.

Secord, D., and Kareiva, P. 1996. Perils and pitfalls in the host specificity paradigm. *BioScience* 46:448–53.

Simberloff, D., Schmitz, D. C., and Brown, T. C., eds. 1997. *Strangers in Paradise: Impact and Management of Nonindigenous Species in Florida*. Washington, D.C.: Island Press.

Simberloff, D., and Stiling, P. 1996. Risks of species introduced for biological control. *Biological Conservation* 78:185–92.

———. 1998. How risky is biological control? Reply. *Ecology* 79:1834–36.

Simmonds, F. J., and Hughes, I. W. 1963. Biological control of snails exerted by *Euglandina rosea* (Ferrussac) in Bermuda. *Entomophaga* 8:219–22.

Sinclair, A.R.E., Olsen, P. D., and Redhead, T. D. 1990. Can predators regulate small mammal populations? Evidence from house mouse outbreaks in Australia. *Oikos* 59:382–92.

Singer, P. 1997. Neither human nor natural: Ethics and feral animals. *Reproduction, Fertility, and Development* 9:157–62.

Smith, H. S. 1919. On some phases of insect control by the biological method. *Journal of Economic Entomology* 12:288–92.

———. 1935. The role of biotic factors in the determination of population densities. *Journal of Economic Entomology* 28:873–98.

Souder, W. 2000. *A Plague of Frogs: The Horrifying True Story*. New York: Hyperion.

Stiling, P. 1990. Calculating the establishment rates of parasitoids in classical biological control. *American Entomologist* 36:225–30.

Stiling, P., and Simberloff, D. 2000. The frequency and strength of nontarget effects of invertebrate biological control agents of plant pests and weeds. In *Nontarget Effects of Biological Control*, ed. P. A. Follet and J. J. Duan, 31–43. Boston, Mass.: Kluwer Academic Publishers.

Strand, M. R., and Obrycki, J. J. 1996. Host specificity of insect parasitoids and predators. *BioScience* 46:422–29.

Strong, D. R. 1997. Fear no weevil? *Science* 277:1058–59.

Strong, D. R., and Pemberton, R. W. 2001. Food webs, risks of alien enemies, and reform of biological control. In *Evaluating Indirect Ecological Effects of Biological Control*, ed. E. Wajnberg, J. K. Scott, and P. C. Quimby, 57–79. Wallingford, U.K.: CABI Publishing.

Sutherst, R. W., Floyd, R. B., and Maywald, G. F. 1995. The potential geographical distribution of the cane toad, *Bufo marinus* L. in Australia. *Conservation Biology* 9:294–99.

Sutherst, R. W., and Maywald, G. F. 1985. A computerized system for matching climates in ecology. *Agriculture, Ecosystems, and Environment* 13:281–99.

Syrett, P., Briese, D. T., and Hoffmann, J. H. 2000. Success in biological control of terrestrial weeds by arthropods. In *Biological Control: Measures of Suc-*

cess, ed. G. Gurr and S. Wratten, 189–230. Dordrecht, Netherlands: Kluwer Academic Publishers.

Syrett, P., Smith, L. A., Bourner, T. C., Fowler, S. V., and Wilcox, A. 2000. A European pest to control a New Zealand weed: Investigating the safety of heather beetle, *Lochmaea suturalis* (Coleoptera: Chrysomelidae) for biological control of heather, *Culluna vulgaris. Bulletin of Entomological Research* 90:169–78.

Tenner, E. 1996. *Why Things Bite Back: Technology and the Revenge of Unintended Consequences.* New York: Knopf.

Todd, K. 2001. *Tinkering with Eden: A Natural History of Exotics in America.* New York: W. W. Norton .

Tothill J. D., Taylor T.H.C., and Paine R. W. (1930). *The Coconut Moth in Fiji: A History of Its Control by Means of Parasites.* London, U.K.: Imperial Bureau of Entomology.

Tyndale-Biscoe, C. H. 1995. Vermin and viruses: Risks and benefits of viral-vectored immunosterilization. *Search* 26:239–44.

Van Driesche, R. G. 1994. Classical biological control of environmental pests. *Florida Entomologist* 77:20–33.

Van Driesche, R. G., and Bellows, T. S., Jr. 1993. *Steps in Classical Arthropod Biological Control.* Lanham, Md.: Entomological Society of America.

———. 1996. *Biological Control.* New York: Chapman and Hall.

Van Driesche, R. G., Heard, T., McClay, A., and Reardon, R. 1999. Proceedings of Session: Host Specificity Testing of Exotic Arthropod Biological Control Agents—The Biological Basis for Improvement in Safety. Morgantown, W.Va.: USDA Forest Service Forest Health Technology Enterprise Team.

Van Driesche, R. G., and Hoddle, M. S. 1997. Should arthropod parasitoids and predators be subject to host range testing when used as biological control agents? *Agriculture and Human Values* 14:211–26.

———. 2000. Arthropod biological control: Measuring success step by step. In *Biological Control: Measures of Success,* ed. G. Gurr and S. Wratten, 39–75. Dordrecht, Netherlands: Kluwer Academic Publishers.

Van Driesche, J., and Van Driesche, R. 2000. *Nature Out of Place: Biological Invasions in a Global Age.* Washington, D.C.: Island Press.

Vitousek, P. M., D'Antonio, C. M., Loope, L. L., and Westbrooks, R. 1996. Biological invasions as global environmental change. *American Scientist* 84:468–78.

Walton, W. E., and Mulla, M. S. 1991. Integrated control of *Culex tarsalis* larvae using *Bacillus sphaericus* and *Gambusia affinis:* Effects on mosquitoes and nontarget organisms in field mesocosms. *Bulletin of the Society of Vector Ecology* 16:203–21.

Williams, C. K. 1997. Development and use of virus-vectored immunocontraception. *Reproduction, Fertility, and Development* 9:169–78.

Williamson, M., and Fitter, A. 1996. The varying success of invaders. *Ecology* 77:1661–66.

Wilson, E. O. 1997. Foreword. In *Strangers in Paradise: Impact and Management of Nonindigenous Species in Florida,* ed. D. Simberloff, D. C. Schmitz, and T. C. Brown, ix–x. Washington, D.C.: Island Press.

Withers, T. M., Browne, L. B., and Stanley, J. 1999. *Host Specificity Testing in Australasia: Towards Improved Assays for Biological Control.* Coorparo, Queensland: The State of Queensland, Department of Natural Resources.

Policy-Related Matters

14 Overview

Anthony C. Steyermark

Careful consideration of the best available scientific data is supposed to be a central feature of the decision-making processes relating to the protection and management of endangered, threatened, or exotic species. The major laws governing these matters in the United States require this consideration. For example, the federal Endangered Species Act of 1973, with amendments, mandates that "The Secretary [of the Interior] shall make determinations [of endangered species and threatened species] solely on the basis of the best scientific and commercial data available" and "shall designate critical habitat . . . on the basis of the best scientific data available and after taking into consideration the economic impact, and any other relevant impact, of specifying any particular area as critical habitat" (16 U.S.C. 1533).

Successful policy and management decisions relating to major aspects of floral and faunal conservation within the United States are therefore legally required to integrate scientific understanding of biology with such things as features of habitats, economics, patterns of land use, attention to indigenous peoples, societal attitudes, and even military needs. On this basis, it is not surprising that, because of often competing or conflicting interests, conservation plans are not always easy to develop, and success is never guaranteed. Because of the uncertainty in both desired and undesired outcomes, conservation plans can be treated as real-world experiments.

Conservation-plan experiments usually have multiple tiers ranging from the level of deciding which questions need to be asked and answered to best help given species, to the level of deciding how best to protect entire ecosystems that span international boundaries. Such a variety of approaches toward making and implementing policy are illustrated in this section. The chapters describe, with examples, how policy decisions themselves are experiments that usually involve multiple disciplines, multiple organizations, even multiple countries. The scientific databases relevant to the examples chosen all include information deriving from experimental studies. The general theme is that successful conservation policies are truly integrative efforts, where outcomes are far from certain, and approaches are imaginative and varied.

Integrative policy experiments are neither easy nor simple. They represent forward thinking in conservation problem-solving, based on the notion that conservation issues are solved not by biologists and other scientists alone but rather by societies as a whole. The key is identifying and communicating with other stakeholders or potential contributors to conservation plans, be they scientists from other biological disciplines or other countries, government officials, or members of the private sector.

Although missed opportunities in policy-setting can both be disappointing and harmful to endangered species, James Spotila and Harold Avery show that such setbacks teach lessons well worth learning. Compromise can often be unsatisfying, leaving none of the parties happy. However, when cooperation and compromise fail, the result is often worse than it would have been with the compromise.

Such was the result when the United States Army determined that it was necessary to expand its National Training Center at Fort Irwin, California, which also happens to be the habitat of the western Mojave population of the threatened desert tortoise. A deadlock developed concerning interpretations of scientific data, and scientists missed the opportunity to have a major impact on the expansion process. Spotila and Avery, in a valuable list of lessons learned, point out that species and ecosystem conservation can be accomplished only in the real world. Just as one learns from scientific experiments that do not work, one must learn from policy efforts that do not succeed.

Integration of diverse agendas and expertise into a complete conservation management policy is also the main theme in John Rodger's chapter. Although good science can aid in forming conservation policies, resource management cannot be accomplished by biologists alone. Scientific approaches to conservation issues are often fragmented, lacking critical components needed to ensure practical outcomes. Rodger shows that successful resource conservation requires the integration of scientific data, local and national needs, and industry resources. Taking a page from the construction and engineering industries' model for integrating expertise, resources, and organizational agendas, the Australian government's Cooperative Research Centres (CRC) Programme aims to link national research efforts with the needs and expertise of industries and with specific environmental needs at the national level. Rodger highlights several case studies from the work of the Marsupial CRC. The successes of the Marsupial CRC emphasize the fact that resource management is a multidisciplinary art and that integration of science, industry, and government needs and strengths can lead to significant conservation accomplishments.

Experimental science and multidisciplinary research efforts are key elements in the cases presented by David Wildt. Wildt argues that research strategies in wildlife biology need to be prioritized toward endangered species; hypothesis driven and experimentally based; maximized through interdisciplinary efforts; and connected to capacity building, or the strengthening of science relative to wildlife issues. Successful captive-reproduction efforts for cheetahs, black-footed ferrets, and giant pandas emphasize the roles of integrative research efforts and experimental biology in the conservation of endangered species. Wildt's description of the breeding program for giant pandas demonstrates how the integration of experimental methods, multidisciplinary research efforts, policy decisions by both the Chinese and the American governments, and international partnerships have led to successful captive breeding of this flagship species.

Finally, Leo Braack's chapter details an experimental approach toward ecosystem management on a grand scale, with a case study of African international park management initiatives. Good fences may make good neighbors, but political boundaries often do not make for healthy ecosystem function. In Africa, political, social, and economic concerns have often superseded environmental ones. The result has been the devastation of whole ecosystems. Braack integrates the history of conservation in Southern Africa, the biological aspects of ecosystem function, and the social, economic, and political issues involved in species conservation. The result is a unique look at the formation of transfrontier parks in Africa. Because of often disparate political agendas, maintaining the success of such cooperative conservation areas requires "sustained nurturing." However, these bold experiments in conservation policy may be the last hope for the protection of whole ecosystems, not just the species within them.

15 The Army and the Desert Tortoise

Can Science Inform Policy Decisions?

James R. Spotila and Harold W. Avery

SUMMARY

The United States Army wishes to expand its National Training Center (NTC) at Fort Irwin, California, in the interests of national security. Environmentalists opposed the expansion because it would destroy habitat in the western Mojave Desert and because it would impact the threatened desert tortoise. After 14 years of deadlock between the Army, the Department of the Interior, and environmentalists, the Army undertook a new strategy. It attempted to tie the expansion of Fort Irwin to the recovery of the desert tortoise. Instead of treating the tortoise as an obstacle, the Army sought to aid in its recovery. When the Army and the Department of the Interior reached a deadlock in negotiations, they turned to a panel of scientists and desert managers to provide recommendations. The panel supported the expansion on the condition that specific measures be taken in the western Mojave Desert and at Fort Irwin to protect the tortoise. The specific recommendations were rejected by user groups, the Department of the Interior, and Congress, in part because the panel was not unanimous and because the actions of its members undermined its findings in public. Congress mandated expansion of the NTC and adopted legislation to implement it. Scientific input became irrelevant when the panel failed to provide clear and unambiguous recommendations to policymakers, politicians, and the public. Nevertheless, the panel recommendations may be useful in mitigating the effects of the expansion.

INTRODUCTION

One of the most important questions in conservation biology is Can science inform policy decisions? or perhaps more honestly, Why doesn't science inform policy decisions? This chapter addresses these questions in the context of the controversy over the U.S. Army's proposed expansion of Fort Irwin in the Mojave Desert of California and the protection of the desert tortoise, *Gopherus agassizii*, a species listed as threatened by the U.S. Fish and Wildlife Service (USFWS). We bring unique insights to this issue because one of the authors (Spotila) served as chief environmental scientist of the Department of the Army (DA) from 1998 to 2000 and the other author (Avery) was a biologist with the Bureau of Land Management (BLM), the National Biological Survey and Service (NBS), and the Biological Resources Division (BRD) of the United States Geological Survey (USGS) from 1990 to 2001. Spotila participated in many of the science discussions and meetings with biologists from the Department of the Interior (DOI), the Army, the Department of Defense (DOD), and academia as well as in policy meetings and negotiations at the highest levels within the Army and between the Army and the DOI. Avery was a leading tortoise biologist, conducting research with the BRD and working with DOI managers, biologists, and environmentalists concerned with protection of the tortoise. He also was involved in environmental education with tortoise and turtle clubs and enthusiasts.

The Army's position is that the expansion of the National Training Center (NTC) at Fort Irwin is essential to maintaining operational readiness for national security. The NTC is the only instrumented training area in the world that is suited for live training of heavy brigade-size task forces and that provides maximum possible testing of what the Army calls its "vision for the brigade transformation." This transformation is the change from a division fighting as a series of battalions to a self-contained brigade (made up of three battalions of tanks and mechanized infantry) fighting as a unit more like the armored combat commands used in Europe in World War II (Houston 1997) than the heavy divisions that would have defended Germany from an attack by the former Soviet Union in the 1970s and 1980s.

The NTC further believes that it is a good environmental steward of its lands because it has strongly supported the Endangered Species Act (ESA) and has spent more than $1.5 million a year in studying and protecting the desert tortoise, the Lane Mountain milkvetch, and other species. In addition, it has a strong program of environmental protection that includes pollution prevention, compliance with all environmental laws, restoration of damaged sites (through its Integrated Training Area Management [ITAM] program), conservation, and its occupational health and safety and explosives safety programs (U.S. Army 1992).

The environmental community contends that the desert tortoise is declining rapidly and that the expansion of Fort Irwin will impact some of the best habitat in the Mojave Desert that contains some of the highest-density populations of this species. Opponents of expansion use the decline of the tortoise, the fact that the proposed expansion area is described as critical habitat in the Desert Tortoise Recovery Plan (Fish and Wildlife Service 1994), the destruction of habitat in training areas of Fort Irwin (Krzysik 1997), and the demise of the Soviet Union as reasons why the expansion should not take place.

In this chapter we review the history of the expansion of Fort Irwin, the basis for the Army's need for expansion, the claim that the desert tortoise is declining, and the history of the recent government actions toward expansion. In addition, we discuss the roles of science, politics, and environmental action in the formation of policy in this matter. We believe that the Army does need to expand Fort Irwin to carry out its mission of preparing for the next war or, even better, of being prepared to prevent the next war. We also believe, despite the limited and sometimes anecdotal nature of the data, that the desert tortoise is in decline and that the Mojave Desert is being negatively impacted by human activities. The real issue, therefore, is whether the expansion of Fort Irwin should be viewed as a problem for the tortoise or as an opportunity for its recovery. Is it a matter of the Army or the desert tortoise, or a matter of the Army and the desert tortoise?

TRAINING AND READINESS

Training and readiness is the mantra of the U.S. Army. The United States, like many nations, has a long history of on-the-job-training for its military, especially the Army. This approach cost money and lives in the past, as can be illustrated by a few examples. The greatest loss of American lives in war occurred in the Civil War, in part because neither side in the conflict had adequate strategic or tactical training. The infantry tactics handbook was out-of-date for the increased power of the weapons available to the armies, and the officers and men were ill trained. The famous battles of Antietam and Gettysburg may be familiar to some, but less familiar battles are also instructive. At Stones River, a superior Union army was almost defeated owing to a tactical error made by a commander before the battle. The battle was won only after several determined defensive stands by well-trained soldiers under General Sheridan and artillery directed by Captain Mendenhall (Cozzens 1990). Similar examples occurred at the Battles of Iuka and Corinth, Chickamauga, and Lookout Mountain and Missionary Ridge (Cozzens 1997, 1992, 1994).

The United States again learned the value of training in World Wars I and II. Concerning the battles in North Africa in 1942 and 1943, General Eisenhower wrote: "This was the great Allied lesson of Tunisia; equally important, on the technical side, was the value of training. Thorough technical, psychological, and physical training is one protection and one weapon that every nation can give to its soldiers before committing them to battle, but since war always comes to a democracy as an unexpected emergency, this training must be largely accomplished in peace. Until world order is an accomplished fact and universal disarmament a logical result, it will always be a crime to excuse men from the types and kinds of training that will give them a decent chance for survival in battle. Many of the crosses standing in Tunisia today are witnesses to this truth" (Eisenhower 1948).

Only a few years later, the United States again learned the value of training, this time in Korea. The first troops, Task Force Smith, sent to stop the invasion of the North Koreans, were wiped out, which led to the Army mantra "No More Task Force Smiths." Commenting on the Korean War, Fehrenbach (2000) stated: "There had been many brave men in the ranks, but they were learning that bravery of itself has little to do with success in battle. On line, most normal men are afraid, have been afraid, or will be afraid. . . . Only when superbly trained and conditioned against the shattering experience of war, only knowing almost from rote what to do, can men carry out their tasks come what may. And knowing they are disciplined, trained, and conditioned brings pride to men—pride in their own toughness, their own ability; and this pride will hold them when all else fails."

In Vietnam, the U.S. Army relearned the value of training yet again. This long history supports Dunnigan's (1982) conclusion that armies remember no more of the past than do their oldest members and that peacetime armies fight each other in tragicomic bloodbaths until one side relearns the lessons of warfare before the other and eventually wins. Thus, the U.S. Army established the NTC at Fort Irwin in 1981 to ensure that its soldiers were always trained and ready to fight with the latest equipment, strategy, and tactics so that they could fight and win the next war with minimum loss of life. This approach was proven correct in the Gulf War in 1991 when soldiers reported that their training at Fort Irwin had been tougher than real battle and had prepared them for victory (Gordon and Trainor 1995; Clancy and Franks 1997, 356–59; H. R. McMaster, personal communication, 1998).

THE REASON FOR EXPANSION OF FORT IRWIN

The NTC at Fort Irwin (Figure 15.1) is the only Army installation with the size and instrumentation capability to train large units (brigade-size) to

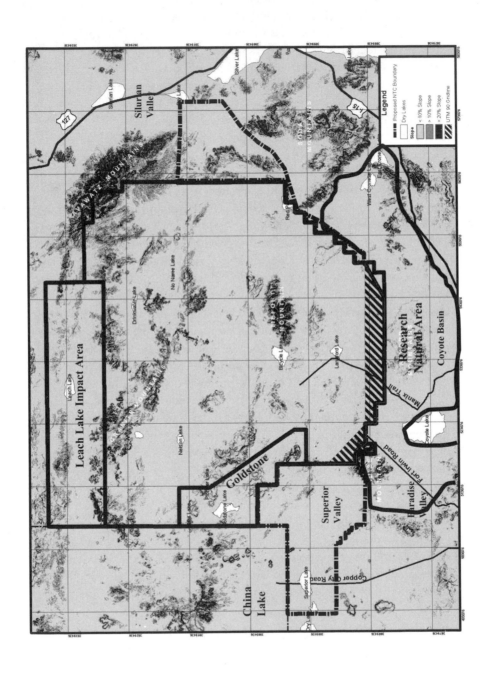

operate with all assigned equipment over realistic distances. The following descriptions are based on documents provided by Fort Irwin. Additional information is available on its Web site (http://www.irwin.army.mil/). The NTC provides challenging, realistic combined arms and joint air–ground training under conditions that mirror combat conditions. Originally the NTC trained Army units to fight Soviet-style Cold War threats. Now it trains tank units in armored brigades and light infantry as well as special operations units on flat terrain and in mountains and gorges. Heavy vehicles are restricted to areas with slopes of less than 20% and without steep gorges and drop offs.

By using global positioning system transponders to track all vehicles, using multiple integrated laser engagement system training equipment to target hits on vehicles and personnel, and recording all radio transmissions, the training staff can record all aspects of practice battles and play them back to soldiers at each unit level (platoon to brigade) for discussions of the lessons learned from each battle. Training units rotate into Fort Irwin for 35–58 days and fight battles in training scenarios that take place 24 hours a day for the entire period. The opposing force, the 11th Armored Cavalry Regiment, is thoroughly trained to operate at any level using enemy organizations, tactics, and doctrine (theory of warfare and method of operation). Observer controllers spend time with rotational units, observing, analyzing, and counseling on unit performance throughout the preparation and execution of all missions. The controllers then advise rotational units on how to improve their battlefield operations. This system, which works like an oral exam for a graduate student, takes place over several weeks and includes the assessment of simulated casualties from mines, artillery, air strikes, tanks, rifles, and chemical weapons. In the

Figure 15.1 *(opposite)*. The National Training Center (NTC) at Fort Irwin, California. The NTC occupies 260,106 ha (642,730 acres) in the western Mojave Desert north of Barstow, Calif. The area below the 90-grid line (hatched area) on the map is off-limits to the Army for training owing to the presence of the desert tortoise. The area just below the 90-grid line shows the approximate boundaries of a research natural area established in the authorizing legislation for the withdrawal of training land for Fort Irwin (H.R. 4577). China Lake is the Navy's air-warfare training station. Goldstone is the NASA deep-space tracking station and is off-limits to the Army for training. Leach Lake is an impact area for artillery and tank shells as well as bombs ranging in size from 500 lbs to over 2000 lbs. It is not available for training owing to the danger of unexploded ordnance. Black lines represent roads.

last part of field training, units conduct live-fire exercises against computer-controlled targets in the northern portion of Fort Irwin. The use of live ammunition makes some areas (e.g., the Leach Lake area) unsuitable for on-the-ground training because they contain large amounts of unexploded ordnance.

Owing to new requirements to train larger units (7000 soldiers) and multiple units simultaneously and to train for new operational environments (e.g., nonlinear, noncontiguous threats; elite irregular forces; operation in built-up areas), Army units now need to train across larger areas. Two land-use requirement studies indicated a shortfall of approximately 193,000 acres for training at Fort Irwin because of changes in doctrine, equipment, speed, and tactics. The Army proposed to expand Fort Irwin to meet this need.

THE DECLINE OF THE DESERT TORTOISE

After the publication of Woodbury and Hardy's (1948) classic study of the natural history and ecology of the desert tortoise, there was a hiatus of about 30 years in which little research occurred on this animal. In the 1970s and 1980s the BLM, the California Energy Commission, and Southern California Edison supported research into the biology and life history of the desert tortoise (Turner 1986; Pearson 1986). Twenty-seven permanent study plots were established in the Mojave Desert after 1970. The BLM supported research at these sites that determined the density, size-class structure, age-class structure, and sex ratios of tortoise populations and attempted to determine cause of death for animals that were found dead (Berry and Nicholson 1984). These sites continue to be monitored occasionally by USGS scientists, and their data suggest that the desert tortoise is in serious decline in the Mojave Desert (K. H. Berry, letter to Molly Brady, BLM, Riverside, Calif., 1996, exhibit 21, *Ranchers vs. BLM 2001*). Unfortunately many of these plots are surveyed only once every 3 to 5 years, and therefore there is little information about the demographic processes at work in these populations. Data indicate only changes in population size on the sites. Berry (1984) edited a report to the USFWS on the status of the desert tortoise throughout its range in the United States, and Berry (1986) summarized research on the desert tortoise in California. Berry and Nicholson (1984) developed maps of tortoise density based on counts of tortoise sign (live individuals, carcasses, scat, cover sites, tracks, burrows, drinking sites, and courtship rings) along line transects. These maps indicated that high densities of desert tortoises still existed in some areas of the Mojave Desert. However, the maps indicated low densities of tortoises in and around Fort Irwin, including the Superior and Paradise Valleys. These results confirmed those of Luckenbach (1982).

In 1990 the USFWS listed the Mojave Desert population of the desert tortoise as a threatened species under the ESA because desert tortoise numbers were declining precipitously in many areas (55 *Federal Register* 12178–12191). The declines were attributed to direct and indirect human-caused mortality coupled with the inadequacy of existing regulatory mechanisms to protect the species and its habitat (Fish and Wildlife Service 1994). These conclusions were based on the best available scientific information at that time, but Bury and Corn (1995) have noted that the information was anecdotal and dependent on circumstances: there were no published or unpublished data that supported claims of long-term tortoise decline in the entire Mojave Desert. Although there are data (mostly unpublished) that do indicate recent declines in some populations (especially in the western Mojave Desert), assigning a cause to that decline is often impossible. Some researchers (Berry 1990; Jacobson 1994) have stated that upper-respiratory-tract disease is a major cause of decline. Others have proposed drought as the cause (Corn 1994). Unfortunately, there have been no long-term (20+ years) demographic or life-history studies of desert tortoise populations. Demographic and life-history studies on other turtles (Frazer, Gibbons, and Greene 1990; Congdon, Dunham, and van Loben Sels 1993, 1994) have provided important insights into factors that control their populations. Such information will be needed if an adaptive management approach is to be used to achieve recovery of the desert tortoise.

Since 1990 considerable effort has gone into surveys, using the Berry and Nicholson (1984) or BLM transect methodology, to determine the density of desert tortoises in and around Fort Irwin as well as in the rest of the western Mojave Desert. The BLM conducted these surveys with funding from the NTC. Opponents of expansion used data from these surveys to claim that the expansion would eliminate a substantial portion of the desert tortoise population in the Mojave Desert. However, there are serious problems with the transect method, and density estimates obtained with it are not valid (Turner et al. 1982; Freilich et al. 2000). The BLM (Berry and Nicholson 1984) transect method gives information on the presence or absence of tortoises but not on the densities, or even relative densities, of tortoises. Therefore, claims of negative effects from expansion were controversial, and scientific input into the policy process was needed.

EXPANSION: THE EARLY YEARS

In 1985 the NTC began to plan its expansion. In 1988 it signed a memorandum of agreement with the BLM to allow BLM to be the proponent of the environmental impact statement for the expansion. At that time, the Army did not anticipate that any serious environmental issues would prevent expansion. In 1989 the NTC proposed a broad expansion to the south (Figure 15.2A). However, in 1990 the USFWS listed the desert tortoise as threatened in the Mojave Desert and in 1991 issued a draft jeopardy opinion for the southern expansion. It also placed 23,600 acres along the southern portion of Fort Irwin off limits to training. In 1993 the NTC proposed an eastern expansion into the Silurian Valley, and in 1997 the BLM issued a draft environmental impact statement stating that the Army preferred the eastern expansion. At that time, user groups (primarily off-highway vehicle users) objected to that plan because of the loss of recreational land that would have resulted. The BLM responded in late 1997 by proposing a modified southern expansion that included the land that was off-limits for training, as well as land to the south (Coyote Basin) and east (small part of Silurian Valley) of Fort Irwin. In response, in 1998 the NTC proposed a modified southern expansion that avoided most of the Coyote Basin, which was not suitable for training, but included the small portion of the Silurian Valley, the off-limits land, and the Paradise and Superior Valleys to the west (Figure 15.2B).

In 1998 and early 1999, there was extensive discussion within the Army and within the DOI concerning this latest proposal, but the two were not talking to each other. At best they were talking past each other. After several briefings and informal discussions with staff of the House of Representatives and Senate, the Army and the DOI were still unable to reach agreement. In March 1999 Congressman Jerry Lewis, chairman of the military subcommittee of the House Appropriations Committee, expressed concerns to the secretary of the Interior and the secretary of the Army about the slow progress of the expansion process (his congressional district includes Fort Irwin).

Until that time the Army took a reactive approach to concerns about the desert tortoise. The attitude of some in the Army could be characterized as "Tell us what it will cost for mitigation. Let us expand, and we will pay the bill. Let Fish and Wildlife figure out the problem. That is their expertise; ours is to train soldiers." As a result the Army was always in a defensive posture and could not see how to move the issue toward closure. At the same time the attitude in the Department of the Interior was essentially "You can't trust the Army. All they want to do is run tanks. The expansion

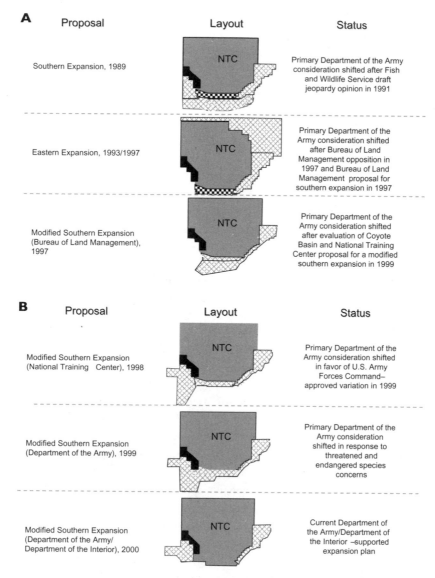

Figure 15.2A and B. The history of the expansion proposals for the National Training Center (NTC) at Fort Irwin. Crosshatched areas represent the proposed expansion land; the black area is the NASA Goldstone Tracking Station, which is off-limits to army training; and the checkered area represents the portion of Fort Irwin that was off-limits to training because of the tortoise.

area is critical to the desert tortoise, and the tortoise is in steep decline. There is nothing we can do about it, so there is nothing the Army can do to mitigate the effects of its expansion."

EXPANSION: A CHANGE IN APPROACH

During 1998 and 1999, there was a dramatic change in the Army's approach toward resolving the expansion impasse. Brigadier General Lovelace, the director of training in the Office of the Deputy Chief of Staff for Operations and Plans (ODCSOPS) of the Army at the Pentagon, and Brigadier General Webster, the commander of the NTC at Fort Irwin, requested and received a series of briefings on the biology of the desert tortoise, its past decline, and its current status. The generals also supported an ecosystem approach (Leslie et. al 1996) to the management of the desert tortoise at Fort Irwin that had been developed by staff members in the Office of the Assistant Secretary of the Army for Installations and Environment (OASAIE). That plan was based on discussions with a team of biologists who had several years of experience with desert tortoise research and conservation. Discussions of mitigation within the Army focused on both off-site and on-site mitigation measures, including research on the biology of the desert tortoise, an understanding of which would be critical to reversing the tortoise's decline. The Army planned to spend up to $35.4 million on this effort, which included the purchase of land for the tortoise. At about the same time, Ray Clark, principal deputy assistant secretary in the OASAIE, directed that the expansion plan include an effort to link expansion to the recovery of the desert tortoise. That is, mitigation should be tied to improving the status of the desert tortoise and not viewed simply as payment of damages. He insisted on a more creative and comprehensive approach to both expansion of Fort Irwin and recovery of the desert tortoise.

It is important to note that the Army staff, led by the chief of staff of the Army (a four-star general), directs the activities of Army units and installations: that is, the uniformed portion of the Army that actually carries out the duties of the Army. The Army Secretariat is the civilian staff of the secretary of the Army and provides civilian oversight for the actions of the uniformed staff. The secretary, assistant secretaries, and principal deputy assistant secretaries are political appointees and ensure that the Army follows the policies of the administration. The new attitude of the uniformed leaders of the Army staff and strong leadership (oversight) by the Army Secretariat led to a more active approach to the expansion, and this approach was then fully supported by the secretary of the Army.

In April 1999 the Department of the Army (DA) proposed a modified southern expansion that included the land in the 1998 NTC proposal along with a larger portion of the slope of the Coyote Basin to allow more room for vehicle maneuvers (Figure 15.2B). Discussions with the DOI focused on boundaries, mitigation, and language in the proposed expansion legislation. Despite frequent and frank discussions between the agencies, the DA and DOI were unable to agree on boundaries for expansion. The main problem was the DOI's claim that Paradise Valley and the slope of the Coyote Basin were the best tortoise habitat in the western Mojave Desert and that, therefore, no mitigation would compensate for the loss of those areas. The DA claimed that not many tortoises were involved, that they could be relocated or left in place, and that mitigation should focus on areas in which tortoise densities were still high but threatened by human activity.

EXPANSION: CONGRESS PUSHES THE PROCESS

In July 1999 Congressman Lewis held up funding ($30 million) in the House Appropriations Committee for land buy-outs in the western Mojave Desert and requested draft legislation from the DA for the expansion of Fort Irwin. He also added $19 million to the 2000 budget for the purchase of land for the expansion. In September 1999 the BLM released information on the densities of tortoises (California State Director to Director of BLM, letter including two maps, in author's personal collection) in the expansion areas based on line transect counts of tortoise sign). The BLM claimed that these counts provided estimates of tortoise densities and indicated the importance of the expansion area for the tortoise in the western Mojave Desert. Unfortunately these data were interpreted incorrectly and released prematurely, in violation of a DA–BLM agreement that the DA would review the data before public release. The data on the maps were presented as total corrected tortoise sign and interpreted to mean densities of tortoises. However, as discussed above (see Turner, et. al. 1982; Freilich et al. 2000), the BLM transect method does not give reliable data on tortoise densities; it indicates only the presence or absence of tortoises. On September 1, 1999, Senator Dianne Feinstein of California sent a letter to the president objecting to the expansion of the NTC on the basis of information released by the BLM. These events breached the trust that had been built up between the DA and DOI over the previous year of negotiations and led to a crisis. The expansion was held up, as was the money for land buy-outs in the Mojave Desert.

In the fall of 1999, the DA and DOI held a series of meetings to resolve this latest impasse. Direction from the highest levels of the DOD and DOI was to find a solution to expansion and the protection of the western Mojave Desert. The DA took the approach advocated by Principal Deputy Assistant Secretary Clark to expand the NTC while recovering the desert tortoise. The DA proposed tortoise reserves, tortoise relocation or resettlement, vaccine development for the upper-respiratory disease afflicting tortoises, reconstruction of Fort Irwin road to include protection of tortoises and establishment of a National Academy of Sciences review to reassess the entire situation. However, the Army medical laboratories determined that a vaccine was impractical, and the DA and DOI determined that a National Academy of Sciences review was also impractical in the time frame available to resolve the issue. In November, Congressman Lewis and Senator Feinstein agreed to link funding for land purchases in the Mojave Desert to expansion of Fort Irwin. Congress appropriated $10 million for land purchases, and Senator Feinstein agreed that the DA and DOI should find a way to expand Fort Irwin in an environmentally sound manner.

During all of this time, a habitat conservation plan for the western Mojave Desert (West Mojave Plan) under Section 10 of the ESA had been under development. The West Mojave Plan was supposed to protect the desert tortoise and other threatened, endangered, and sensitive species while allowing for development of the region. Although local and state government, user groups, and others were involved in the development of the plan, the military installations in the western Mojave were not, because they operate under Section 7 of the ESA and are not required to obtain a permit for their actions under Section 10. However, since the expansion of Fort Irwin would affect the land available for mitigation in the West Mojave Plan, Congressman Lewis and Senator Feinstein encouraged the Army to coordinate its expansion of Fort Irwin with the development of the West Mojave Plan as much as possible.

EXPANSION: A CHANCE FOR SCIENTIFIC INPUT

In December 1999 the DA and DOI met to discuss the expansion and agreed to form a scientific panel to obtain advice on the impact of expansion on the desert tortoise. With input from the DA, the BLM organized the panel, which comprised three members from BLM, two scientists from the USGS, one manager from California Fish and Game, two civilians from the Army, and four scientists from academia. The panel was charged with answering two questions: Given the proposed expansion of Fort Irwin, can the desert tortoise be recovered in the western Mojave Desert? And if so, what actions are needed to accomplish that recovery? In January 2000 the panel met for two days and concluded that, yes, the tortoise could be recovered, but the required actions would be extensive and very costly.

The panel concluded that the expansion area could be devoted to Army maneuvers, and the tortoise could still be recovered if there was full implementation of the panel's conservation strategy with expansion to the south. The panel reached several additional conclusions. The tortoise was further from recovery than when it had been listed in 1990. Delays in implementing the Desert Tortoise Recovery Plan of 1994 had contributed to the decline of the tortoise. If no action was taken, the tortoise would continue to decline. The expansion of Fort Irwin with no mitigation would jeopardize the recovery of the tortoise. The expansion areas (Paradise and Superior Valleys) were relatively pristine, undisturbed habitat that was essential to the survival of the tortoise. However, if the recovery plan was implemented or if the panel's conservation strategy was implemented, then recovery could occur with expansion. This conclusion was key because for the first time a panel of scientists had concluded that expansion could take place and the tortoise could be recovered in the western Mojave Desert. These findings provided a basis for mitigation of the expansion.

The panel also made several general recommendations. The cost of the proposed mitigation measures should be analyzed. The four desert wildlife management areas (approximately 1.4 million acres) proposed in the recovery plan should be set aside as reserves for the desert tortoise. The reserves should be managed for the immediate recovery and long-term conservation of the desert tortoise. All private land in the reserves (approximately 460,000 acres) should be purchased. There should be a 1% development cap in the reserves. All public land in the reserves should be withdrawn from mining. No new utility corridors should be established in the reserves. More BLM rangers should be hired to patrol the reserves. A route network consistent with tortoise recovery should be designated, and all vehicle use should be limited to designated routes. Parking and camping in the reserves should be restricted to designated areas. Hunting should be allowed in the reserves, but not vehicle races. The panel also made specific recommendations for each reserve and for the Army.

For the Army, the panel recommended that there be no vehicular traffic within 500 m of dry lakes in the expansion area. There should be no off-road access through the reserve south of Fort Irwin along the Manix Trail (Figure 15.1), which is used to move vehicles from a railroad siding to Fort Irwin. The NTC should implement best practices for training to eliminate adverse impacts of wind-blown dust and obscurant smoke on the organisms and environment of the reserve. The NTC should establish an independent advisory group of scientists and stakeholders for peer review of research and conservation activities on Fort Irwin. The NTC should fund several research studies vital to improving the understanding of desert tortoise biology. These studies should include comprehensive demographic and health surveys of tortoises in all proposed expansion areas, a 25-year study of the life history and demography of the desert tortoise for populations in the expansion areas and other reference areas, a 10-year study of the physiology, diseases, and pathology of the tortoise for populations in the expansion areas and other reference areas, a 10-year study of the epidemiology of disease in the tortoise in the western Mojave, and a pilot translocation study to determine the efficacy of relocating healthy tortoises from the proposed expansion area. The panel recommended that NTC remove tortoises from the expansion lands if studies indicated that this was necessary (or establish a provisional off-limits area on expansion lands that would remain in force until the independent advisory group resolved the future of land use). The NTC should establish a desert tortoise study area on or adjacent to Fort Irwin, construct barrier fences along the common boundary between Fort Irwin and the reserve areas, and establish a cooperative desert tortoise adoption center and health program in Barstow, California.

After the panel completed most of its work in January and formulated its recommendations, some members engaged in a series of e-mail exchanges that questioned the value of the recommendations, suggested that the Army should pay nearly $400 million for mitigation, and raised doubts about the effectiveness of the measures already agreed upon. This exchange led to a reconvening of the panel in February to revisit the recommendations. The panel reaffirmed its findings and recommendations. After that meeting, there were additional e-mail exchanges again reconsidering portions of the report. In the end, nine of the twelve panel members agreed to sign the report (LaRue 2000).

In March 2000 the Army accepted the report and proposed buying 153,000 acres of land as mitigation for expansion ($70 million) and spending approximately $2 million per year for 10 years on the required studies. In March 2000 the BLM and USFWS (letters to DOI in the author's personal collection) rejected the conclusions of the report, claiming that there was disagreement among the panel members about its conclusions and that since only nine of twelve members signed the report there was no consensus as to the recommendations. The rejection letters were quickly placed on Internet sites. In April, news articles appeared in local newspapers (e.g., in the *San Bernardino Sun*, April 26, 2000) misrepresenting the report and overstating the requirements of the conservation measures. DOI employees and consultants who were members of the panel and had signed the report disavowed it, stating to the media that the requirements were too strong.

Between April and June 2000, the DA and DOI attempted to draft administration legislation for the expansion that included as many of the panel's conservation measures as possible. In June final efforts were made to formulate an administration proposal. Most of the language was completed, but the DOI calculated that the purchase of all the private land in the proposed reserves would cost about $400 million. The DOI wanted the DA to find the money. The Army did not believe that it was responsible for all the problems of the desert tortoise in the western Mojave Desert, but was willing to discuss some amount over $70 million. Estimates to buy "substantially" all the land as required by the panel report were closer to $100 million than to $400 million, since land in the suburbs of cities and towns and in the expensive western portion of the desert would not have to be purchased. On June 27, 2000, Congressman Lewis and Senator Feinstein stated that the proposal to establish a tortoise reserve of 1 million acres was both unrealistic and unworkable and that a mitigation ratio of 1 to 1 should be used for the expansion (i.e., 1 acre of land should be purchased for the reserve for every 1 acre of land withdrawn for expansion). On July 26, a "sense of the Congress resolution" passed, stating that "prompt expansion of the National Training Center is vital to the national security interests of the United States." The recommendations of the scientific panel were no longer relevant, and the opportunity for scientists to have a major impact on the expansion process was lost.

EXPANSION: A MISSED OPPORTUNITY

Although the scientific panel missed an opportunity to define the conservation measures for the desert tortoise, the panel did an excellent job of evaluating the available information about the biology of the desert tortoise and the threats to its recovery in the western Mojave Desert, given its short time for action. The panel's report was not perfect, but it provided a good basis for development of mitigation for the expansion of Fort Irwin and set a standard that the West Mojave Plan will have to reach or surpass to establish mitigation measures for completion of a habitat conservation plan for development in the desert. However, the failure of the scientists and managers on the panel to broker a deal—that is, find a conservation solution that was both scientifically sound and politically acceptable—made the panel report irrelevant in the process of mitigating the impacts of the expansion of Fort Irwin.

The panel spent too little time debating its findings and recommendations in the on-site meetings. It also failed to go to the actual expansion sites and instead relied on data from the BLM transects around Fort Irwin and on statements of BLM fieldworkers who claimed that the expansion area was pristine habitat and full of tortoises. This failure caused the panel to recommend stringent mitigation measures for expansion. If panel members had visited the sites, they would have discovered that the habitat was not pristine and that it supported very few tortoises. It is also interesting that the panel stated that an evaluation of the available data indicated that the tortoise was in such peril that the measures recommended by the panel were needed even without an expansion of Fort Irwin. The major problem was that without that expansion, there was no mechanism to pay for even a small part of those recommendations.

After the completion of the panel report, the minority claimed that a nine to three vote was not sufficient support for the report and that therefore there was no consensus. However, 75% of the members did support the report, and this figure meets most definitions of a consensus. In addition, two of the three dissenting members were regulatory biologists for the state of California and USFWS and should have been *ex officio* members. They should have abstained from voting since voting could prejudice their regulatory function. Abstention would have resulted in a vote of nine to one. Some members of the panel failed to recognize that they were translating scientific information into policy recommendations. Leaks, inappropriate comments to the media, and second thoughts transmitted by e-mail (thus providing a permanent electronic record) publicized the apparent doubts of the panel members and weakened the conclusions of the report. Some panel members apparently thought that if implemented, the panel report would affect the deliberations of the West Mojave Plan participants. What the panel members took as scientific discourse the media took as evidence for the lack of scientific support for the policy recommended by the panel. Finally, the panel's inability to determine the cost–benefit ratio of the panel recommendations weakened the scientific arguments of the panel. However, although the panel's report was shelved by policymakers, it still affected the expansion process because it kept pressure on the agencies by establishing guidelines for protection of the tortoise during the expansion.

EXPANSION: RACING TOWARD A CONCLUSION

In August 2000 the effort to draft administration legislation failed again, over the details of the language of who controlled the expansion land before training began, environmental provisions, water rights, and other important matters. The Council on Environmental Quality (CEQ) in the Executive Office of the President required agreement on environmental provisions, and the DA and DOI could not agree. In September, at the request of Congressman Lewis, the Army provided draft legislation to his office for the expansion. The Army viewed this as a normal response to a congressional request. The CEQ and the Office of Management and Budget in the Executive Office of the President viewed this action as an end run around proper administrative channels. This difference in views created a crisis. With the approach of the 2000 presidential election, advocates of expansion and advocates for the tortoise in the administration and Congress now agreed that this issue needed to be resolved.

In early October 2000 Secretary of Defense William Cohen and Secretary of the Interior Bruce Babbitt directed their agencies to find a solution to the expansion problem. They established a top-level task force that agreed to hold another meeting of policymakers, scientists, and managers at Fort Irwin later that month. At that meeting General Webster, now the director of training at ODCSOPS, made a dramatic offer. He suggested that the Army would give up expansion into Paradise Valley and the Coyote Basin if the rest of the expansion was approved (Figures 15.1 and 15.2B). This gesture was extraordinary because it meant that the Army was willing to accept less land for training in order to respond to concerns about the desert tortoise. It also meant that DOI could move forward with consideration of mitigation measures that were less stringent than those proposed in the scientific panel's report. Webster's gesture raised the question of whether the Army really needed all the land it had been asking for over the past 15 years. General Webster's answer was yes. However, he calculated that training on part of the expansion land would be better than having expansion blocked for a number of years. He also hoped that an agreement could be made with the Navy to use the southeast portion of the China Lake facility for a staging area like that planned for the eastern portion of the expansion in the Silurian Valley. Such an agreement would free up a larger area of the Superior Valley for battle training. That hope remains a calculated risk because at the time of this writing, no agreement had been reached with the navy. General Webster was the only participant to risk his career to accomplish the twin goals of training and recovery of the tortoise. If the compromise did not work, then he would face retirement. It remains to be seen whether his approach will be sufficient to meet the training needs of the Army. It certainly reduced the impact of the expansion on the desert tortoise and freed up considerable resources for tortoise conservation.

At the October 2000 meeting, the Army proposed that the land south of Fort Irwin that would no longer be in the expansion area should be turned into a reserve for the desert tortoise. At first BLM representatives stated that that land was not particularly valuable for the tortoise. However, after discussion about reallocating the land for expansion of the NTC, they agreed that a tortoise reserve was needed to provide a corridor for tortoises between the western and the eastern Mojave Deserts. That area was then set aside as a research natural area.

Between October and December, the DA and DOI formulated legislation that the administration forwarded to Congress. In December 2000 Congress passed the Miscellaneous Appropriation Act of 2000 (Public Law 106–554), which authorized expansion of Fort Irwin and established a timetable and tasks for the expansion to be completed. President Clinton signed that act into law on December 21, 2000. The act directed the DA and DOI to develop a plan for the expansion of Fort Irwin. In 45 days, a key elements report was due to Congress outlining the major facets of the plan. The USFWS was to provide a preliminary review of the key elements report in another 45 days. Thirty days later, another report to Congress and draft withdrawal language were due. Thirty months were allowed for environmental and endangered species (ESA) actions. The act also authorized up to $75 million for mitigation and $5 million for environmental studies and administrative activities during the planning process.

Between January and July 2001, the reports were produced, meetings took place, and draft legislation was approved by the administration and sent to Congress. In December 2001, the legislation was approved as part of the military appropriations bill. Training on the expansion land will not take place until the Section 7 consultation is completed. The mitigation funds will then be available and tortoise reserves established. A working group from the DA and DOI will be established to recommend to USFWS what mitigation measures should be supported with the $75 million. In addition, it is anticipated that Fort Irwin, in its biological assessment of the training activity, will propose other actions needed to assess the impact of training on tortoises in the expansion area. The process is expected to be completed within the time allowed in Public Law 106–554. There will still be an opportunity for scientific input into the process in the working group deliberations.

LESSONS LEARNED

One of the important things that the Army does after any action is to assess the lessons learned from the action. That process is the basis of training at

Fort Irwin and is used throughout the Army system. Other agencies engage in similar activities. What lessons were learned from the expansion process at Fort Irwin?

1. All conservation is local. No matter what the overall policies of government agencies are, it is the activities of the local facility or local agency office that determine whether conservation will work or not.

2. Although Army commanders knew that conservation of the desert tortoise was important, they initially viewed the tortoise as a problem. Later, when they learned that the Army could improve training for soldiers and simultaneously help in the recovery of the tortoise, they were more willing to work with the DOI to find a solution.

3. Policymakers and politicians tend to dismiss scientists because scientists do not distinguish between their need to discuss scientific uncertainty and their need to provide clear answers within the deadlines set for policy decisions and elections. Because the desert tortoise scientific panel did not sharply define and then stick to its policy recommendations, policymakers and politicians discounted its recommendations. In addition, the pursuit of individual agendas after the panel met undermined the positive work done by the panel under difficult time constraints.

4. Conservation biologists often state that conservation will succeed only if the needs of indigenous peoples are met (Schwartzman, Moreira, and Nepstad 2000; Schwartzman, Nepstad, and Moreira 2000; Colchester 2000; Chicchon 2000). This statement holds true not only in Latin America and other tropical areas but also in California. The desert tortoise cannot be saved unless the people who live in the desert support that goal. They need to be convinced of the value of saving their piece of biodiversity. However, clear laws and regulations based on science are required in order for conservation to be effective (Chaves et al. 1996; Terborgh 1999, 2000; Redford and Sanderson 2000). In the case of the Fort Irwin expansion, the local residents were engaged not as active participants in the process but rather through their congressman.

5. Protection of endangered species and ecosystems can be accomplished only in the real world. Scientists interested in influencing conservation policy need to be sure that they have the best possible

Figure 15.3. *(A)* Degradation of tortoise habitat in the western Mojave Desert. Off-highway vehicle (OHV) use, dumping, urban encroachment, and other related impacts reduce habitat quality within much of the western Mojave Desert Wildlife Management Area (Photo taken by Harold W. Avery near Helendale, California, September 2001).

data, that they interpret that data clearly for the nonscientist, that they make clear recommendations, and that they accept the fact that those recommendations will be imperfect. Second thoughts are not useful in the policy process. Brokering deals for conservation is both an art and a science (Richter and Redford 1999). The United States is a democracy, and all social activities involve compromise. The conservation scientist's job is to obtain the best compromise that protects species and ecosystems.

6. At first, the opportunity to enhance the recovery of the western Mojave population of the desert tortoise using Army-funded mitigation (resulting from Fort Irwin expansion) may not be apparent. However, much of the designated critical tortoise habitat in the western Mojave desert wildlife management area is broken up

Figure 15.3. *(B)* Free-living transmittered male tortoise using degraded habitat in the western Mojave Desert (note gunshot holes in surrounding debris). Despite being heavily impacted by human use, some degraded habitat in the western Mojave Desert is still occupied by high densities of tortoises. Without mitigation funds to purchase private land and provide habitat protection and restoration, such tortoise populations will eventually perish (Photo taken by Harold W. Avery near Helendale, California, September 2001).

by private inholdings, which produce a checkerboard pattern of land ownership. Besides being unmanageable, these areas are typically affected by off-highway vehicle use, dumping, urban encroachment, and other sources of habitat degradation (Figure 15.3A). Despite the litany of human disturbances affecting tortoise habitat in the western Mojave Desert, some of these areas still have significant densities of tortoises that may greatly exceed tortoise densities in proposed expansion areas. Tortoises still occupy degraded habitat (Figure 15.3B), but without immediate habitat protection and restoration, these tortoise populations will ultimately perish. Mitigation funds used to buy private inholdings, and fund protection and restoration of critical habitat, will be vital to tortoise survival and recovery in the western Mojave Desert.

REFERENCES

Berry, K. H. 1984. The status of the desert tortoise in the United States. In *The Status of the Desert Tortoise in the United States*, ed. K. H. Berry, Desert Tortoise Council Report to U.S. Fish and Wildlife Service, order no. 11310–0083–81, Sacramento, Calif.

———. 1986. Desert tortoise *(Gopherus agassizii)* research in California 1976–1985. *Herpetologica* 42:62–67.

———. 1990. Amended 1997. Draft Report to the Fish and Wildlife Service, Region 1, Portland, Oregon.

Berry, K. H., and Nicholson, L. L. 1984. The distribution and density of desert tortoise populations in California in the 1970s. In *The Status of the Desert Tortoise in the United States*, ed. K. H. Berry, 26–60. Desert Tortoise Council Report to U.S. Fish and Wildlife Service, order no. 11310–0083–81, Sacramento, Calif.

Bury, R. B., and Corn, P. S. 1995. Have desert tortoises undergone a long-term decline in abundance? *Wildl. Soc. Bull.* 23:41–47.

Chaves, A., Serrano, G., Marin, G., Arguedas, E., Jimenez, A., and Spotila, J. R. 1996. Biology and conservation of leatherback turtles, *Dermochelys coriacea*, at Playa Langosta, Costa Rica. *Chel. Cons. Biol.* 2:184–89.

Chicchon, A. 2000. Conservation theory meets practice. *Cons. Biol.* 14:1368–69.

Clancy, T., and Franks, F. M., Jr. 1997. *Into the Storm: A Study in Command.* New York: G. P. Putnam's Sons.

Colchester, M. 2000. Self-determination or environmental determinism for indigenous peoples in tropical forest conservation. *Cons. Biol.* 14:1365–67.

Congdon, J. D., Dunham, A. E., and van Loben Sels, R. C. 1993. Delayed sexual maturity and demographics of Blanding's turtle *(Emydoidea blandingii)*: Implications for conservation and management of long-lived organisms. *Cons. Biol.* 7:826–33.

———. 1994. Demographics of common snapping turtles *(Chelydra serpentina)*: Implications for conservation and management of long-lived organisms. *Am Zool.* 34:397–408.

Corn, P. S. 1994. Recent trends of desert tortoise populations in the Mojave Desert. In *Biology of North American Tortoises*, ed. R. B. Bury and D. J. Germano, 85–93. Fish and Wildlife Research 13, U.S. Department of the Interior, National Biological Survey, Washington, D.C.

Cozzens, P. 1990. *No Better Place to Die: The Battle of Stones River.* Urbana: University of Illinois Press.

———. 1992. *This Terrible Sound: The Battle of Chickamauga.* Urbana: University of Illinois Press.

———. 1994. *The Shipwreck of Their Hopes: The Battles for Chattanooga.* Urbana: University of Illinois Press.

———1997. *The Darkest Days of the War: The Battles of Iuka and Corinth.* Chapel Hill: University of North Carolina Press.

Dunnigan, J. F. 1982. *How to Make War: A Comprehensive Guide to Modern Warfare.* New York: William Morrow and Co..

Eisenhower, D. D. 1948. *Crusade in Europe.* Garden City, N.Y.: Doubleday and Co.

Fehrenbach, T. R. 2000. *This Kind of War: The Classic Korean War History.* 50th anniversary ed. Dulles, Va.: Brassey's.

Fish and Wildlife Service. 1994. *Desert tortoise (Mojave population) Recovery Plan.* U.S. Fish and Wildlife Service, Portland, Oregon.

Frazer, N. B., Gibbons, J. W., and Greene, J. L. 1990. Life tables of a slider turtle population. In *Life History and Ecology of the Slider Turtle,* ed. J. W. Gibbons, 183–200. Washington, D.C.: Smithsonian Institution Press.

Freilich, J. E., Burnham, K. P., Collins, C. M., and Garry, C. A. 2000. Factors affecting population assessments of desert tortoises. *Cons. Biol.* 14:1479–89.

Gordon, M. R., and Trainor, B. E. 1995. *The Generals' War: The Inside Story of the Conflict in the Gulf.* Boston, Mass.: Little, Brown and Co.

Houston, D. E. 1977. *Hell on Wheels: The 2nd Armored Division.* Novato, Calif.: Presidio Press.

Jacobson, E. R. 1994. Causes of mortality and disease in tortoises: A review. *J. Zoo Wildl. Med.* 25:2–17.

Krzysik, A. J. 1997. Desert tortoise populations in the Mojave Desert and a half-century of military training activities. In *Proceedings: Conservation, Restoration, and Management of Tortoises and Turtles—an International Conference,* ed. J. Van Abbema, 61–73. New York: New York Turtle and Tortoise Society.

LaRue, E. 2000. Results of the Fort Irwin Tortoise Panel meeting of 18–19 January and 18 February 2000, BLM CA 062.97, March 15, Barstow Calif.

Leslie, M., Meffe, G. K., Hardesty, J. L., and Adams, D. L. 1996. *Conserving Biodiversity on Military Lands: A Handbook for Natural Resource Managers.* Arlington, Va.: The Nature Conservancy.

Luckenbach, R. A. 1981. Ecology and management of the desert tortoise (*Gopherus agassizii*) in California. In *North American Tortoises: Conservation and Ecology,* ed. R. B. Bury, 1–37. Fish and Wildlife Research 12, U.S. Department of the Interior, Fish and Wildlife Service, Washington, D.C.

Pearson, D. C. 1986. The desert tortoise and energy development in southeastern California. *Herpetologica* 42:58–59.

Redford, K. H., and Sanderson, S. E. 2000. Extracting humans from nature. *Cons. Biol.* 14:1362–64.

Richter, B. D., and Redford, K. H. 1999. The art (and science) of brokering deals between conservation and use. *Cons. Biol.* 13:1235–37.

Schwartzman, S., Moreira, A., and Nepstad, D. 2000. Rethinking tropical forest conservation: Perils in parks. *Cons. Biol.* 14:1351–57.

Schwartzman, S., Nepstad, D., and Moreira, A. 2000. Arguing tropical forest conservation: People vs parks. *Cons. Biol.* 14:1370–74.

Terborgh, J. 1999. *Requiem for Nature.* Washington, D.C.: Island Press.

————. 2000. The fate of tropical forests: A matter of stewardship. *Cons. Biol.* 14:1358–61.

Turner, F. B. 1986. Foreword: Management of the desert tortoise in California. *Herpetologica* 42:56–58.

Turner, F. B., Thelander, C. G., Pearson, D. C., and Burge, B. L. 1982. An evaluation of the transect technique for estimating desert tortoise density at a prospective power plant site in Ivanpah Valley, California. In *The Desert Tortoise Council: Proceedings of the 1982 Symposium*, 134–53. Long Beach, Calif.: Desert Tortoise Council.

U.S. Army 1992. *U.S. Army Environmental Strategy into the 21st Century.* Atlanta, Ga.: Army Environmental Policy Institute.

Woodbury, A. M., and Hardy, R. 1948. Studies of the desert tortoise, *Gopherus agassizii. Ecol. Monogr.* 18:145–200.

16 Integrating Experimental Research with the Needs of Natural-Resource and Land Managers

Case Studies from Australia and New Zealand

John C. Rodger

SUMMARY

Finding effective ways to link the skills and expertise of researchers, especially in academia, with the needs of industry and the nation has been recognised as a major problem in many countries, including Australia. Such linkage is also needed in areas not traditionally seen as commercial, such as natural-resource management, which includes conservation. In the early 1990s, the Australian government set up a completely new national competitive programme to encourage such multi-institutional, multidisciplinary research and technology development called the Cooperative Research Centres (CRC) Programme.

The Marsupial CRC comprises core member organisations from across Australia and New Zealand: universities, state government agencies, and relevant industries (particularly in agriculture and forestry). It also has strategic international alliances with public-interest groups operating in areas such as animal welfare and the environment to build its skills base and networks and to ensure that its work is understood and supported by the broader community.

To illustrate this interaction between researchers and managers, this paper draws on examples from the Marsupial CRC's research and communications programmes. Areas covered include the application of a variety of laboratory and field disciplines of experimental biology to (1) the conservation of endangered marsupials and (2) the development of fertility-based population management tools for the control of exotic marsupial pests in New Zealand and abundant marsupials in Australia.

INTRODUCTION

The Theme Is Integration of Skills and Resources

Integration means more than simply the incorporation of scientific data into policy and field applications. An integrated approach to conservation is required from research-project design to eventual application of the results. Integration also includes breaking down the barriers between scientific disciplines and, in particular, between field- and laboratory-based scientists. However, integration also requires linking this research capacity more directly with the expertise and perspectives of public and commercial natural-resource managers and industry, especially major stakeholders and land managers. Collaboration of this type is not unknown, but it tends to depend on the interests and personalities of individuals and is thus ad hoc and patchy. The integration envisaged here would also operate at the level of organisations, so that major and complementary human, financial, and physical resources of diverse stakeholders would be brought together to work toward shared goals. Only by moving to this more inclusive approach can we expect to find meet challenges facing the world in the conservation of biodiversity and the management of abundant, and pest, species.

Thinking Bigger

My vision is of integrated wildlife science projects that bring together all the expertise, resources, and organisational agendas needed to identify needs, undertake the research, and then turn the research outputs into solutions to problems on the ground. This model of integration is familiar in the construction industry and engineering but not in the environmental and life sciences. The latter have traditionally focused on individually resourced "microprojects." Linking of these projects to deliver significant outcomes has been overlooked, or mechanisms to achieve this linkage have been absent. An outcome of this fragmented approach has been that, almost inevitably, critical components needed to complete the whole or to ensure that research is turned into practical outcomes are lacking. This is largely because scientific justification for funding focuses on component projects individually not as a linked whole. The result is that "exciting" projects get funded and less exciting (or less fashionable?) but nonetheless critical projects do not.

By way of analogy, consider the construction of an office building. There is an original vision, and there is a design. Funding is obtained, many professions come together to deal with the details, and then a managed team of tradespeople and labourers execute the plan in a coordinated manner. It is inconceivable that only some parts of this plan, such as exciting new information technology facilities, would be implemented, and that the less exciting parts, like the plumbing, would not. The key implications of this approach are the following:

1. The need for project management to coordinate design and effort, integrate components and disciplines, and communicate between the diverse elements to ensure individual components combine in a timely way to meet the goals of the project.

2. The need for appropriate funding and management structures to support such integrated, multidisciplinary, multiorganisational, goal-oriented research and development.

In the early 1990s, the Australian government set up a completely new national competitive programme to encourage such multi-institutional, multidisciplinary research and technology development called the Cooperative Research Centres (CRC) Programme. The CRC Programme is a national experiment in research funding that attempts to meet this challenge across a broad range of science and national need. The CRC Programme is a useful model through which public-policy-makers can examine the benefits and limitations of such a big-picture approach and identify generic components of potential relevance to their own national systems.

The Australian Situation

Despite its relatively small population (now approaching 20 million), Australia has an outstanding record of international-standard publicly funded research in universities, the Commonwealth Scientific and Industrial Research Organisation (CSIRO), major hospitals, federal and state government agencies, and research institutes. The record of industry-funded research in agriculture and mining, often in partnership with government, is also strong. However, the record in industry-funded research in manufacturing and value adding to primary products is relatively poor. Where Australia has a problem, and this is not unique, is in capturing the economic and national benefits of its publicly funded research. This, and the growing recognition that successful national economies in the future will be global and knowledge-based, led the Australian government to establish the CRC Programme in 1990. The programme aims to better link national research capacity and effort with the needs and applications expertise of existing and emerging industries and with national needs such as the environment and sustainable development.

THE AUSTRALIAN CRC PROGRAMME

The CRC Programme uses a carrot and not a stick as the instrument of policy, and it does not set out to pick winners. On a 2-year cycle, open competitive bid rounds are held and new or existing CRCs are funded for a fixed term. The potential cash funding from the programme for any individual CRC ranges from A$2–4 million per annum for up to 7 years. CRCs must be multiorganisational and involve at least one university and one industry, or research user organisation. In practice most CRCs involve many more organisations, commonly around ten as core members plus associated supporting or collaborating organisations and individuals. Government funding is pooled with contributions (mainly in-kind: staff time, facilities, etc.) from the member organisations such that the CRC Programme funds generally make up one-third to one-fourth of the total resources of any CRC. The programme has been reviewed very favourably twice since its inception (Myers 1995; Mercer and Stocker 1998) and now has strong bipartisan political support.

At any one time, there are around sixty-five CRCs. They cover the areas of mining, information technology, manufacturing, biomedicine, agriculture, and the environment. CRCs are actual organisations (some fully or partly incorporated), not simply collaborations. Whatever their structure (and the system is quite flexible), CRCs operate like small research companies with a group of owners (the member organisations), a board, a CEO, and an executive. CRCs use standard business practices such as strategic and operational planning, project-based budgeting, and project management tools. CRCs undergo stringent review at years 2 and 5 and can rebid for a further 7 years of funding, but this is in open competition with all other bids in that round.

There are around twelve environmental sector CRCs at any one time. They deal with such national issues as water resources, water salinity, land degradation, conservation of biodiversity, and the Great Barrier Reef. The major industry or research users tend to be government natural-resource agencies that are seeking robust scientific input to policy development or decision-making systems or practical tools for innovation in environmental management.

Competing CRC bids are judged against the following nine selection criteria:

1. Well-defined objectives that address a specific community or industry need
2. Outcomes that will make a significant contribution to Australia's sustainable economic and social development

3. A high-quality research programme, with well-defined, relevant, clear, and achievable outputs

4. A utilisation or commercialisation strategy that is well structured and practicable

5. Graduate education and training oriented to user and industry needs and job opportunities

6. Collaborative arrangements with strong commitment to building internal and industry–user linkages to enhance research and education programmes

7. Budgeted resources (cash, in-kind, time) that demonstrate commitment and support for research and education programmes

8. Effective management structure (financial, operational, research) to ensure that objectives are realised

9. Performance monitoring and evaluation strategy for internal assessment of research and education programmes that also meet reporting requirements of the CRC Programme

To illustrate the interaction between researchers and managers at work, this paper draws on examples from the research and communications programmes of the Cooperative Research Centre for the Conservation and Management of Marsupials, which is involved in developing innovative approaches to the conservation of threatened marsupials and the management of abundant marsupials and their negative impacts.

THE MARSUPIAL CRC

The Marsupial CRC encompasses a diverse range of scientific fields, including reproductive biology, conservation genetics, immunology, ecology, behaviour, and molecular biotechnology. The CRC began with five founding members: two universities, the University of Newcastle and Macquarie University; a New Zealand government–owned research company, Landcare Research; and two industry–user participants, Perth Zoo and the Queensland State Department of Primary Industries. The participation of these five organisations means that the CRC has nodes spread from the west coast to the east coast of Australia and in New Zealand. As the CRC has grown and developed, we have extended our links to include the expertise of another six Australian universities and a second New Zealand research company (AgResearch). We have also made strategic international alliances with organisations in the United States (Boyce Thompson Institute, University North Carolina, and UCLA) and in Austria (University of Vienna). The CRC is also engaged in constructing networks with public-interest groups operating in areas such as animal welfare and the environment to ensure its work is understood and supported by the broader community.

In addition to increasing research expertise and capacity, the CRC has been building industry and research user links to facilitate technology transfer or commercialisation of its research, primarily through alliances with state natural-resource and conservation agencies in Australia and possum pest-management interests in New Zealand. In addition, in late 2002, a spin-off company was established to support product development and commercialisation of CRC-developed fertility-control vaccines for possum control in New Zealand.

To illustrate the interaction between researchers and managers at work, I draw on three examples from the Marsupial CRC's research and communications programmes: conservation projects in Western Australia, fertility control for possums, and the marsupial night-stalk.

MARSUPIAL CRC CASE STUDIES

Conservation in Western Australia

The state of Western Australia (WA) has unique flora and fauna that are quite different from those of eastern Australia, and the state has many conservation issues around threatened marsupials. The Marsupial CRC, working through Perth Zoo, the major state zoo, has fostered close collaboration between the Department of Conservation and Land Management (the responsible state agency) and two of the local universities (University of Western Australia and Murdoch University). The CRC has funded some key staff and supported research costs, especially of graduate students, to facilitate the collaboration (see Fletcher and Morris 2003 for a more detailed discussion).

Some of the highlights have been in the application of molecular genetics to conservation questions. Out of this has come policy advice on management of the critically endangered Mala (rufus hare wallaby, a rabbit-sized macropod) based on microsatellite study of genetic diversity of island and captive populations. CRC support of reproductive research, captive breeding, and postrelease monitoring has assisted in the recovery of the threatened chuditch, or western quoll (a catlike marsupial carnivore) and the state emblem, the numbat (a highly specialised marsupial termite eater). Recently the CRC has commenced a new collaboration on the captive breeding and biology of Gilbert's potoroo (another rabbit-sized macropod), which is the most critically endangered Australian marsupial species. The only known population of Gilbert's potoroo (around thirty individuals) has a poor breeding record. CRC expertise in assisted reproduction (artificial insemination in particular) is expected to be needed to improve breeding performance.

Fertility-Control Tools for Possums in New Zealand

The Australian brushtail possum and several wallaby species were introduced to New Zealand in the nineteenth century to establish a fur industry (New Zealand has essentially no native land mammals). Unfortunately the possum is now New Zealand's major vertebrate pest with serious conservation and animal health impacts (see Montague 2000 for a detailed coverage of the possum in New Zealand). Traditional possum control has been based on poisoning. This technology is now highly efficient, but only a limited area can be controlled. In addition, there are serious community concerns about the technology, and there is a recognised need for alternative cost-effective control methods for the longer term. An important goal has been to find ways to control possum fertility, and the CRC has focused on contraceptive or sterilising vaccines targeting the egg coat and sperm surface. This work is a good example of highly integrated science at work. It involves many scientific disciplines, significant research infrastructure, and many organisations; and the individual elements must be highly coordinated. Expertise includes reproduction, molecular biology, immunology, reproductive technology, animal husbandry, biotechnology, vaccine technologies, ecology, behaviour, population biology, and population modeling. The team includes universities (Australia, New Zealand, United States, Austria), research organisations (New Zealand, United States), and New Zealand industry organisations, government agencies, and commercial investors (see Figure 16.1). CRC research has focused on proof-of-concept using crude and recombinant vaccines through to 2000; testing of the effectiveness of oral delivery for the first-generation vaccines should be complete by mid-2002 (see Mate and Hinds 2003; Cowan, Pech, and Curtis 2003 for more detail on this work). The CRC and its commercial partners have set themselves the goal of having a product ready for limited field trial by 2005.

The Great Australian Marsupial Night-Stalk

Public understanding and support are key elements of all wildlife conservation and management, especially when native animals, new technologies, or invasive or "undesirable" procedures are involved. The CRC has established links and lines of communication with key community groups and non-

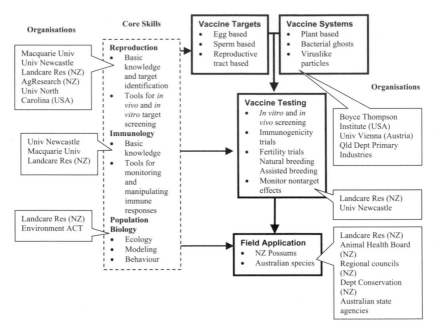

Figure 16.1. Marsupial CRC fertility-control vaccine strategy.

governmental organisations focused on animal welfare and conservation. The CRC recognised, however, that it needed to reach the community more broadly. Our tool to do this has been the Great Australian Marsupial Night-stalk, an annual national community-participation spotlight survey (of native and pest species) in local neighbourhoods. The CRC has provided the staff to coordinate the programme, but its success has been based on community interest (from schools, youth clubs, nature groups, etc.), support from state government agencies, the media, and funding from the federal government's National Science Week programme. In 1999 the night-stalk was largely limited to Western Australia, from where it was coordinated. By 2002 it had grown to cover 50% of the Australian bioregions and essentially all the major areas of human population. In addition to providing a great experience and developing local capacity, night-stalk is also proving to be a useful source of information on marsupials for the agencies. The information is mainly qualitative, but thousands of enthusiastic eyes spread across Australia are proving an invaluable source of information on the distribution of species, in some cases even possible sightings of threatened species in new areas. In the longer term, the closer links between natural-resource agencies and their communities, which the night-stalk is encouraging, may well be its most significant contribution. For more details on the night-stalk, visit the Perth Zoo Web site for a copy of the newsletter *Spotlight* (www.perthzoo.wa.gov.au).

THE CHALLENGES FACING CRCS

Thus far I have focused on the positive outcomes of CRC-stimulated collaboration and integration, but what about the problems? Any innovation brings challenges, and the CRC Programme's radical new vision for how science is done is no exception.

Building a Joint Venture Organisation

A highly integrated, multiorganisation, multinode research "center" (almost a nonsense word in many cases) will not function on goodwill alone. Formal structures are required. CRCs are bound together by a formal agreement that describes the rights and obligations of the partner organisations and the obligations of the staff they have seconded to the CRC as in-kind contributions (usually a significant percentage of their total time, 20–50%). The agreement covers such issues as management structure, principles of operation, intellectual-property protection and ownership, and commercialisation strategy. In practice CRCs operate as research companies with a CEO reporting to a board of management made up of nominees of the member organisations and independents. In many cases, individual CRCs have been established as actual companies with boards of directors and all the normal rights and obligations of corporations. The CEO, with an executive of programme managers and leading researchers, and the support of a small administration office, is responsible for the day-to-day operations of the center.

Changing Culture and Building Commitment

Because CRCs are formed from multiple organisations with different goals, structures, and practices, building a new organisation requires considerable work for everyone involved. Some of the tasks are intangible, such as developing a shared vision for and allegiance to the CRC and its goals among staff who remain within and loyal to their parent organisation. A major factor in achieving this shared vision is alignment between the parent organisation's goals and CRC goals. Without alignment, conflict is inevitable. A key issue that many organisations involved in CRCs have difficulty in grappling with is that they are the CRC and that its success depends on the synergy they achieve with the other partner organisations.

The more tangible challenges of building a CRC are putting in place effective communication and management structures and procedures to make the distributed and essentially "virtual" organisation operate like a more conventional organisation. There is no single formula to address these challenges, but all CRCs and other multiorganisation groups must have effective strategies to address them. In practice, the Marsupial CRC has found that the most important factor has been commitment of individuals to the CRC goals. Significant problems have occurred only when this commitment was lacking. A factor that contributes to this problem is the fact that a significant proportion of each CRC's budget (in particular staff time) are in-kind contributions from the party organisations. This means that a large part of the CRC budget has limited flexibility and that the CRC has no direct employer control over in-kind staff should problems arise.

Developing New Skills

Experimental biologists, particularly those working in academic environments, normally manage quite small groups, and their staff are often literally in the room next door. Larger research groups do exist, of course, and in the commercial environment, outcomes-focused research management is the norm. However, these groups rarely include staff at multiple and different types of organisations or staff for which the research leader has no line management responsibility or control. To address these management needs, most CRCs have adopted a mixed strategy of on-the-job training (e.g., adopting project and intellectual-property management practices and reporting regimes) and formal management training through universities or business support organisations such as the Australian Institute of Management (e.g., team leadership, project management, media and communications). New skills development also involves working with professionals like patent attorneys and commercial lawyers on issues like patent applications, research contracts, and intellectual-property licensing agreements. In practice, it has been the Marsupial CRC's experience that the vast majority of its scientists accept this challenge and find their new skills rewarding and valuable for their career development. Such extra skills development also extends to graduate students who develop their science within this outcomes-focused research-management environment (e.g., Marsupial CRC graduate students working in relevant areas have had experience in developing patent applications based on their work).

Measuring Outcomes

Traditionally the productivity of scientists is measured by the number of peer-reviewed publications. This may be a blunt instrument, but it is well understood and broadly accepted. CRC staff and CRCs do publish in peer-reviewed journals; but much of a CRC's activity is not directed to this end, and this has been a problem for the external assessment of CRCs and their staff. CRCs have a good idea of individual performance because of their active research-management practices. However, all CRCs have an agreed-upon set of performance indicators in their contracts with the government. Unfortunately these indicators focus on easily quantified outcome such as numbers of publications, patents, and graduate student completions. These metrics are important, but they tell only part of the story for an organisation whose success must ultimately be judged by its performance against its agreed-upon research and educational targets and the commercial or public-good outcomes of the application of its research. Finding more effective ways to measure CRC performance remains a major issue and is currently under review.

Another problem is that current performance indicator setting is from the perspective of the funding body and is largely an administrative exercise. The most critical perspective is arguably that of the industry or government agencies that need the CRC's outcomes for their business. In this model, the end users of the CRC's research would have a large role in setting the standards by which the CRC is judged and in the performance assessment processes.

Broadening What Scientists Do

Much has changed in science and its management over the last two decades. When I began my training, accountability issues, such as animal care and ethics, were virtually unheard-of but are now an accepted part of an experimental biologist's life. I would argue that experimental biologists need to rethink the paradigm of how they work. Funding structures may still favour the lone investigator with a team of willing students and assistants, but many of the challenges we face require a far broader and multidisciplinary approach. In addition, society has views about what experimental biologists should do and how their research should be applied. They are no longer unquestioned experts who know what's best (consider, for example, the controversies over animal cloning and genetic modification of food plants). Thus scientists need to be much more involved in public issues and in communicating what they do and the practical outcomes of their research. For some, this means developing new skills in areas like dealing with the media. I know that some of my colleagues are uncomfortable with this changing and demanding environment. However, I am encouraged that so many, especially among the young, see that the new relevance of experimental biology offers great opportunities and that their science can be not only exciting but useful.

CONCLUSIONS

Opportunities flow from the integration of multidisciplinary science with the needs and expertise of resource managers, industry, and the broader community. This model is based on large projects that seek to bring together all the expertise and resources needed to solve problems, rather than the traditional model of disconnected "micro" life-science projects. Key implications of this approach are project management, people management, and organisational management, which means new skills are needed beyond traditional science discipline training.

The Australian CRC Programme has shown that collaborative, multiorganisational research with clear goals can work. It has also shown that the management challenges are not as great as we might have imagined. However, none of the projects discussed above would exist, and none of the critical linkages would have been developed, without the funding structure of the CRC Programme, which not only recognised the need but also provided long-term resources for collaborative outcomes-focused research. This innovation in funding policy was the critical incentive to draw together disparate elements into collaborative teams. It is extremely unlikely that the big-project integrated research that I advocate will occur without such dedicated funding.

ACKNOWLEDGMENTS

The Marsupial CRC is supported by the Australian government's Cooperative Research Centres Programme. I am indebted to my colleagues Drs. David Wildt and Bill Holt for their valuable advice and criticism.

REFERENCES

Cowan, P. E., Pech, R., and Curtis, P. 2003. Field applications of fertility control for wildlife management. In *Reproductive Sciences and Integrated Conservation,* ed. W. V. Holt, A. R. Pickard, J. C. Rodger, and D. E. Wildt, 305–18. Cambridge: Cambridge University Press.

Fletcher, T., and Morris, K. 2003. Captive breeding and predator control: A successful strategy for conservation in Western Australia. In *Reproductive Sciences and Integrated Conservation,* ed. W. V. Holt, A. R. Pickard, J. C. Rodger, and D. E. Wildt, 232–48. Cambridge: Cambridge University Press.

Mate, K. E., and Hinds, L. A. 2003. Contraceptive vaccine development. In *Reproductive Sciences and Integrated Conservation,* ed. W. V. Holt, A. R. Pickard, J. C. Rodger, and D. E. Wildt, 291–304. Cambridge: Cambridge University Press.

Mercer, D., and Stocker, J. 1998. *Review of Greater Commercialisation and Self Funding in the Cooperative Research Centres Programme.* Department of Industry, Science, and Tourism, Canberra, Australia.

Montague, T. L. 2000. *The Brushtail Possum: Biology, Impact, and Management of an Introduced Marsupial.* Christchurch, N.Z.: Manaaki Whenua Press.

Myers, R. 1995. *Changing the Culture. Report of the CRC Programme Evaluation Committee.* Australian Government Printing Service, Canberra, Australia.

17   Making Wildlife Research More Meaningful by Prioritizing Science, Linking Disciplines, and Building Capacity

David E. Wildt

INTRODUCTION

Research strategies for wildlife biology are packaged in many ways. There are those scientists whose first priority is the *in situ* approach—understanding habitats (and, secondarily, the complex species interactions within). Others trudge through landscapes to study a single species, population, or even individual. Some focus on animals living *ex situ* (in zoos or breeding centers) or sometimes simply on animal "parts" (germplasm, blood components, tissues, cells, DNA) in laboratories. Still others never study the wild animal at all but rather concentrate on taxonomic relatives as "models" for new information (e.g., the domestic cow for wild antelopes, the domestic cat for tigers).

Having been involved in wildlife research for more than a quarter century, I have used all these strategies at one time or another. And, surely I have experienced most of the arguments for and against the intrinsic values of various research strategies for enhancing wildlife management and conservation. I appreciate the work of ecologists for whom habitat monitoring is a priority, as well as the work of the most esoteric of my colleagues, who study cellular enzyme kinetics. However, within these two radically different approaches—monitoring versus experimental studies of biodiversity—can one be better or more useful? Certainly surveying and monitoring nature is essential, but it is not sufficient. Scrutinizing habitats is noble and exciting, but it often simply documents growing degradation, fragmentation, and biodiversity loss while offering no scientific understanding or conservation solutions. Conversely, laboratory studies seem to attract substantial funding while generating massive arcana of little use to holistic biology, let alone species management or protection.

If the goals of wildlife research are to understand and save biodiversity, it would be useful if such research

- was a priority;
- was always based on sound science;
- maximized scholarly output to improve multidisciplinary applications; and
- was inextricably linked to strengthening science and decision-making abilities—capacity building.

Attending to these general principles could eliminate some of the tired controversies about how best to do wildlife research. If these principles were followed, the issue of *in situ* versus *ex situ* research would become less relevant because the likelihood of generating useful scholarly knowledge would increase, regardless of approach. Adherence to these standards also would help eliminate wasteful efforts to promote one biological field over another. Disciplinary chauvinism only perpetuates territoriality, distrust, and glacial progress. Most importantly, a more inclusive paradigm that also was concerned about capacity building would ensure a solid future for wildlife research. Our generation is struggling to save biodiversity, even in the developed world. Biodiversity hotspots, with high levels of endemism, are occurring in less developed regions that desperately need more local, competent scientists (who know the scientific method), informed decision-makers, and a public that appreciates its wildlife heritage.

PRIORITIZING RESEARCH

Historically, most wildlife research was carried out by academicians who investigated a question or a species out of personal inquisitiveness. For example, an ecologist may have studied white-tailed deer because of a fundamental interest rather than because this species, now often considered a pest, ruins gardens and causes traffic mishaps. From George Schaller's studies of giant pandas and snow leopards to E. O. Wilson's examinations of insect societies, the literature is filled with wonderful illustrations inspired by simple curiosity about nature.

However, there is growing discomfort within the conservation community about abstruse research of little applied value. Ironically, many of us were attracted to science by the promise of self-determination, the opportunity to discover new knowledge no matter how far-reaching or obscure the purpose. Given the scope of biodiversity loss, however, now is the time to change attitudes and actions. We need more biologists working for wildlife and clearer priorities for action because literally thousands of species have yet to be studied.

My colleagues and I recently reviewed the number of published papers from 1999 through mid-2000 that centered on the reproductive biology of wildlife species (Wildt et al. 2003). Reproduction obviously regulates a species' ability to survive. There are thousands of professional reproductive biologists on the planet. The Society for the Study of Reproduction (a largely U.S.-based organization) lists almost 3,000 members; the Society for the Study of Reproduction and Fertility (United Kingdom–based) has more than 1,000 members. But few of these individuals are working for the preservation of biodiversity. An analysis of ten top-rated reproductive biology journals from the United States, the United Kingdom, and Australia revealed abysmally few wildlife-related manuscripts. More than 90% of all journal space was devoted to 14 common species (including human, cow, pig, dog, and mouse). By far, most reproductive investigations were directed at mammals, with a total of 256 species studied. However, 75% of these (192 species) were represented by three or fewer publications. Only 3 species on the IUCN Red List of Threatened Species (Asian elephant, African elephant, cheetah) were studied extensively (i.e., with ten or more publications). We concluded that the reproduction of only 84 "wildlife" and 14 "common" (including human) mammals has been adequately studied . Since there are an estimated 4,629 mammalian species on Earth, a rather amazing 97.9% of all mammals have gone unstudied in this rather popular scientific discipline. Even more alarming, this percentage will be higher for birds, reptiles, amphibians, and fish, which are less charismatic and generally receive less attention than mammals.

Thus, nearly all experimental biology is disproportionately directed at already well-studied species, while the majority, many of which are sliding toward extinction, are being ignored. The excuses may be financial (lack of research funds), logistical (how does one study polar bears, for example), ignorance (of the number of species in trouble), or lack of initiative (once you are a rat biologist, it is easier to study rats than another species). Overcoming these hurdles is not impossible. Funds can be found (our laboratory has secured millions of dollars for such research), and research animals are accessible if one makes the appropriate contacts, that is, becomes a part of the "network." Probably the greatest impediment is that colleagues are unaware of the desperate need to extend their professional talents to benefit wildlife. Would it not make more sense for the entrepreneurial biologist who investigates the well-studied laboratory baboon to refocus efforts on the declining lemurs of Madagascar, the last mountain gorillas of Rwanda, or the nonhuman primates of Borneo (where more than 80% of species are under threat)? Should not the scientist studying the turkey, rabbit, or cow place a higher priority on the biological mysteries of the rare Attwater's prairie chicken, the riverine rabbit, or the roan antelope?

The difficulty of finding new investigators seems related to a lack of information on how one really begins—how do we prioritize species and studies? With a few exceptions, research priorities have not been effectively addressed in wildlife research, probably because so many species need attention in virtually all scientific areas. So little effort has been expended toward what needs to be done first. There is consensus that too little is known about what is out there in the natural world and its status (thus, conservation's mantra emphasizing survey and monitoring). But it is important to remember that this priority is largely observational. It is data acquisition, often not bona fide generation of knowledge through experimentation, that contributes to understanding biological mechanisms.

Although little attention has been given to identifying exactly what kinds of research are important, there is a growing body of literature that identifies species deserving research emphasis. The Red List classifies rare species into three categories: critically endangered (at least 50% probability of extinction within 10 years or three generations, whichever is longer), endangered (20% probability of extinction within 20 years or five generations), or vulnerable (at least 10% probability of extinction within 100 years) (http://www.redlist.org). The Red List now categorizes on the basis of information from which declines in numbers of individuals, the geographic areas occupied by the species and the number of populations and even breeding individuals can be observed, estimated, inferred, or suspected.

Although the Red List provides an overview of threatened animal (and plant) species, more-specific prioritization sometimes is generated through taxonomic specialist groups within the IUCN Species Survival Commission (SSC). For example, the IUCN Cat Specialist Group has recognized the Iberian lynx as the most vulnerable to extinction of the 37 extant felid species (Nowell and Jackson 1996). Likewise, the Canid Specialist Group has identified the Ethiopian wolf as the most threatened and deserving canid for research (Sillero-Zubiri and Macdonald 1997). The specific studies that should be undertaken are sometimes elucidated in the specialist groups' action plans, which are published under the aegis of the IUCN. Unfortunately, priorities often are set by a few specialists and can be biased by personal agendas. More effective than the specialist groups have been stakeholder-based workshops comprised of scientists representing different specialties as well as wildlife authorities, community leaders, sociologists, and educators. The Conservation Breeding Specialist Group (www.cbsg.org; Ellis and Seal 1995) has pioneered these participatory workshops. For example, the general issues affecting a species or habitat are identified in an early plenary session and are then collapsed into themes and dealt with in small working groups. Over the course of several days, problem-solving strategies are identified, recommendations are made, unambiguous actions with time lines are called for, and people are given responsibility for particular tasks. These blueprints for conservation generally point out the need for more research often in explicit areas that might involve *ex situ* as well as *in situ* animal populations. Often, though, recommendations, actions, and time lines cannot be met, because of lack of resources, including biologists willing to take on the challenge. However, there are stellar examples of how combining a prioritization scheme with organizational partnerships has resulted in milestones in conservation. The Florida panther, black-footed ferret, and the giant panda are three examples addressed later in this chapter.

The facilitated-workshop approach to identifying conservation priorities is not the exclusive domain of the IUCN. Others, such as Conservation International (a nongovernmental organization), use a similar approach for priority setting, including identifying urgently needed conservation activities with the involvement of a wide range of local stakeholders as well as scientific experts (e.g., Supriatna et al. 1999). One component of Conservation International's appeal is accountability; substantial emphasis is placed on securing funding to allow its scientists (or appropriate collaborators) to follow up on recommendations.

Zoos also are beginning to help prioritize species and have the advantage of access to millions of zoo-goers who can help rally financial support for research and conservation action. For example, for the wild cats, there is a North American Felid Taxon Advisory Group (TAG) under the umbrella of the American Zoo and Aquarium Association (AZA). The Felid TAG's primary responsibility is to provide zoos and breeding centers with a regional collection plan, a set of recommendations regarding the number of animals in specific species to be maintained in North American zoos for exhibition, education, and research and as a hedge against extinction. However, the Felid TAG also explicitly provides priority areas of research study (e.g., in animal health, bioinformatics, genetics, genome resource banking, behavior, nutrition, animal husbandry, and *in situ* surveys (www.felidtag.org). Each TAG also stands as a group of specialists willing to evaluate proposals, assist in securing research funding and endorsing projects of highest priority. Information on the Felid TAG Web site is valuable for the novice as well as the experienced biologist, because scientifically sound prioritizations can assist with funding. More organizations (including foundations) are demanding that their valuable funds be allocated only to high-priority projects. Endorsements by IUCN specialist groups and nongovernmental organizations (including those that are zoo based, such as the AZA and TAG), serve this function.

Finally, it is worth noting that the groups creating these priorities (whether they be species-based or discipline-based within species, whether they originate within the IUCN, wildlife nongovernmental organizations, or regional zoo associations) are a means for linking researchers to the wildlife world. These vast networks can assist the biologist in identifying research priorities. However, these groups also can offer partnership opportunities that, in turn, can lead to animal access and to more effective, multidisciplinary research.

SOUND EXPERIMENTAL SCIENCE

Wildlife research has not been historically associated with sophisticated technologies; standing around in short pants on the African plains watching animal behaviors through binoculars appears pretty low-tech no matter how rigorous the study. So in certain arenas, wildlife research has a soft-science image. Conducting statistically solid *in situ* studies is a challenge because the environment cannot be controlled, and because often there are too few research animals that are of appropriate age and sex ratio and that are located in one place and uninfluenced by zoo visitors (for *ex situ* studies). *In situ* and *ex situ* research also has been constrained by welfare arguments requiring that studies be noninvasive, or hands-off. If this shortsighted philosophy had been successful, much of what we now know about wild-animal physiology (including significant strides made in wildlife anesthesia, restraint and disease surveillance/prevention) would never have been discovered.

The point is that wildlife studies always should be hypothesis driven, experimentally based, and rigidly tested statistically. The fact that wildlife species generally are timorous, excitable, and dangerous is no excuse for a half-hearted approach. However, these traits do require caution during all the steps of a successful investigation, from early planning through raw-data interpretation. Wildlife studies also are usually executed differently than traditional academic (i.e., laboratory) studies. Rarely can wildlife investigators initiate a project in isolation—partners are crucial, especially wildlife authorities (whether field managers or zoo curators) and veterinarians (especially when direct animal handling is necessary). For *ex situ* studies, partners usually come from different institutions, because one zoo rarely holds adequate numbers of animals to allow statistically valid studies.

There also is mounting pressure for projects to be "applied," that is, to directly affect species management or conservation. The distinctions between applied and basic research can be ambiguous, and it always is dangerous to restrict studies to only those that are perceived to have an immediate, practical value. Wildlife managers often fail to appreciate the value of basic research. At the same time, it is a researcher's responsibility to explain the merit of all studies, from the practical to the arcane, even if pragmatic value will not be apparent until far into the future. The most important objective should be generating the scholarly knowledge most needed to enhance the management and conservation of a priority species. Therefore, anesthetizing an African elephant to measure the length of the auditory canal out of curiosity will seem frivolous compared to anesthetizing one to assess a new diagnostic test for tuberculosis. Auditory morphometrics may someday allow us to learn more about elephant communication and, thus, herd socialization and well-being. Nonetheless, it is important that animal managers perceive the research as valuable. It is the researcher who must demonstrate the value of basic research to those who control animal access.

Finally, it is foolish if not obtuse, to assume that applied wildlife research will result in quick-fix solutions (Wildt, Ellis, and Howard 2001). My personal example is our studies of the cheetah initiated now more than two decades ago. As reproductive biologists, we became enchanted with the idea of routinely propagating cheetahs by artificial insemination (AI) or *in vitro* fertilization (IVF) and embryo transfer. We presumed that these methods would rapidly solve the problem of cheetahs failing to breed in zoos. Armed with established technologies developed in cattle, we directly applied ungulate methods to this specialized carnivore. Our repeated failures to produce AI or IVF cheetah offspring pointed to our naiveté—how could we have believed that methods accommodating the physiology of the herbivorous, domesticated cow would be appropriate for the predatory, carnivorous cheetah? This flash of discovery, a belated appreciation for physiological specialization, changed our way of doing science. We began a 15-year systematic and experimental examination of the cheetah's specific reproductive biology. No doubt we tested the patience of our cheetah curatorial partners with seemingly inane basic studies (e.g., pituitary–gonadal responsiveness to exogenous releasing hormones; Wildt et al. 1983). But a decade's worth of research in gamete biology, endocrinology, and embryology in zoo-held versus free-living cheetahs resulted in new information on species biology.

Publishing positive (as well as negative) data in scientific journals enhanced the research's value and our laboratory's credibility. Results also had a direct application. Because our studies were geared to fertility of the whole animal, the program and the findings were widely accepted and valued, even by managers. More important, knowledge was used to reliably produce cheetah cubs by AI. One study relied upon thawed sperm imported from wild cheetahs in Africa to produce cubs in zoos, thereby bolstering the genetic diversity of an *ex situ* population without removing animals from nature (Wildt, Ellis, and Howard 2001). Meanwhile, parallel studies of more than twenty-five other felid species affirmed the wisdom of a species-specific, basic-research approach. Physiological variations among cat species in the family Felidae were remarkable (Wildt, Brown, and Swanson 1998). None of this could have been understood or appreciated without serious hypothesis-driven research.

Thus, an orthodox experimental approach was effectively applied to a rare wildlife species. Some studies were rather invasive, involving risky, manipulative studies under field conditions (Wildt, O'Brien, et al. 1987). Yet no animal was harmed, and the studies maximized knowledge of cheetah reproductive biology. Nonetheless, even this approach may not be bold enough.

MAXIMIZING INFORMATION THROUGH MULTIDISCIPLINARY RESEARCH

My own first experience mixing scientific disciplines to generate more meaningful information involved the cheetah. In the early 1980s, geneticist Stephen O'Brien (National Cancer Institute) was launching studies of felid phylogeny. By serendipity and largely as a professional courtesy, we provided his laboratory "extra" cheetah blood samples collected during some of our original reproductive studies. O'Brien's provocative findings that the cheetah expressed low genetic diversity led to a series of publications on population bottleneck theory and the impact of lost genetic variation on health and reproductive fitness (see O'Brien, Wildt, and Bush 1986). The discovery of lack of genetic variation was intriguing given our simultaneous observation of high levels of malformed sperm in the cheetah's ejaculate, and other observations of disease susceptibility in captivity. Other disciplines (husbandry, ecology, behavior, and most recently stress physiology; Wielebnowski et al. 2002) merged to study the species cooperatively and intensively. The product was not simply the acceptance or rejection of any single hypothesis. Rather, over time, the biology of an entire species and how it "ticks" (or doesn't tick) in zoos and nature was better understood (see Wildt, Ellis, and Howard 2001). Not all questions were answered, and perhaps more new mysteries emerged than were resolved. Nonetheless, this multidisciplinary experiment convinced us that wildlife research is best and most efficiently conducted among collaborators who can cooperatively blend diverse specialties. There simply is too little time and too few resources to do otherwise.

The cheetah experience was not a fluke. Our interest in the relationships between genetic diversity, reproductive traits, and disease susceptibility was extended to include the lion and Florida panther. In both cases, experimental studies were possible because we had access to different types of populations living in nature. For example, we studied free-living lions in the Serengeti (controls), lions isolated in an extinct volcanic caldera (the Ngorongoro Crater of Tanzania), and lions in a wildlife sanctuary in India (where the population exceeded habitat capacity) (Wildt, Bush, et al. 1987). We compared isolated Florida panthers living in south Florida's Big Cypress Swamp to puma cousins living throughout North, Meso-, and South America. Partnering among scientists again mingled the disciplines to more rigorously understand the consequences of lost genetic diversity in small populations (review, Roelke, Martenson, and O'Brien 1993). For the Florida panther, useful preresearch organization occurred during CBSG stakeholder workshops that identified priorities. Having wildlife managers involved in the planning process facilitated collaboration and prevented the data from disappearing into the scientific literature. Findings were valued and used to create novel *in situ* strategies, including genetic augmentation of the depauperate Florida panther genotype by selectively introducing pumas of a more robust subspecies.

Another success story from our experience, involving basic and applied research, has involved the black-footed ferret, once believed to be extinct. This mustelid of the western plains of the United States suffered from loss in habitat and prey (owing to expanded agriculture and government-sponsored extermination of the prairie dog, its primary prey). Rediscovery of a founder population of eighteen black-footed ferrets in the 1980s led to a series of CBSG workshops that were followed by intensive studies in small-population biology, husbandry, genetics, disease, and, in our laboratory, reproductive biology (see Wolf et al. 2000a, b). The challenge was to determine how to breed this remnant population in captivity, preferably rapidly, to expand the population for potential reintroduction. At the time, our laboratory had no experience with mustelids and little direct access to the rare founder population, which understandably was under a hands-off management protocol to promote natural breeding. While others toiled at basic husbandry, genetics, nutrition, and disease susceptibility, we began studying fundamental gamete biology in the taxonomically related European and Siberian ferrets. Once we adequately understood sperm biology and ovulation induction in these species, we quickly succeeded in developing ferret AI. Eventually and not without controversy (Wildt, Ellis, and Howard 2001), we applied the AI technique successfully to the black-footed ferret, the technique being especially useful for propagating underrepresented genotypes within the narrow species lineage. Today, AI, along with other breeding strategies, is used to more effectively produce black-footed ferrets on a multi-institutional scale. This, in turn, is contributing to species reintroduction onto the prairies of the United States and Mexico. Again, interdisciplinary research and partnerships were key to linking new knowledge from *ex situ* studies to benefit an endangered species *in situ*.

ANOTHER MORE DETAILED EXAMPLE OF INTEGRATIVE RESEARCH AND CAPACITY BUILDING

Although making interdisciplinary research work in the developing world is a bigger challenge, these regions are where the action is, where there is the most biodiversity requiring urgent attention. In the developing world, most problems are landscape based and related to the needs of local people who are striving simply to survive. Wildlife often is not valued, except perhaps as a source of food. Changing this attitude is perhaps the biggest challenge to biodiversity conservation today. But because of a recent unique opportunity, I have begun to believe that we may be able to partially meet this challenge through flagship species such as the giant panda. China as a nation values the giant panda and recognizes its international significance as a conservation symbol. However, the status of giant pandas in nature is precarious. The species now is isolated in more than forty fragments of mountainous bamboo forest with no corridors for genetic exchange. No accurate population count is available, and nothing is known about the self-sustainability of these small populations. The challenge of managing giant pandas living in the wild, where status information is unreliable and resources are poor, is staggering. The remote, rugged habitat rarely allows even the most experienced biologist a peek at this elusive bearlike herbivore. Populations in those wobbly fragments indeed are at risk of dying out. And if the wild giant pandas disappear, other wildlife also will be lost because there will be little incentive to preserve the remaining forests from human encroachment.

Perhaps because effective *in situ* conservation of giant pandas seems so overwhelming, the *ex situ* panda population in China receives a great deal of attention from zoos as well as reserve authorities. Giant pandas in zoos and breeding centers in China may well be the key to the future of their wild counterparts. The *ex situ* population is a powerful force for educating the public about the plight of wild pandas. This charismatic animal inspires remarkably effective fund-raising, while simultaneously raising ethical issues about the impact of exportations of giant panda for loans to zoos. Could the popular giant panda be exhibited to death if its self-sustainability is compromised by the lures offered by Western zoos? This controversy was effectively quashed by a 1998 U.S. Fish & Wildlife Service policy that mandated that such loans to U.S. zoos first prove that compensation would benefit giant panda conservation in China, mostly *in situ*. The die was cast—giant pandas would enter U.S. zoos only if strict conservation guidelines could be met.

While these political debates swirled, there were issues of biology and management facing giant pandas living in zoos and breeding centers in China. In 1996, I was privileged to participate in a CBSG workshop in China that addressed common problems in managing giant pandas *ex situ*. Workshop participants, most of whom were Chinese (there were only six Western participants), first identified a laudable goal: "that the captive population be self-sustaining to support giant pandas living in nature." That itself was a radical mission statement in that Chinese zoos had commonly plucked giant pandas from the wild to support captive breeding programs. That single phrase represented the first commitment to development of an effective breeding program. However, if the population was not self-sustaining, how could it ever support wild giant pandas, especially given that a compensatory loan program to foreign zoos would further deplete an already shaky captive population? Intensive workshop discussions led to one conclusion—before the *ex situ* giant panda population could be managed and increased in size, it needed to be understood. That is, it was necessary to understand its population demographics, health, and reproductive fitness. This finding led to a biomedical survey of more than sixty giant pandas, a binational collaboration of twenty North American scientists and more than fifty Chinese counterparts representing the fields of genetics, population biology, reproductive physiology, endocrinology, behavior, nutrition, and veterinary medicine (Wildt, Ellis, and Howard 2001; Wildt et al. 2003). Side-by-side, scientists collected historical information and conducted intensive physical examinations of anesthetized pandas. Massive amounts of data were collated, interpreted, and shared. Because a multidisciplinary approach was used, factors that may have escaped the notice of, say, a nutritionist might be captured by a geneticist; those missed by a reproductive biologist might be discovered by a veterinarian. The biomedical survey identified more than a half dozen variables that were limiting reproductive and management success. Details are in other publications (Wildt, Ellis, and Howard 2001; Wildt et al. 2003), but three illustrations are briefly mentioned here. For example, the supplemental feeding of a high-protein concentrate revealed that pandas were receiving inadequate dietary fiber, which accentuated poor health and reproduction in some individuals. Use of multiple males during breeding produced many offspring with unknown paternities, a problem that was resolved by developing a laboratory at one panda facility with microsatellite analysis capability. The discovery of testicular hypoplasia in males from another facility raised concerns about the potential of a heritable defect that might even occur in the wild population.

These integrative, cross-cultural collaborations produced significant new data on giant panda biology. New knowledge and new technologies also improved captive breeding success (e.g., eighteen cubs were produced at the three largest facilities in 2000). But the most interesting story is less about biology or *ex situ* cub production than about linkage and spin-offs. Working together in teams with transparent sharing of information led to significant trust building. Giant pandas historically have been managed under competitive systems, including individual and federal centers. Facilities managing giant pandas under different "flags" gradually began communicating and working together. The introduction of new technologies during the survey also wetted appetites for advanced training, which led to courses in veterinary medicine and genome resource banking of biomaterials.

Because of respect for CBSG and the value of facilitated workshops, we also were invited to help identify *in situ* priorities in partnership with the State Forestry Administration. Training in all field-related sciences to allow better monitoring of wild giant pandas and their habitat was needed. These independent efforts in China almost seamlessly linked themselves to a giant panda loan that was being developed simultaneously by my home institution, the Smithsonian's National Zoological Park. In exchange for the 10-year exhibition of two giant pandas in Washington D.C., the National Zoo agreed to contribute more than $1.3 million annually to giant panda research, personnel training, and infrastructure development, most of it in China. Funds were donated by dedicated people and corporations. Most importantly, achieving the training priorities identified by our Chinese colleagues now became possible. For example, recent workshops sponsored by the Smithsonian's National Zoological Park have included training in geographical information systems, small-mammal surveys, and wildlife management. These courses were geared not to zoos but to on-the-ground staff from some of the most remote giant panda reserves. One of the most recent workshops involved training both zoo and reserve staff in small-population biology and genetic management with a goal of linking the growing *ex situ* panda population to wild populations, perhaps eventually through reintroduction. In a way, it was a celebratory workshop illustrating how far we had progressed. The partners had helped to stabilize and then expand the *ex situ* population as well as plan a future when the Chinese could achieve their original 1996 goal—using the captive giant panda population to support its counterparts in nature.

The crucial ingredients of this anomalous mix have been diverse biologists, wildlife authorities, husbandry specialists, zoos, corporations, and private citizens. While they all contribute in different ways, the common mandate is an improved understanding and conservation of the giant panda. The goal has not yet been entirely met, but a foundation has been laid on the basis of partnerships, people, and process. What remains unclear is how can we extend the giant panda model to the thousands of less charismatic (but no less needy) wildlife species deserving attention. Perhaps it is as simple as latching the less electrifying species onto flagship species like the giant panda. For example, as well-funded multidisciplinary research is established for pandas, why not set up parallel studies for other rare Chinese mammals, birds, reptiles, amphibians, and fish? Only marginal additional resources may be needed to begin rigorous studies of such species.

THE FUTURE: CAPACITY BUILDING AS A KEY

Over time I have become even more convinced that our most intensive and novel approaches in contemporary conservation science are still inadequate. Certainly we can continue to advocate a paradigm that focuses on saving habitat, on monitoring space and taxa. But in the absence of serious scientific inquiry and experimental approaches, I suspect that in the end we will lose. The victims will be not only species but also intellectual capital—that is, we will fail to understand the complex mechanisms of adaptation, fitness, and survival relevant to conservation and, perhaps, to the preservation of humankind.

To meet this challenge, the scientific method needs to be applied to priority conservation issues in every country. This will literally require a global army of experimental biologists studying biodiversity (not the current handful of Westerners often focused on a single species and a one-dimensional, basic research approach; Wildt et al. 2003). Of course, this army in turn will require capacity building, especially in the developing world. Training is hardly an original idea, and many institutions (including the Smithsonian's National Zoological Park), nongovernmental organizations, and universities are committed to this endeavor. However, obviously we are not near perfecting capacity building because there are way too few local people in biologically rich regions leading conservation programs. For example, I always have been dismayed by the lack of black Africans in conservation science; Africa is the wondrous continent that inspired my initial devotion to wildlife biology. Likewise I have come to realize that the educators of children do not know how to teach the value of wildlife heritage, which is indeed perhaps a community's best chance for a better socioeconomic future. Ironically, it is these teachers who are on the frontline of conservation. It is they who need to instill a love and respect for nature (and even scientific study) in students, and hopefully the next generation of conservationists.

The value of having scientists working interactively with teachers to translate real scientific stories into useful and fun classroom activities and materials has been discussed in one of our recent papers (Wildt et al. 2003). Such materials can illustrate basic biological principles (e.g., the insidious consequences of population bottlenecks, using the cheetah example) as well as promote scientists as role models. Colleagues have developed several effective curriculum modules in partnerships with teachers. The teachers share these resources with their colleagues in formal training courses, thus creating a multiplier effect. Although we are relatively new to this type of training, the program has been enthusiastically received by teachers and students alike and in both Western and developing countries.

Even with teachers beginning to "mass motivate" students to consider careers in wildlife biology, we cannot wait another generation for action. These teachers offer hope for the future, but the need for more professional capacity building in adults is now urgent. The need for continuous, long-term training in developing countries is recognized, but not well organized or funded. Ideas advocated by others have included the need to (1) commit to regions and locales with repeated courses over years rather than "one-off" experiences, (2) ensure that courses are adequately organized and translated (if necessary) and involve weeks of learning, not days, (3) identify "stars" that can receive advanced training and perhaps become future trainers themselves, and (4) provide physical resources and incentive grants after the course ends to motivate as well as seed new projects undertaken by trainees. We also have initiated a "conservation diplomacy" program whereby a qualified candidate is brought to our institution for short-term (6 weeks to 1 year), intensive training. The opportunity is advertised and touted as prestigious, and an official certificate of accomplishment (diploma) from the Smithsonian Institution is awarded. This certificate often is all that is needed for the trainee to secure a respectable conservation position upon return to the home country.

In the end, all the prioritization of conservation needs and buy-in on the value of linking disciplines are of little use without resources. Funding conservation science is neither within the scope of this chapter nor is it of significant personal anxiety. If the guiding principles are in place—*conducting high priority, scientifically sound, integrative research with the ever present parallel goal of training others*—then the programs are likely to sell themselves. Within this context, we as biologists should be more concerned about advocating the value of both basic and applied research to all ignored species and as quickly as possible. Finally, it would be in the best interest of biodiversity if we realized that our lasting legacy as scientists was to commit to training a worldwide cadre of biologists, especially in the developing world, who are keen about wildlife and schooled in the scientific method.

ACKNOWLEDGMENTS

The author thanks Susie Ellis for critical comments on an early draft of the manuscript. Stephen O'Brien and Mitch Bush were partners to realizing the value of multidisciplinary research. JoGayle Howard led the described black-footed ferret research. Giant panda studies involved more than 100 dedicated colleagues, but special acknowledgment is extended to Susie Ellis, Don Janssen, JoGayle Howard, Mabel Lam, Lucy Spelman, Stephen O'Brien, Zhang Anju, Zhang Zhihe and Zhang Hemin and the Smithsonian's National Zoological Park and its Friends of the National Zoo, CBSG, Chengdu Base of Giant Panda Breeding and China Research and Conservation Center for Giant Panda. Jennifer Buff was instrumental in developing new capacity building tools for teachers.

REFERENCES

Ellis, S., and Seal, U. S. 1995. Tools of the trade to aid decision-making for species survival. *Biol. Cons.* 4:553–72.

Nowell, K., and Jackson, P., comps. 1996. *Wild Cats: Status Survey and Conservation Action Plan*, 106–110. Gland: IUCN.

O'Brien, S. J., Wildt, D. E., and Bush, M. 1986. The cheetah in genetic peril. *Sci. Am.* 254:84–92.

Roelke, M. E., Martenson, J. S., and O'Brien, S. J. 1993. The consequences of demographic reduction and genetic depletion in the endangered Florida panther. *Curr. Biol.* 3:340–50.

Sillero-Zubiri, C., and Macdonald, D., comps. and eds. 1997. *The Ethiopian Wolf: Status Survey and Conservation Action Plan,* 123. Gland: IUCN.

Supriatna, J., Talbot, K., Thomson, J., and Mittermeier, R. 1999. *The Irian Jaya Biodiversity Conservation Priority-Setting Workshop,* ed. J B. Burnett, 71. Washington, D.C.: Conservation International.

Wielebnowski, N. C., Ziegler, K., Wildt, D. E., Lukas, J., and Brown, J. L. 2002. Impact of social management on reproductive, adrenal, and behavioural activity in the cheetah *(Acinonyx jubatus). Anim. Cons.* 5:291–301.

Wildt, D. E., Brown, J. L., and Swanson, W. F. 1998. Reproduction in cats. In *Encyclopedia of Reproduction,* ed. E. Knobil and J. Neill, 497–510. New York: Academic Press.

Wildt, D. E., Bush, M., Goodrowe, K. L., Packer, C., Pusey, A. E., Brown, J. L., Joslin, P., and O'Brien, S. J. 1987. Reproductive and genetic consequences of founding isolated lion populations. *Nature* 329:328–31.

Wildt, D. E., Chakraborty, P. K., Meltzer, D., and Bush, M. 1983. Pituitary and gonadal response to luteinizing hormone-releasing hormone administration in the male and female cheetah. *J. Endocrinol.* 101:51–56.

Wildt, D. E., Ellis, S., and Howard, J. G. 2001. Linkage of reproductive sciences: From "quick fix" to "integrated" conservation. *J. Reprod. Fertil. Suppl.* 57:295–307.

Wildt, D. E., Ellis, S., Janssen, D., and Buff, J. 2003. Toward more effective reproductive science in conservation. In *Reproductive Sciences and Integrated Conservation,* ed. W. V. Holt, A. Pickard, J. Rodger, and D. E. Wildt, 2–20. Cambridge: Cambridge University Press.

Wildt, D. E., O'Brien, S. J., Howard, J. G., Caro, T. M., Roelke, M. E., Brown, J. L., and Bush, M. 1987. Similarity in ejaculate-endocrine characteristics in captive versus free-ranging cheetahs of two subspecies. *Biol. Reprod.* 36:353–60.

Wolf, K. N., Wildt, D. E., Vargas, A., Marinari, P. E., Kreeger, J. S., Ottinger, M. A., and Howard, J. G. 2000. Age dependent changes in sperm production, semen quality, and testicular volume in the black-footed ferret *(Mustela nigripes). Biol. Reprod.* 63:179–87.

Wolf, K. N., Wildt, D. E., Vargas, A., Marinari, P. E., Ottinger, M. A., and Howard, J. G. 2000. Reproductive inefficiency in male black-footed ferrets *(Mustela nigripes). Zoo Biol.* 19:517–28.

18 African National Parks under Challenge

Novel Approaches in South Africa May Offer Respite

Leo Braack

INTRODUCTION

Africa is well known for the diversity and abundance of its wildlife, particularly within its savanna areas, and the combination of game and scenic wilderness remains a prime attraction for tourists from all over the world. Yet these areas and the fascinating diversity of life they contain are but remnants of what was once present, and even these remnants are now seriously threatened.

Following millennia of evolution and adaptation, in a relative blink of the eye over the last three centuries, colonizing powers carved up the continent into political blocks (Packenham 2000), opening the doors to a steady stream of adventurous missionaries, miners, farmers, and industrialists who, with gunpowder, rapidly reduced the previously widespread wildlife into diminishing pockets (Robertson 1978). The introduction of medicines, public-health measures, and transport systems boosted the sparse indigenous population into burgeoning abundance and, unfortunately, over time also disrupted traditional value systems and contributed to chronic poverty and dependence (Jeal 1996). To stem the destruction, wildlife preserves were independently established in virtually every country, but usually in leftover areas of low economic potential. These national parks and other conservancies were then developed as tourism destinations for a relatively affluent minority and were managed by generations of primarily white specialist staff. These conservancies failed to maintain or instill in traditional cultures an appreciation of the benefits and value of wildlife and wilderness. By the time the colonial powers retreated, the indigenous nations understood the tourism and revenue potential of conservation areas but had largely lost their cultural closeness to and appreciation of their natural resources. This lack of appreciation led to poaching, mismanagement, and widespread loss of wildlife and decline in conservation standards. These may indeed be strong words, but by and large they reflect the situation in most African countries today. At no previous time in its history has the continent faced such an array of challenges on such a large scale.

The obstacles that confront wildlife conservation in Africa can be grouped into three main categories, each with a number of subcomponents: financial constraints, social factors, and habitat limitations. Various innovative and experimental developments have emerged in recent times, many of which promise to alleviate some of these problems. Among these developments is the recent trend toward the creation of transfrontier parks, which, although not novel elsewhere, are new in the African context and hold potential for clear benefits that Africa needs.

OVERVIEW OF CONSERVATION DEVELOPMENT IN AFRICA

European travelers in Africa during the seventeenth, eighteenth, and nineteenth centuries unanimously described the abundance and widespread range of the continent's wildlife (Burchell 1822; Harris 1947, 1986; Lichtenstein 1812–1815). Vast, seemingly inexhaustible herds of migrating antelope numbering in the millions of individuals were features of the landscape in southern Africa (Cronwright-Schreiner 1925; Harris 1947, 1986). Yet with the increasing number of Europeans dispersing into the hinterland, the increased availability of firearms, improved transport, and pressure to "subdue" the land for farming purposes, the impact on wildlife reached such proportions that it can only be described as wholesale slaughter (Pringle 1982). Elephants *(Loxodonta africana)* were shot for ivory; hippos *(Hippopotamus amphibious),* buffalo *(Syncerus caffer),* and antelope for meat and their skins; and lions *(Panthera leo)* and other predators because they posed a threat to sheep, cattle, and horses (Carruthers 1990). Animals were shot because they carried disease to domestic animals and people (McKelvey 1973; Robertson 1978), because they damaged agricultural crops, and also simply as a recreational pastime (Carruthers 1990). This slaughter reduced this wildlife cornucopia to isolated small populations, endangered the status of species such as bontebok *(Damaliscus dorcas dorcas)* and mountain zebra *(Equus zebra zebra),* and led to the complete extinction of some species, such as the bloubok (blue antelope, *Hippotragus laucophaeus*) and quagga *(Equus quagga)* (Pringle 1982).

The southern tip of Africa was the first to be extensively colonized by European powers—in 1652 by the Dutch—and by the early nineteenth century, the massive herds of springbuck *(Antidorcas marsupialis)* and other wildlife had been practically annihilated within the Cape Colony. Now the hinterland and areas further north were under siege. The rapid diminishment of wildlife caused increasingly vocal concern among many people and led to formal government attempts to limit further excessive reductions. The first tactic was the introduction of legislation to control hunting throughout all areas of the colonial sphere of influence (at the national level) so as to achieve an improvement throughout an entire region, but this legislation proved impossible to enforce. Setting aside areas dedicated for the conservation of wildlife was more practical, and the first of these embryonic state game reserves appeared in the Cape Colony as early as 1822. However, the first truly significant areas set aside for conservation were the Pongola Game Reserve, established in 1894; the Hluhluwe and Umfolozi Game Reserves established in 1897; and the Sabi Game Reserve—precursor of the present-day Kruger National Park—established in 1898 (Carruthers 1995).

Events elsewhere in the world did not escape the notice of African conservationists. In particular, the creation of Yellowstone National Park as a symbol of national pride and government responsibility did much to motivate and ensure the successful proclamation in 1926 of Africa's first equivalent, the Kruger National Park (Stevenson-Hamilton 1937).

It is important here to state clearly that, at the time, the colonial powers controlled the entire process of legally designating national parks, managing and policing them, and promoting tourism. National parks largely represented a Eurocentric set of values and methods for conserving wildlife. In maintaining these values, the colonial or white-controlled governments heavily subsidized these national parks and ensured effective management (Carruthers 1995).

Political independence arrived for most African countries in waves during the late 1950s and early 1960s, although there was some delay for Mozambique (1974) and Zimbabwe (1980). The new independent governments, led by indigenous black people, identified new national priorities to address the long-standing neglect of poverty, education, health, and other social issues. The net result, although never intentional, was that conservation agencies in all African countries were faced with diminishing resources and support.

Given the almost global realization of the serious pressures facing our planet's biodiversity, given the commitment of most governments to formally engage in actions to conserve at least representative samples of such natural heritage, and given the fact that much of that national action is represented in the form of national parks, many African countries are now faced with the reality that prevailing social and economic pressures prevent them from adequately servicing that national commitment.

In a number of southern African countries, a high percentage of land—in some cases approaching the percentage held in the country's national parks—is being conserved through private initiative. By and large these private reserves are effectively, profitably, and sustainably managed, and in many cases they retain the full spectrum of species richness historically present, which of course prompts the question of why private enterprise can achieve this but state or parastatal agencies cannot. With few exceptions, these private reserves are relatively small and fenced, and they are often run according to their own particular profit-motivated management objectives. They have even less capacity than the national parks to allow ecological processes such as drought, disease, and migration to shape genetic fitness of wildlife populations and so ensure appropriate adaptive conditioning. Conservation of true biodiversity, which comprises not just species richness but also diversity in structure and function,[1] therefore continues to remain in the state-controlled national parks, imperfect as these areas may be as representations of their broader ecosystems.

The problems that African countries are faced with, and how they deal with them, can be reflected by examining the South African situation in greater detail. While South Africa does represent somewhat of a special case—its infrastructure and skills base are greater than those of most other African countries, and its economy is more advanced—it has retained its focus on African values and in many respects is viewed as a leader and role model within the continent. For all these reasons, therefore, it is appropriate to focus on how South Africa is dealing with the conservation challenges that, to greater or lesser degree, are common to all African countries.

South Africa's system of nineteen national parks (3,583,734 ha), despite its many flaws and the criticisms that have been leveled at certain management policies, remains the most effectively managed system in Africa. The parks are run by South African National Parks (SANParks), a parastatal organization supervised by a statutory board appointed by Parliament. Although the organization has the responsibility to be the ultimate conservator of the nation's natural heritage and is doing so on state land, the organization receives only about 15% of its annual financial needs (approximately U.S.$40 million) from government; the remainder it has to source itself (SANParks 2000/2001). Of the nineteen national parks, only two are usually profitable; and of these, it is the Kruger National Park that generates the revenue necessary to cross-subsidize and allow the survival of the national park network.

PRESSURES ON PARKS

Diminishing Financial Support from Government

Although SANParks receives only approximately 15% of its annual expenditure from government, the reality is that the government expects SANParks to further reduce its dependence on this subsidy. Already in recent years the organization has lost the annual roads subsidy from government (about U.S.$1 million), which in effect was a partial return of taxes that SANParks paid to the central government. To be fair, the South African government was generous with disaster relief during the 2000/2001 financial year to help SANParks recover from the substantial effects of the subcontinental floods experienced in February 2000. The government also contributed substantially to enable a major staff reduction and an organizational restructuring exercise during 2001, after it became apparent that SANParks was experiencing severe financial problems brought about by excessive human-resource costs coupled with declining income.

South Africa is by no means the only country suffering from reductions in government funding. The government of Kenya, for example, has other priorities that prevent it from adequately financing biodiversity conservation in the country, and there is no prospect of the situation improving. Recognizing this fact, Richard Leakey, the former head of Kenya Wildlife Service, believes it is now time for the international community to take up the responsibility for the conservation of natural resources, and he supports the establishment of a U.S.$300 million endowment for the protection of savanna species in East Africa.[2]

Following suit, on November 2, 2001, South Africa announced, at the Seventh World Wilderness Congress in Port Elizabeth, South Africa, the launch of an ambitious program called My Acre of Africa. This program will be marketed worldwide in collaboration with prominent South African business partners and with Nelson Mandela as patron-in-chief. With the goal of establishing a conservation trust governed by a board of trustees composed of prominent South African leaders, the program will call on individuals and organizations to "sponsor an acre of Africa."[3]

Elsewhere, other institutions have stepped in to assist languishing conservation programs. For example, the World Bank in recent years has committed an initial U.S.$5 million toward the planning and implementation of conservation areas in Mozambique, and U.S.$60 million toward the revitalization of national parks in Zimbabwe. Some progress has been achieved in Mozambique, in particular on the planning of transfrontier parks, but huge deficits in staffing and resources exist. A grant of DM 12 million by the

German Kreditanstalt fur Wiederaufbau in 2001 has enabled substantial development to commence within the Limpopo National Park, which was proclaimed in November 2001 in Mozambique. With only a small proportion used, the U.S.$60 million intended for Zimbabwe has been suspended because of the political turmoil within that country.[4]

Declining Profitability of Parks

The South African economy, which is far more stable and resilient than the economies of most other African countries, has experienced inflation of 6–15% in the preceding decade. SANParks has had to absorb these increases (especially in the cost of fuel and maintenance materials) in its operational expenditure, but the organization has not been able to pass on these rising costs to tourists, who are reluctant to pay higher accommodation and entry fees. With increasing expenditures and reduced income, the organization has been forced into the unsustainable situation of trying to compensate by giving smaller annual salary increases to staff, cutting down on maintenance costs, and postponing the replacement of worn-out vehicles and equipment. The consequences of these actions have manifested themselves in many ways. Poor customer service has led to customer dissatisfaction. Staff morale is low, and the skilled workers who have been lost cannot be replaced at equivalent levels of remuneration. The parks cannot compete with the multitude of alternative tourist accommodations around the periphery of the park, and long-standing traditional South African visitors feel that standards within the national parks are dropping.

To address this untenable situation, SANParks engaged in a landmark exercise that represents a complete shift from previously entrenched dogma and cherished conservation principles—at least in the African context. Using the International Finance Corporation (IFC) as its lead advisor, in the late 1990s SANParks embarked on a commercialization program to allow private operators to build and operate tourism facilities within the national parks. Previously, all operations within the parks—from building and maintaining roads, bridges, and buildings to operating shops, restaurants, and laundries and providing security (only airport operations at Skukuza, commercial vehicle rental, and police, bank, and postal services were privately run)—were done by SANParks employees. Starting in 2000, private operators in SANParks land were granted contractual rights to use defined areas of land and any buildings on that land, over a specified period, in return for concession fees. To date, thirteen such concession areas, averaging about 10,000 ha in size, have been granted, including eight within the 1,878,682 ha Kruger National Park.

In return, the concessionaires pay SANParks a total guaranteed minimum income (net present value) of well in excess of U.S.$30 million over a 20-year period. However, there is a sliding scale of profit-related commission, which means that the actual fees paid by concessionaires are likely to be considerably higher than forecasted. Taking the commercialization program one step further, in an attempt to outsource all functions not deemed to be related to core conservation activities, SANParks during 2001 also gave rights to private operators to take over shops and restaurants within many of the national parks, including the Kruger National Park, on conditions similar to those granted to land-concession operators. All these operators are contractually bound to a set of obligations regarding financial terms, environmental management, social objectives (e.g., major percentages of shares must be held by historically disadvantaged individuals),[5] and so on (SANParks 2001).

The commercialization of national conservation areas—public property—has not been without major concern, although the public increasingly accepts that properly managed commercialization could contribute to more-efficient services and greater sustainability of the national parks network, without compromising biodiversity conservation. Richard Leakey has publicly expressed reservations about allowing private concessionaires to operate within those national conservation areas that represent the nation's ultimate commitment to biodiversity conservation, citing concern that the need for such operators to show profits may require previously unacceptable compromises to be made in favor of such operators.[6]

Loss of Skilled Staff

For various reasons, skilled staff have trickled from the services of SANParks, and although in some cases the departing workers have been replaced by staff of excellent quality, often there have been either no replacements or the replacements have been of poorer quality. The reasons that staff members have left are varied but are generally associated with financial pressures facing the organization, and associated low salaries, as well as with social and political changes taking place in the country. There is no doubt that the size of the Kruger National Park's staff during the mid-1980s to mid-1990s (nearly 3,000) represented inefficient use of human resources and justified some restructuring. The restructuring in early 2001 within the Kruger National Park reduced the staff size to 1,987 through natural attrition as well as extensive retrenchment. It is also true that the composition of staff up until the time of the 2001 restructuring, especially within management ranks, was heavily weighted in favor of whites, reflecting apartheid appointments made prior to the 1994 democratic elections. Pressure to move toward having a staff composition that reflects the demographic composition of the general South African population, which is overwhelmingly black, has meant that new appointments tend to be black, and advancement to senior management also favors previously disadvantaged groups.

As alluded to earlier in this chapter, formal wildlife conservation in colonial Africa was largely the domain of white people, and training and employment opportunities for black people in skilled posts or even lower-management posts were virtually non-existent. Coupled with that has been the fact that posts in wildlife conservation have traditionally received poor salaries, but the white people who took these posts often remained in service as a lifelong career or at least for extended periods, building up essential skills and experience. Because black people were deprived of quality jobs for decades, indeed centuries, and had no incentive to train for wildlife positions, there is now a relative dearth of appropriately skilled blacks from which to recruit. The current approach therefore is to employ black candidates with basic qualifications and potential and to provide on-the-job training, but even this approach suffers because most high-quality candidates are drawn to better-paying posts in the private sector. This all adds up to difficulty in effectively replacing the experienced old hands who are rapidly being lost through retirement, retrenchment, and resignation. The collective corporate experience, or "corporate memory," has become impoverished. Many of the new staff say that the loss of the old guard is in fact a good thing that will better position the organization in the new social circumstances in South Africa. Only the passage of time will reveal the benefits or demerits of the decisive changes in recent time.

Uncompetitive Remuneration Packages Leading to Poor Service

A long-standing and widespread complaint among virtually all staff from the lowest ranks to midlevel management is that salaries and benefits are poor relative to those received in the private sector or even other parts of the public sector. Several surveys in recent years have shown that, on average, remuneration packages for SANParks positions fall within the lowest one-third of the spread of remuneration associated with the same or similar positions outside the organization.[7]

Although it would not be fair to tar all staff with the same brush—many employees show great loyalty, productivity and diligence despite salary dissatisfactions—such salaries offer no real incentive to existing staff to perform above the basic minimum expected of them, nor do these salaries attract high-quality candidates when vacancies arise. Many applicants lose interest when informed of the remuneration package.

The unfortunate result of this situation of course is that the organization, and the visiting public, receive average service at best and often substandard service. Numerous travelers to the Kruger National Park and other national parks in the late 1990s wrote to newspapers to air their dissatisfaction.[8] The commercialization program currently under way within SANParks, discussed elsewhere in this paper, together with other revenue-generating initiatives, has the potential to improve income levels and thereby the organization's capacity to better compensate employees and compete with the private sector for skilled staff.

Changing Social Expectations of Product Quality

The standard accommodations and amenities in many African national parks still offer the spartan style offered half a century ago. Rooms do not provide the comfort most visitors have become accustomed to, and facilities are lacking. Restaurants provide average meals, and there is little for young children or teenagers to amuse themselves with. Many park managers realize the disincentives this represents, but entrenched policies, insufficient decision-making authority, and budgetary pressures allow them little leeway to make meaningful changes. As a result, many families having small or teenage children take their main annual holiday away from the national parks.

In another landmark change within SANParks, the management style within the Kruger National Park was dramatically changed during 2001, when the park was compartmentalized into four business units, each headed by its own business-oriented manager having far greater freedom to make decisions relating to operational priorities, budget expenditure, and packaging of tourism opportunities. This business restructuring has been hailed by SANParks authorities as representing a fundamental change aimed at financial viability without compromising biodiversity objectives, principally achieved through targeting customer needs and improved customer satisfaction.

Land Claims by Descendants of Evicted Residents

The phenomenon of land claims made by descendents of evicted residents is evident in several African countries, with major consequences for national park systems. In South Africa, even before the introduction of the apartheid homelands system,[9] with its consequent mass relocation of people, removal of rural tribal people to accommodate the plans of town design, agricultural schemes, national parks, and so on frequently meant such people had little choice but to follow official instructions. With implementation of fully democratic government in 1994 in South Africa, land restitution became a major government commitment, and it has led to numerous land claims by the descendants of people who were forcibly removed from ancestral land. National parks have not escaped this process: portions of the Kalahari Gemsbok Park (now part of the Kgalagadi Transfrontier Park) and the Kruger National Park (the Makuleke region) have been returned to tribal claimants. In both these cases, SANParks successfully negotiated for these areas to remain incorporated as "contractual parks," meaning that they will remain as conservation areas fully integrated with the adjoining national park, jointly managed by the national park authorities and the land owners (de Villiers 1999a). The land making up the Richtersveld National Park is entirely owned by local people and is also managed as a contractual park by a joint park committee of landowners and SANParks authorities. Joint management and co-ownership have frequently resulted in serious differences between the stakeholders or members of the joint management boards and have complicated the process of managing wildlife. Communities' desire to see an economic return from their land has led to complete divergence from previous conservation dogmas, including allowing limited and controlled hunting in some of these contractual parks.

On occasion, attempts to establish new parks have failed, to the detriment of wildlife conservation. To address a shortcoming in conservation of representative habitat types, SANParks during the 1990s identified a large area to be purchased and designated as the Highveld National Park. During the process of planning and purchasing this land, claims instituted for an extensive portion of the area finally led to SANParks abandoning all efforts to designate the area as a national park.

Nevertheless, the willingness of SANParks authorities and private-land owners to engage in partnerships under clear contractual conditions has meant that substantial additional land areas could be linked to national parks, with major benefits to wildlife and ecological processes. Not only did the Makuleke region remain an effective part of the Kruger National Park, but considerable additional areas in private game reserves along the western boundary of Kruger were also joined and managed compatibly with the wildlife management policies of the park. No fences exist between any of these areas, which increases the area under effective conservation from the 1,878,682 ha represented by the core Kruger National Park (state land) to the more than 2,176,235 ha of the Greater Kruger National Park.

Ecosystem Limitations

At nearly 19,000 km^2, the Kruger National Park is one of the largest protected areas on the continent and is probably the most intensively studied. Yet even this park is clearly not extensive enough to accommodate a number of key ecological processes that historically have shaped gene pools or habitat structure and composition (Braack 1997). These processes include seasonal migration to access food and water during periods of scarcity, endemic disease outbreaks, water catchment and quality maintenance, fire, and the impact of large herbivore population cycles on vegetation. This ecosystem limitation is an outcome of the fact that wildlife protection areas in Africa—and probably on most other continents too—were declared not on the basis of ideal ecosystem requirements but rather on the basis of what land was left over after prime agricultural and forestry land had been allocated, mining areas had been identified, and other economic or strategic concerns had been satisfied.

Migration   Fortunately, fencing of wildlife areas is not a general feature in all countries of Africa, but it is common in the more affluent countries that export beef or other products from domestic stock. Because of the widespread and endemic nature of wildlife diseases such as foot-and-mouth virus, brucellosis, east coast fever, African swine fever, and others, veterinary authorities in countries like Botswana, Namibia, South Africa, and Zimbabwe generally strongly insist that wildlife areas be fenced off to prevent disease transmission to domestic animals. These wildlife pathogens cause few or no symptoms in their wildlife hosts but have severe consequences in unadapted stock. Conservation authorities are also often pressured to use fences to minimize the impact of wildlife predation (by lions, leopards, and so on) on farming stock and also crop damage by elephants.

Complying with these regulations has meant fencing along the unnaturally defined conservation borders, has interfered with previous wildlife movement patterns, and has led to animal population declines (Berry 1980; Williamson and Williamson 1985; Whyte and Joubert 1988). Attempts to partially redress the situation have brought about some unusual partnerships in recent years: fences have been dropped between many private game farms, private game reserves, and even private game reserves and adjoining national parks. The next step being taken is between nations, to link conservation areas across international boundaries and so create international parks or transfrontier parks. Transfrontier parks have considerable potential to contribute to the long-term viability and sustainability of African conservation areas.

Disease Although the countries that engaged in extensive fencing may have had to contend with particular consequences, countries that did not do so have in some cases had to deal with their own problems. Confirmed outbreaks of rabies and canine distemper are believed to have caused the extinction of wild dogs *(Lycaon pictus)* within the Serengeti–Masai Mara ecosystem; these diseases were probably contracted from known endemically infected domestic dogs kept by pastoral people adjoining the unfenced park area (Woodroffe et al. 1997). In a similar example that dates back to the late 1950s, before fencing was commenced, the mingling of domestic stock and wildlife near the southern border of the Kruger National Park is now believed to have caused the introduction of exotic bovine tuberculosis (BTB) from cattle to buffalo. This insidious disease has now spread through almost the entire park—with prevalence levels of BTB exceeding 60% in many buffalo herds—and has raised major concern, at least in some quarters, about the long-term impact on buffalo populations and, particularly, on the far more vulnerable lion populations (Bengis 1999; De Vos, Raath, and Bengis 1995; Keet, Kriek, and Penrith 1996). Quite aside from the direct pathogenic effect of the disease on its host, the mere presence of the disease within Kruger and the fear of transmitting it to other areas of South Africa—where the disease has since been eradicated— have led to the threat of veterinary restrictions on the movement of rhinos and other wildlife species that are part of reintroduction programs from Kruger to restock new and existing parks.

As a general rule, the smaller a conservation area is, the more intensively it needs to be managed and the greater the need for frequent human intervention. Disease outbreaks are a case in point. For example, anthrax is considered endemic in the far northern areas of the Kruger National Park, which unfortunately coincide with the preferred distribution of rare species such as the roan antelope *(Hippotragus equinus)*, tsessebe *(Damaliscus lunatus)*, Lichtenstein's hartebeest *(Sigmoceros lichtensteinii)*, eland *(Taurotragus oryx)*, and others. Anthrax smolders at low prevalence in most years, with low annual mortality—fewer than ten individuals, mostly among kudu *(Tragelaphus strepsiceros)* and buffalo *(Syncerus caffer)*—but epidemics occasionally occur during very dry spells and kill many hundreds of animals. Kruger is by no means a small park, but even here the risk is great that a single outbreak of this disease could wipe out the entire populations of roan antelope or other rare species (De Vos 1989).

During earlier decades, great effort and expense was incurred during outbreaks to laboriously track down dead animals, burn the carcasses, and disinfect watering points that had became contaminated by vultures. Since the mid-1990s, the policy has shifted, so that anthrax is accepted as an endemic disease and active intervention is limited to immunization of a core number of individuals from rare species by firing vaccine into these animals from a hovering helicopter. Such operations are costly in time, finances, manpower and other resources. Dwindling budgets place pressure on park managers, who have to balance the various needs of the park. The development of a transfrontier park with Mozambique and Zimbabwe to add larger refuge areas and also support larger populations of rare species should assist in relieving the need for large-scale intervention.

However, not even innovative linkage of conservation areas is a panacea for the infrequent catastrophic events that occur on a continental scale. The ravages that a single outbreak of a virulent exotic disease can wreak are illustrated by the disastrous effects of the rinderpest virus epidemic that swept through Africa between 1889 and 1903. Strongly suspected of entering Africa in 1889 with cattle imported from India and Aden to provision the Italian army in Somaliland, the disease then moved southward throughout the continent. The disease killed up to 90% of the cattle—causing death by starvation of many thousands of tribal people—and decimated susceptible wildlife populations, mainly buffalo, eland, warthog *(Phacochoerus aethiopicus)*, and bushpig *(Potamochoerus porcus)*. Much of the current discontinuous and isolated distribution of certain wildlife species is attributed to the dramatic impact of this epidemic (Spinage 1962; Walsh 1987).

Water Catchment and Quality Maintenance   With few exceptions, African countries tend to have little water. The six major rivers that flow through the Kruger National Park—which is long and narrow—do so from west to east directly across its narrow axis and are critically important in supporting a high proportion of the biodiversity in the park. On the western side of South Africa, the catchment areas of each of these six rivers have been dramatically affected, so that reduced water quantity and quality now threaten the rich aquatic life historically present in these rivers. Dams have stabilized and slowed the average flow of water and thus produced extensive silt deposition with consequent growth of extensive and thick reed-beds *(Phragmites australis)*. Stable flow means no periodic floods to scour and clean deep hippo pools, which also become silted up; and rocky as well as sandy habitats become diminished, to the detriment of fish and invertebrates that prefer such areas. Huge plantations of water-loving eucalypts and other exotic trees in the catchment areas have steadily reduced the supply of water to rivers such as the Sabie and pose a major threat to this river, which is widely acknowledged to be the richest in the country in terms of biodiversity. The combined effects of dams, plantations, and agricultural irrigation have changed the status of at least two rivers from permanently flowing to only seasonal flow, and some forms of aquatic life have been lost. Irrigation of cane fields adjoining rivers has leached fertilizers into the river, and the resulting eutrophication and build-up of dense blankets of aquatic weeds again have unnaturally modified habitat. Copper and sulfate mines occasionally spill high loads of sediment and toxic chemicals into the Olifants River and cause severe die-off of fish and other organisms.

Although the average width of the Kruger National Park (about 60 km) does allow a certain amount of water purification to take place, the mosaic of suitable habitats to sustain the historic range of aquatic organisms is threatened. Concern about this threat to the aquatic biodiversity of South Africa's premier national park led to the establishment in 1988 of the multidisciplinary, multiorganizational Kruger National Park Rivers Research Programme, which over a period of 12 years has provided critical insights into the dynamics of aquatic systems. This research program has resulted not only in a far wider awareness of the scale and nature of the problem but also, more meaningfully, in a wide range of monitoring actions, collaborative catchment management groups, and changes in management approaches that now are starting to yield dramatically improved protection of aquatic systems (Breen et al. 2000; Rogers and Biggs 1999).

Fire   Fire has had a major role in shaping terrestrial ecosystems in Africa, and fires are a regular seasonal feature in virtually all savanna systems on the continent. Humans in Africa, the "cradle of mankind," have been a part of these savanna systems for many millennia and, together with lightning, are the main sources of ignition of such fires (Van Wilgen et al. 1997).

However, the impacts of humans at the time when these savanna ecosystems experienced their main evolutionary development were very different from the impacts of humans in the last two centuries. People are now far more widely distributed, in far greater densities, and their cultural practices have greatly modified the historical fire regimes to which these ecosystems are primarily adapted. National parks and other conservation areas may be well managed inside their boundaries, but they are nevertheless affected in a major way by what happens outside their borders. High densities of impoverished rural people living adjacent to such national parks have greatly modified that bordering landscape. Some areas have been denuded

through overstocking and overgrazing and therefore contain less com-
bustible material and experience fewer fires. In other areas, deliberate firing
of moribund crops or pasture to promote new growth has led to more fires,
all of which means that fire patterns differ substantially from those that
existed historically. Fires that escape these areas and enter the national park
have a different seasonality and frequency than historically, and if sustained
they will almost certainly change the vegetation composition and structure
of the landscape, with concomitant changes in associated animal life.

Recognizing the importance of fire as an evolutionary tool and habitat
determinant, the Kruger National Park management team in 1996 adopted
a policy of allowing, within specified limits, all lightning fires to burn
unchecked and of confining human-ignited fires to the smallest possible
area because too many of them were being ignited. However, the reality has
been that since 1996 the area burned inside Kruger by accidental human
fires—most often ignited by job-seeking illegal migrants crossing from
Mozambique to South Africa—far exceeds the areas burned by lightning,
and the entire policy is now under review. Fire management in national
parks in Africa is a major challenge because it is such a key habitat determinant, relatively little is known about true ecological requirements, and conservation managers have little control over the majority of fires. By enlarg-
ing these conservation areas in the form of transfrontier parks, managers
can achieve greater resilience and accommodate habitat mosaics to satisfy
the needs of the broad spectrum of biodiversity present within the region.

Impact of Large Herbivores   Africa has a range of "mega-herbivores" such
as elephant, black *(Diceros bicornis)* and white rhinos *(Ceratotherium
simum)*, buffalo, hippo, and giraffe *(Giraffa camelopardalis)*. Given their
voracious appetites, it stands to reason that at least some of these species
require extensive areas within which to forage for preferred foodplants. If
particular species have a tendency toward destructive feeding (such as elephants, which routinely break branches from, debark, and not infrequently
mortally ring-bark favored trees or uproot trees to access tasty underground
parts) and if such species under modern conditions are permanently confined to specific areas, then the potential for an unsustainable relationship
becomes high. Changes in habitat structure or composition are fine if
brought about by "normal" evolutionary processes, but they are generally
unacceptable if induced by human actions (artificial boundaries) that favor
one species (elephants) and affect all other elements of biodiversity in that
system (Whyte 2001).

Elephants have been shown to have the capacity to dramatically alter the habitat in which they occur, changing forest to more-open environment (Douglas-Hamilton 1987; Dublin, Sinclair, and McGlade 1990; Dublin 1994; Jachmann and Croes 1991; Leuthold 1996), suppressing rejuvenation of selected woodland trees (Viljoen 1988), and in the process significantly reducing the species richness of birds and other taxa (Cumming et al. 1997). So although elephants are an important component of African savanna systems, modern realities of restricted ranges and confined space mean that elephants now survive within artificial systems that many conservationists believe no longer have the capacity to accommodate the potentially destructive impact of elephants (Whyte, van Aarde, and Pimm 1998). Elephants are ill-suited to the modern-day situation, because—as at least some people believe—their population cycles are adapted to operate at scales that are no longer available. Elephants evolved in the pre-gunpowder era, with no real natural enemies or known significant diseases, probably having population build-ups and crashes over time spans approaching a century or more, causing their own decline by destructive consumption of food resources (Caughley 1976). On a subcontinental scale, one would thus have a mosaic of areas, some with moderate numbers of elephants harmoniously coexisting with savanna woodland, some areas with high elephant numbers where woodland was actively being converted to more open grassland, and post-crash, low-density elephant areas where woodland was in recovery (Whyte, can Aarde, and Pimm 1998). If this scenario is valid, then clearly even the largest of conservation areas in Africa today are still too small to accommodate these megascale impacts and cycles of elephants.

Because of massive poaching during the 1970s and 1980s, elephant populations declined dramatically in much of Africa, leading to widespread concern and control measures. However, in some countries such as Botswana, Namibia, South Africa, and Zimbabwe, more-favorable wildlife-protection conditions led to a situation in which wildlife managers felt that elephant populations in parts of these countries had exceeded acceptable thresholds and thus required control.

To limit the habitat-modifying capacity of large numbers of elephants, conservation managers in the Kruger National Park for decades maintained the elephant population at around 7,500 individuals by annual culling of the excess, that is, killing several hundred individuals each year. Public outcries over this organized killing of elephants eventually led SANParks to declare in 1995 a moratorium on culling and to review the elephant management policy. The review concluded that elephant numbers should not be allowed to increase indefinitely, that elephants must be managed so as not to detrimentally affect other components of biodiversity, and that the most practical way of achieving this was by zoning the Kruger National Park to accommodate different levels of elephant impact. To simulate the hypothetical natural situation, two areas would be zoned for high elephant impact, two zones for moderate impact, and another two zones for low numbers of elephants and associated low impact. Through an ambitious monitoring program, the effects of this policy would be tracked until any of a range of predetermined "thresholds of potential concern" were reached, at which point the zones could be switched or other appropriate management options considered (Whyte et al. 1997).

However, this policy still meant that elephant numbers would have to be reduced in the low-impact zones. This is where social perceptions again entered the equation. Politics around countries maneuvering to sell accumulated stocks of ivory and lingering sensitivities to potential public outcries over resumed elephant killing meant the highest-level conservation directors shied away from giving approval to implement the policy. At the time of writing (November 2002), this policy has still not been implemented, and it reflects how even in what many regard as Africa's best-managed park, concerns about political ramifications and public reactions can compromise or even thwart management of these areas.

Elephant control will remain a sensitive and controversial issue in society. Some attempts have already been made to achieve such control by means other than outright killing. Live capture and translocation of entire family groups—several cows with associated juveniles—and fully-grown bulls are now routine operations in South African parks, but the demand for such live elephants to stock newly proclaimed parks or game ranches is far too low to address the perceived growing overpopulation in places such as the Kruger National Park. A more novel approach still being investigated is the possibility of contraception for elephants. Two methods have been tested in the park thus far, the first based on a porcine zona pellucida (PZP) immunocontraceptive vaccine and the second on subcutaneous estradiol-17β (estrogen) implants to achieve hormonal control. Both methods have flaws that are unacceptable to management of SANParks: contraception rates (PZP) are lower than desired, and bulls have been provoked into near-constant harassment of treated cows secreting high estrogen levels, which has led to loss of calves (estrogen method) (Whyte and Grobler 1998). The fatal flaw in these methods, however, is that it has been convincingly shown that to achieve zero population growth in the elephant population of the Kruger National Park, about three-quarters (well over 3000) of the reproductively-active female elephants would need contraceptive treatment, which has to be done several times a year (Whyte and Grobler 1998). This requirement makes the PZP method impossible from a cost and time perspective. It may, however, be a useful method for small parks or game ranches, and investigations are assessing its value in smaller-scale conservation areas.

So, if public opinion prevents culling, live-removal options offer too little relief, and contraception is not practical, the only remaining options are either to allow elephants to increase in numbers indefinitely and thereby allow woodland habitat to be lost on a grand scale, or to expand the area in which elephants can forage. Gross modification of habitat is precisely what is happening in some areas of Africa now. Large areas of Botswana and Zimbabwe have elephant densities that are too high to control by means of conventional culling, and the impact of these high densities concerns local conservationists, who are unable now to do anything about it. However, in certain areas the potential for dropping fences and establishing transfrontier parks offers at least a medium-term reprieve to this problem. The Great Limpopo Transfrontier Park, for example, will allow 1000 elephants to be moved over the next 3 years from the overstocked Kruger National Park to Mozambique, where elephants were eradicated during two decades of civil war.

BOLD NEW INITIATIVES: AFRICAN TRANSFRONTIER PARKS

Transfrontier parks, or rather the concept they represent, have been described as areas in which cooperation to manage natural resources occurs across boundaries (Griffin et al. 1999). This concept is not new, and we see its first effective application in the 1932 agreement that linked Glacier National Park (Montana, U.S.) and Waterton Lakes National Park (Alberta, Canada) (de Villiers 1999b). Since then a variety of terms has evolved to describe this linkage, including peace parks, transboundary natural-resource management areas, transfrontier conservation areas, and transfrontier parks. Each has its own specific emphasis, but they all promote natural resource conservation across international boundaries. In this chapter, I use the term *transfrontier park* to mean a government-supported, contractual linkage between extensive areas that are dedicated to the primary purposes of wildlife conservation and ecotourism and that straddle international political borders where humanmade barriers have been removed to promote ecological integrity. Transfrontier parks differ slightly from transfrontier conservation areas, which may also include extensive areas with rural populations practicing low-impact subsistence or small-scale irrigation farming without the primary purpose of wildlife conservation and ecotourism. Both transfrontier parks and conservation areas can be said to be striving for effective and integrated transboundary natural-resource management, and they may or may not have been established to promote peace in the sense of peace parks in a conflict-prone region.

Although these areas in their different guises may have achieved their objectives elsewhere in the developed world, they represent a new and experimental approach in Africa, where social, economic, and political conditions vary considerably from those in the developed world, where situations are not as predictable and stable, and where resources are not as readily available. Yet the concept of integrating conservation areas across international boundaries, removing previous barriers impeding the flow of wildlife and people, and adopting harmonized management approaches for improved ecological and socioeconomic objectives holds clear and major benefits in a continent where wildlife conservation is under threat and where previous solutions have not always proved sustainable.

ADVANTAGES OF TRANSFRONTIER PARKS

Transfrontier parks, given the fact that they link extensive contiguous areas often previously fenced from each other for national disease control, have the great advantage of promoting the unimpeded influence of previously disrupted ecosystem processes. As referred to earlier, these processes include migration, the creation of additional resource areas for rare species having particular habitat requirements, disease outbreaks, fire, and droughts, all of which affect large-scale population dynamics and long-term genetic fitness. In an intuitive sense, transfrontier parks then promote the long-term viability and sustainability of the area and the wildlife it contains, and reduce the need for regular management intervention to address situations that would have represented a crisis in a smaller area.

Linking different areas into one unit often also increases the range of tourism attractions available, improves marketing potential, encourages donor support, and if properly approached should lead to increased tourism and income associated with it. Increased tourism has major benefits in terms of demonstrating the legitimacy of conservation as an effective, justifiable, and financially productive form of land use and promoting social acceptance—especially among land-hungry impoverished rural communities—of wildlife conservation. Tourism should also lead to improved job opportunities and small-business opportunities for communities adjoining such transfrontier parks.

In the African context, where real and often major disparities exist

TABLE 18.1 *Southern African Transboundary Park Areas already established or in development*

Transboundary Park Areas	Countries Involved	Date Memorandum of Understanding Signed	Formal Launch	Size of Area
Kgalagadi TFP	Botswana, South Africa	April 1999	12 May 2000	37,991 km²
Great Limpopo TFP	Mozambique, South Africa, Zimbabwe	10 November 2000	9 December 2002	35,771 km²
Maloti Drakensberg TFC&DA	Lesotho, South Africa	11 June 2001	22 August 2003	14,744 km²
Chimanimani TFP	Mozambique, Zimbabwe	10 June 2001	?	2,056 km²
Ai-Ais Richtersveld TFP	Namibia, South Africa	17 August 2001	1 August 2003	6,046 km²
Limpopo Shashe TFP	Botswana, South Africa, Zimbabwe	?	?	4,872 km²
Lubombo TFCA	Mozambique, South Africa, Swaziland	signed as part of five protocols in 2000	?	ca. 25,000 km²
Iona Skeleton Coast TFCA	Angola, Namibia	1 August 2003	?	31,540 km²
Nyika TFCA	Malawi, Zambia	?	?	3,243 km²
Vwaza Marsh Lundazi TFCA	Malawi, Zambia	?	?	>984 km²
Kasungu Lukusuzi TFCA	Malawi, Zambia	?	?	5,936 km²

NOTE: TFP = Transfrontier Park; TFCA = Transfrontier Conservation Area; TFC&DA = Transfrontier Conservation and Development Area.

between neighboring countries in terms of resources, opportunities, and skills, such parks promote the transfer of skills, capacity-building, the raising of conservation-management standards, and, through joint management and regular contact, a spirit of cooperation and good neighborliness, all to the benefit of the areas being conserved.

Unlike the social environment in many parts of the more affluent, developed world, the social environment in Africa is not conducive to the setting aside of large areas purely for wilderness management. In Africa there is increasing expectation that conservation areas should compete with other land uses and show an economic return to remain supported and viable. Creating transfrontier parks improves the opportunities and potential to do just that, and having greatly expanded areas also allows for zoning of certain areas as pristine wilderness with minimal human impact because the expanded park is sufficiently large to have substantial areas for tourism.

Africa has an unfortunate history of regular wars and tension among neighbors, but in the recent decade an increasingly globalized world, interdependent economies, and the need for regional cooperation have meant that relative peace and stability have descended over a considerable portion of Africa. Such peace and stability are a prerequisite for the establishment of transfrontier parks and partially explain why such parks have only just started emerging in Africa. Southern Africa has been at the forefront in establishing these transfrontier parks, and eleven such initiatives have been formally established or are being implemented (Table 18.1).

The concept of transfrontier parks has generated an enormous amount of support among politicians and the general public of southern Africa and, indeed, even internationally, as demonstrated by the international media coverage of the release of elephants from the Kruger National Park into Mozambique in October 2001. Such support, as well as the wide acceptance of the transfrontier park idea, has been a much needed boost for conservation in the region and has generated substantial donor commitment to the process. Nevertheless, transfrontier parks remain an experimental phenomenon within Africa, and these parks will need sustained nurturing if they are to realize the many benefits they promise.

NOTES

1. *Biodiversity* in its full sense includes composition, structure, and function and extends from the genetic level to the landscape level. That is, true biodiversity requires variation within and between not only populations but also communities and habitats; landscape heterogeneity; and variation in the multitude

of evolutionary mechanisms that enable adaptation and sustained survival of populations.

2. Richard Leakey, Conservation, species loss, and poverty in the 21st century, public lecture, August 20, 2001, Delta Park, Johannesburg, South Africa.

3. The aim of the program is to generate well in excess of U.S.$100 million by getting people to sponsor "conservation bricks," each costing U.S.$45–450. The bricks will be used to build a large scale model of the Kruger National Park, and the site will incorporate a wilderness conservation and education center.

4. Rod de Vletter, World Bank, Maputo, conversation with author.

5. Members of nonwhite population groups, women, and handicapped persons.

6. Richard Leakey, Conservation, species loss, and poverty in the 21st century, public lecture, Delta Park, Johannesburg, South Africa.

7. Jan Small, General Manager, Human Resources, SANParks, South Africa, conversation with author.

8. Interestingly, despite the poor image of SANParks, the number of visitors to Kruger—which bore the brunt of complaints from a disgruntled public—generally kept rising, so that between 1996 and 2000 the average number of visitors per annum was 927,080, the highest number ever (April 1996–March 1997 = 906,999; 1997/1998 = 954,398; 1998/1999 = 948,732; 1999/2000 = 898,191). There are several reasons for this increase, including the declining value of the South African rand (fewer South Africans take overseas holidays), less disposable income for the average South African (national parks are thus more affordable than many other holiday destinations), and increasing numbers of overseas visitors (South Africa is no longer ostracized by the international community).

9. Tribal reserves are specific, defined areas within South Africa known to have been the core areas of historic habitation of specific tribal groupings at the time Europeans arrived on the subcontinent.

REFERENCES

Bengis, R. 1999. Tuberculosis in free-ranging mammals. In *Zoo & Wild Animal Medicine: Current Therapy*, ed. M. E. Fowler and R. E. Miller, 101–14. 4th ed. Philadelphia, Pa.: W. B. Saunders.

Berry, H. H. 1980. Behavioural and eco-physiological studies on blue wildebeest *(Connochaetus taurinus)* at the Etosha National Park. Ph.D. diss., University of Cape Town.

Braack, L.E.O., ed. 1997. *A Revision of Parts of the Management Plan for the Kruger National Park.* Vol. 3, *Policy Proposals regarding Issues relating to Biodiversity Maintenance, Maintenance of Wilderness Qualities, and Provision of Human Benefits.* Unpublished document. Available at www.parks-sa.co.za as of October 2003.

Breen, C., Dent, M., Jaganyi, J., Madikizela, B., Maganbehari, J., Ndlovu, A., O'Keefe, J., Rogers, K., Uys, M., and Venter, F. 2000. *The Kruger National*

Park Rivers Research Programme. Pretoria, South Africa: Water Research Commission.

Burchell, W. J. 1822. *Travels in the Interior of Southern Africa.* London: Longmans.

Carruthers, J. 1990. Towards an environmental history of Southern Africa: Some perspectives. *South African Historical Journal* 23:184–95.

———. 1995. *The Kruger National Park: A Social and Political History.* Pietermaritzburg, South Africa: University of Natal Press.

Caughley, G. 1976. The elephant problem—an alternative hypothesis. *East African Journal of Wildlife Management* 14:265–82.

Cronwright-Schreiner, S. C. 1925. *The Migratory Springbucks of South Africa.* London: T. Fisher Unwin.

Cumming, D. H., Fenton, M. B., Rautenbach, I. L., Taylor, R. D., Cumming, C. S., Dunlop, J. M., Ford, A. G., Hovorka, M. D., Johnston, D. S., Kalcounis, M., Malangu, Z., and Portfors, C.V.R. 1997. Elephants, woodlands, and biodiversity in Southern Africa. *South African Journal of Science* 93:231–36.

de Villiers, B. 1999a. Makuleke land claim and the Kruger National Park—joint management—a benchmark for conservation areas. *South African Public Law* 1999:309–30.

———. 1999b. *Peace Parks: The Way Ahead.* Pretoria, South Africa: HSRC Publishers.

De Vos, V. 1990. The ecology of anthrax in the Kruger National Park, South Africa. *Salisbury Medical Bulletin* 68 (suppl.): 19–23.

De Vos, V., Raath, J. P., and Bengis, R. 1995. The epidemiology of bovine tuberculosis in the Kruger National Park, South Africa. *Proceedings of the Otago Conference on Tuberculosis in Wildlife and Domestic Animals,* 255–59. Dunedin, New Zealand: University of Otago, Otago.

Douglas-Hamilton, I. 1987. African elephants: Population trends and their causes. *Oryx* 12:11–24.

Dublin, H. T. 1994. Vegetation dynamics in the Serengeti-Mara ecosystem: The role of elephants, fire, and other factors. In *Serengeti II: Dynamics, Management, and Conservation of an Ecosystem,* ed. A.R.E. Sinclair and P. Arcese, 71–90. Chicago: University of Chicago Press.

Dublin, H. T., Sinclair, A.R.E., and McGlade, J. 1990. Elephants and fire as causes of multiple stable states in the Serengeti-Mara woodlands. *Journal of Animal Ecology* 59:1147–64.

Griffin, J., Cumming, D., Metcalfe, S., t'Sas-Rolfes, M., Singh, J., Chonguica, E., Rowen, M., and Oglethorpe, J. 1999. *Study on the Development of Transboundary Natural Resource Management Areas in Southern Africa.* Washington, D.C.: Biodiversity Support Program.

Harris, W. C. 1947. *Wild Sports of Southern Africa.* London: Macmillan.

———. 1986. *Portraits of the Game and Wild Animals of Southern Africa.* Cape Town, South Africa: Sable Publishers.

Jachmann, H., and Croes, T. 1991. Effects of browsing by elephants on the

Combretum/Terminalia woodland at the Nazinga Game Ranch, Burkina Faso, West Africa. *Biological Conservation* 57:13–24.

Jeal, T. 1996. *Livingstone.* London: Pimlico.

Keet, D. F., Kriek, N.P.J., and Penrith, M.-L. 1996. Tuberculosis in buffaloes *(Syncerus caffer)* in the Kruger National Park: Spread of the disease to other species. *Onderstepoort Journal of Veterinary Research* 63:239–44.

Leuthold, W. 1996. Recovery of woody vegetation in Tsavo National Park, Kenya 1970–1994. *African Journal of Ecology* 34B:101–12.

Lichtenstein, M.H.K. 1812–1815. *Travels in Southern Africa, 1803–06.* London: Colburn.

McKelvey, J. J. 1973. *Man against Tsetse—Struggle for Africa.* Ithaca, N.Y.: Cornell University Press.

Packenham, T. 2000. *The Scramble for Africa.* Johannesburg: Jonathon Ball.

Pringle, J. 1982. *The Conservationists and the Killers: The Story of Game Protection and the Wildlife Society of Southern Africa.* Cape Town, South Africa: T. V. Bulpin and Books of Africa.

Robertson, T. C. 1978. *South African Mosaic.* Cape Town, South Africa: C. Struik Publishers.

Rogers, K., and Biggs, H. 1999. Integrated indicators, endpoints, and value systems in the strategic management of the rivers of the Kruger National Park. *Freshwater Biology* 41:439–51.

SANParks. 2000/2001. *Annual Report.* Pretoria, South Africa.

———. 2001. *Information Memorandum on Phase 2 of the Concession Programme.* Unpublished memorandum, Pretoria, South Africa.

Spinage, C. A. 1962. Rinderpest and faunal distribution patterns. *African Wildlife* 16 (1): 55–60.

Stevenson-Hamilton, J. 1937. *South African Eden.* London: Cassell & Company.

Van Wilgen, B. W., Andreae, M. O., Goldammer, J. G., and Lindsay, J. A. 1997. *Fire in Southern African Savannas.* Johannesburg, South Africa: Witwatersrand University Press.

Viljoen, A. J. 1988. Long term changes in the tree component of the vegetation in Kruger National Park. In *Long Term Data Series Relating to Southern Africa's Renewable Natural Resources,* ed. I.A.W. Macdonald and R.J.M. Crawford, 310–15. South African National Scientific Programmes Report no. 157. Pretoria, South Africa: Council for Scientific and Industrial Research.

Walsh, J. 1987. War on cattle disease divides the troops. *Science* 237:1289–91.

Whyte, I. J. 2001. Conservation management of the Kruger National Park elephant population. Ph.D. diss., University of Pretoria, South Africa.

Whyte, I. J., Biggs, H. C., Gaylard, A., and Braack, L.E.O. 1997. A proposed new policy for the management of the elephant population of the Kruger National Park. In *Management Plan for the Kruger National Park,* 8:117–39. Skukuza, South Africa: South African National Parks.

Whyte, I. J., and Grobler, D. 1998. Elephant contraception research in the Kruger National Park. *Pachyderm,* no. 25:45–52.

Whyte, I. J., and Joubert, S.C.J. 1988. Blue wildebeest population trends in the Kruger National Park and the effects of fencing. *South African Journal of Wildlife Research* 18 (3): 78–87.

Whyte, I. J., van Aarde, R., and Pimm, S. L. 1998. Managing the elephants of Kruger National Park. *Animal Conservation* 1:77–83.

Williamson, D., and Williamson, J. 1985. Botswana's fences and the depletion of Kalahari wildlife. *Parks* 102:5–7.

Woodroffe, R., Ginsberg, J. R., Macdonald, D. W., and the IUCN/SSC Canid Specialist Group. 1997. *The African Wild Dog—Status Survey and Conservation Action Plan.* Switzerland: IUCN, Gland.

Systematic Index

Subject Index

Page numbers in *italic* refer to figures and tables.

Compositor:	BookMatters, Berkeley
Text:	10/13 Aldus
Display:	Aldus
Printer and binder:	Sheridan Books, Inc.
Subject index:	Andrew Joron